Historical Technology, Materials and Conservation

SEM and Microanalysis

On behalf of the many scientists, conservators, curators, archaeologists and historians who have worked with him over the years and benefited from his expertise, although this is not an official *Festschrift*, we would like to thank and dedicate this volume to Nigel Meeks.

Historical Technology, Materials and Conservation

SEM and Microanalysis

Edited by Nigel Meeks, Caroline Cartwright,
Andrew Meek and Aude Mongiatti

Archetype Publications

in association with

The British Museum

First published 2012 by Archetype Publications Ltd. in association
with the British Museum

Archetype Publications Ltd
c/o International Academic Projects
One Birdcage Walk
London SW1H 9JJ
www.archetype.co.uk

ISBN: 978-1-904982-65-4

British Library Cataloguing in Publication Data
A catalogue record for this book is available from the British Library.

Designed by Marcus Nichols at PDQ Digital Media Solutions Ltd.
Typeset by PDQ Digital Media Solutions Ltd, Bungay
Printed and bound in Great Britain by Latimer Trend & Co. Ltd., Plymouth

Contents

Foreword

This volume contains written versions of presentations given at a two-day meeting entitled 'SEM and microanalysis in the study of historical technology, materials and conservation' held at the British Museum on 9–10 September 2010, organised by the Department of Conservation and Scientific Research at the British Museum, in association with Hitachi High-Technologies Europe. While this volume will be of relevance to all with an involvement or interest in the application of scanning electron microscopy and microanalysis (SEM-EDX) to the study of materials, manufacturing methods and deterioration processes of objects from ancient through to contemporary cultures, we hope it will also be of wider inspiration to those using these techniques in other fields.

At the British Museum, we have a long and solid tradition of SEM-EDX applications and developments applied to the wide range of materials in the collections of the Museum. SEM-EDX – often used in conjunction with other microscopic and analytical instrumentation in the Department of Conservation and Scientific Research – is one of the core techniques needed to research the collections. In recent times, this work has benefited from the support of the Museum's management and the generosity of the Friends of the British Museum, the John Ellerman Foundation, several private individuals, Oxford Instruments NanoAnalysis and Hitachi High-Technologies Europe in providing funding for two state-of-the-art SEMs with EDX analysers.

The purpose of the meeting was to explore the huge influence that SEM and microanalysis have had over the decades on the understanding of the material technologies and associated cultures that produced such objects, and on the conservation and preservation of these materials for the long-term benefit of world cultural heritage. While broadly falling under three categories, 'materials, technology and manufacturing processes', 'conservation' and 'new applications and instrumental developments', the wide range of topics represented by the 46 papers included in the volume demonstrates the variety of investigations being carried out across the world. They also explore how techniques have been developed to extend their use into new fields and showcase some of the new applications and instrumental developments. We hope that these papers demonstrate the key role that SEM and microanalysis have had and continue to play in the study of historical technology, materials and conservation. The new developments of these powerful techniques are allowing an even wider range of materials, objects and samples to be investigated, opening up whole new areas of research that were not possible with previous instrumentation, promising a very exciting future for those working in these fields.

Nigel Meeks, Caroline Cartwright,
Andrew Meek and Aude Mongiatti

Acknowledgements

The editors would like to thank the other members of the Organising Committee (Eleanor Beyer, Catherine Higgitt, Duncan Hook, Chris Jones, Susan La Niece and Antony Simpson) for their efforts in bringing the meeting to fruition, and the team of helpers and other colleagues at the British Museum who kept things running smoothly during the meeting itself. We also gratefully acknowledge the support offered by Hitachi High-Technologies Europe both for the organisational costs of this conference and the publication of this volume, and through the involvement of Chris Jones (then at Hitachi High-Technologies) as a member of the Organising Committee.

For this publication we are particularly indebted to Antony Simpson who has carried out extensive work on the illustrations. We would also like to thank the numerous anonymous reviewers who undertook the difficult task of reading and commenting on the papers, and the participants themselves for the high quality of the papers submitted and for making the meeting a stimulating and inspirational two days.

The Bedford Lemere collection: investigating degrading glass plate negatives

Sarah Allen, Jenny Hodgson, David Dungworth and Sarah Paynter

ABSTRACT Bedford Lemere & Company was one of the leading architectural photography firms of the late nineteenth and early twentieth century. The English Heritage Archive (formerly the National Monuments Record, the public archive of English Heritage) has the largest extant holding of Bedford Lemere & Co. photographic materials. In November 2009 the English Heritage Archive began a project to conserve, digitise and catalogue the 23,000 large format glass plate negatives that are the largest constituent part of the collection.

The glass plate negatives have many preservation needs. One of the most significant issues in the collection is that of peeling emulsion – a particularly devastating form of deterioration that can range from small blisters to instances where the entire photographic pellicle delaminates from the glass base. This phenomenon tends to be accompanied by the formation of a white crystalline substance and a change in the appearance of the glass. The English Heritage Archive Conservation Department instigated a cross-disciplinary, interdepartmental project in collaboration with the English Heritage Archaeological Conservation and Technology team in order to investigate the mechanisms of deterioration.

Scanning electron microscopy with energy dispersive X-ray analysis (SEM-EDX) was used to investigate the topography of the glass plate, the composition of the glass, deterioration products and other related materials. This paper discusses the results of the analyses.

Preliminary findings indicate that the glass composition matches that expected for mid-to-late nineteenth-century glass. However, the layer of glass directly in contact with the photographic emulsion shows the loss of some sodium and calcium. Analysis indicates that the glass is being de-alkalised by reacting with sulphur-rich compounds, resulting in the formation of alkali sulphate salts. It is thought that this weakens the bond between the photographic emulsion and the glass, ultimately causing the pellicle to delaminate.

KEYWORDS photography, glass, negative, deterioration, conservation

Bedford Lemere & Co. at the National Monuments Record

The English Heritage Archive (EH Archive) is the largest publicly accessible archive dedicated to the historic environment in England, with a purpose-built photographic materials cold store containing over 10 million items. The Bedford Lemere & Co. collection is one of the most important held in the archive. The firm was established in 1867 by Bedford Lemere, though its most prolific period began in 1881 when his son, Henry Bedford Lemere, joined the firm. He cemented the company's reputation as one of the nation's leading architectural photography firms. Although the studio was based in London, the photographers worked nationwide capturing images of a burgeoning industrial Victorian society. Common subjects included exterior architectural shots, opulent interiors, industrial scenes and social history.

The total output of Bedford Lemere & Co. is estimated to have been around 100,000 negatives. The EH Archive has by far the largest extant holding of Bedford Lemere material globally, with approximately 28,000 items dating from the 1870s to the 1920s. This includes about 5000 prints and wet collodion negatives but the vast majority of the collection is made up of gelatine dry plates. These large format, 10 × 12 in, machine-made negatives contain extremely high quality images that capture great detail. The emulsion side of the negatives is almost always varnished with shellac.[1]

In November 2009 the EH Archive began an ambitious project to conserve, digitise and catalogue this exceptional collection, which culminated in a publication and exhibition at the Victoria and Albert Museum (London) in 2011. The aims of the project were to make this resource available to a wider audience, as well as ensuring its long-term preservation.

The conservation project

Prior to digitisation, each negative is examined to determine its specific conservation needs. Any negatives that need stabilisation are treated by the EH Archive Conservation Department. Around 80% of the collection needs some form of interventive conservation treatment. Common problems include heavily soiled surfaces, broken or cracked glass, mould damage and of course, delamination. The phenomenon of delamination is not unique to this collection or this format [1–3]. Delaminating emulsion appears in many different forms and there are

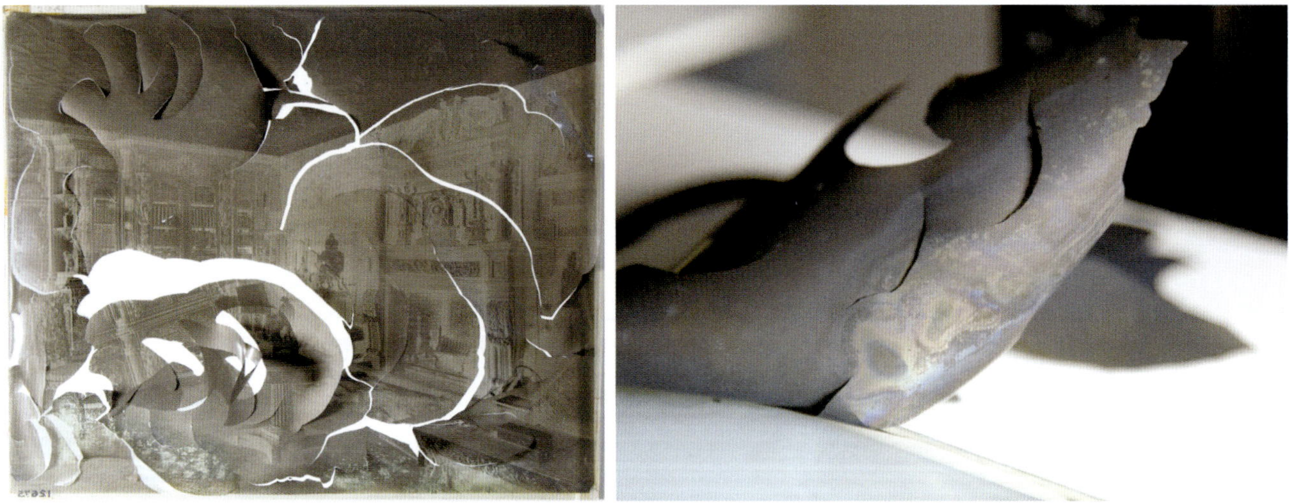

Figure 1 Glass plate negative with delaminated emulsion (left); detail of delaminated emulsion and glass with iridescent appearance.

numerous types of delamination affecting the Bedford Lemere & Co. collection. These are easily characterised visually as each type is very distinctive.

The form of delamination that this paper is concerned with has the following characteristics:

- Sections of emulsion, sometimes in a concentric pattern, are no longer attached to the glass base. This can range from small areas around 50 mm^2 to instances where the entire emulsion has peeled back from the glass.
- The glass underneath the lifted emulsion is matte and slightly translucent due to a white crystalline substance present on the surface.
- The glass has an iridescent appearance when viewed from the non-emulsion side, sometimes also in the areas where the emulsion has not yet delaminated. The emulsion that has been in contact with the glass also has this iridescent appearance (Figure 1).

Approximately 3% of the collection is affected by this type of deterioration. A further 5% of negatives exhibit the distinctive iridescent appearance visible through the non-emulsion side, though the emulsion is still bonded to the glass. It is likely that the iridescent appearance is linked to the subsequent delamination of the emulsion, although this has not yet been confirmed.

Historic sources

The Bedford Lemere & Co. collection not only contains negatives but also administrative archive material including notebooks and photographers' day books. The latter, in particular, are valuable sources of information with regard to this research, as they are a record of exactly when and where each negative was taken. These day books had been transcribed as part of the EH Archive Bedford Lemere project, meaning a database could be built in order to ascertain any patterns in the occurrence of negatives where the emulsion eventually would go on to delaminate. The database was analysed to determine the frequency at which delaminated negatives occurred, at what time of year these negatives were processed and whether or not they occurred in batches or any other type of significant pattern.

The database showed that the spread of delaminated negatives was random but steady, with a few occurring every month since Bedford Lemere began using gelatine dry plates at the beginning of the 1880s. No runs of negatives were affected and no particular year had a significantly higher occurrence rate than any other.

An important discovery, however, was that the phenomenon of this particular form of delamination stopped occurring in any negative processed after 7 July 1899. Delaminated negatives are found regularly up to this date, suggesting that it was not simply an age-related deterioration (i.e. they do not begin to occur less frequently before this date which might be expected if this was the case).

This suggests that there was a significant change in practice or method of manufacture around this date. It is a tantalising clue as to the cause of this type of deterioration, but unfortunately there is no corresponding record documenting any change in practice in the Bedford Lemere & Co. archive material. Thus, material analysis is needed to better understand the mechanisms of deterioration in these delaminated negatives.

Microscope examination

Surface characteristics

A number of samples of glass were removed from delaminated negatives. Though the samples needed were small (1–2 mm^2), it was still necessarily destructive. Therefore, samples were only taken from previously broken negatives. Though this has implications for the statistical validity of the sampling procedure, it was decided that care of the original object was paramount, thus the results of the analysis should be interpreted with this in mind. Glass samples were taken from the least obtrusive image areas possible although, as the emulsion had already delaminated from the glass, none of the image was removed. A range of sample types was removed including glass that had previously been in contact with the emulsion, glass from the centre of the plates and glass from the non-emulsion side.

The samples were examined at a range of magnifications using a scanning electron microscope (SEM) in secondary electron (SE) mode to look at surface topography. In the samples of glass that had previously been in contact with the

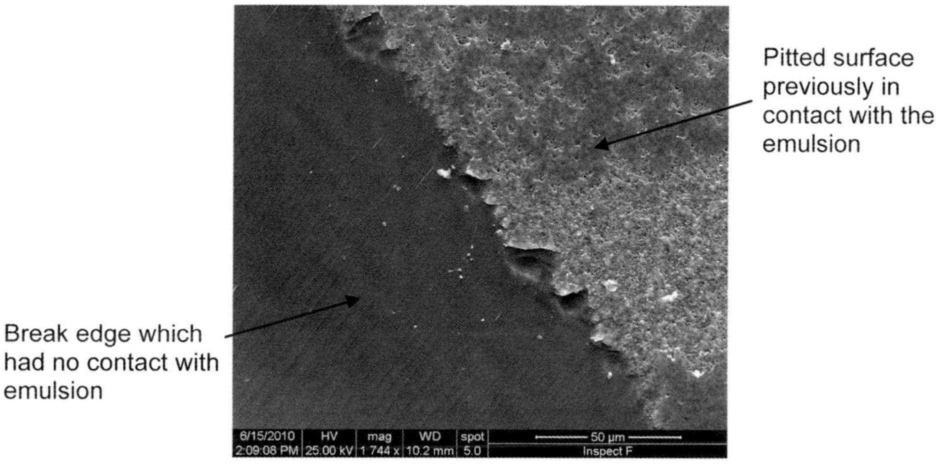

Figure 2 SEM-SE image of the surface of a deteriorating glass plate showing (top right) the pitted surface where the glass was previously in contact with the gelatine emulsion layer and (bottom left) the break edge that had no contact with the emulsion.

emulsion, a pitted, porous surface could be seen at the glass and emulsion interface. In contrast, samples that had come from break edges or the non-emulsion side tended to have a much smoother surface without any pitting. An example of this phenomenon can be seen in Figure 2.

Cross-sections

Cross-sections of glass samples were also examined. This time samples were taken from negatives where the emulsion was still in good condition. Again, samples were no more than 2 mm^2 and were only taken from plates that had already been broken. At this stage one sample was removed that had a small amount of emulsion adhered. This sample was taken from an area at the edge of the image, with no significant detail. The area of the emulsion that was removed was no more than 0.5 mm^2.

These samples were mounted in resin to show the cross-section through the glass plate and then ground and polished to a 1µm diamond paste finish and subsequently examined using the SEM in backscattered electron (BSE) mode, which shows areas with different compositions as varying greyscale shades.

The characteristics observed echoed those that could be seen topographically. Samples from negatives where the emulsion had delaminated showed distinct depressions at the surface, previously the glass/emulsion interface (see Figure 3). In contrast, in the negative where the emulsion was still intact, this interface appeared to have little or no pitting (see Figure 4). This was also the case in samples taken from the non-emulsion side of delaminated negatives – the outer surface of the glass did not have any signs of pitting.

The micro-porous, pitted layer, present only on delaminated negatives could be the cause of the iridescent appearance referred to earlier, due to interference of light reflected from the numerous layers.

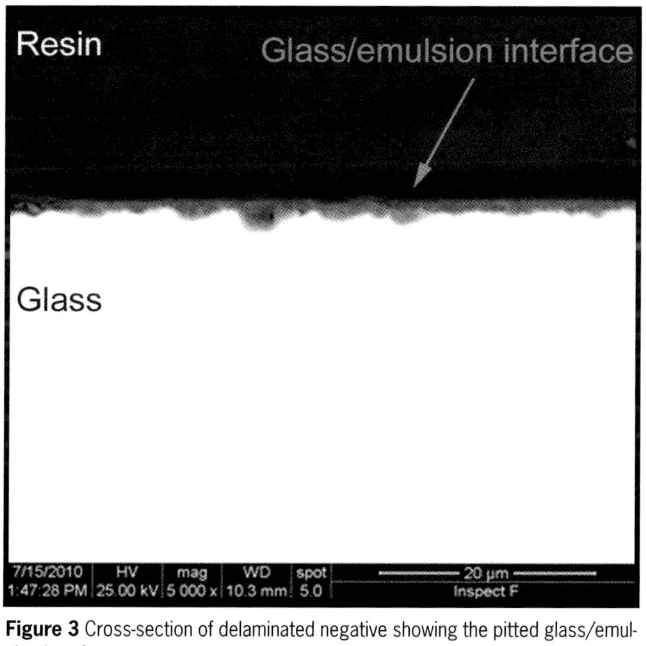

Figure 3 Cross-section of delaminated negative showing the pitted glass/emulsion interface.

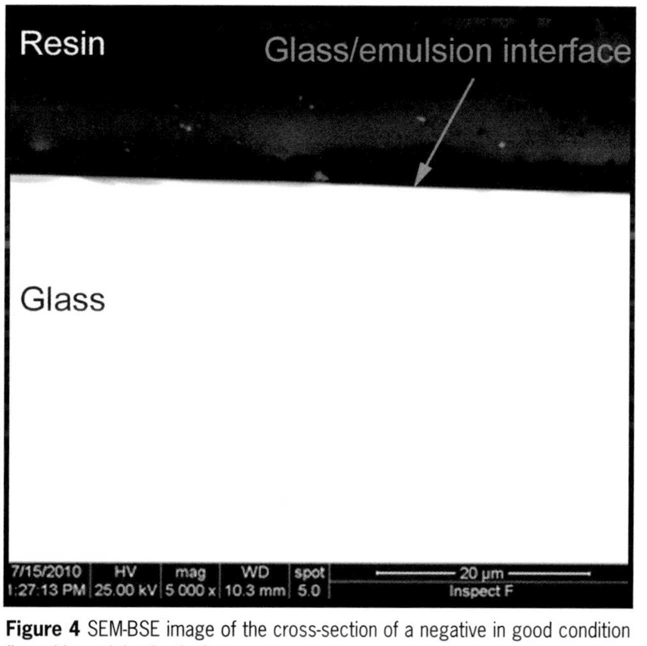

Figure 4 SEM-BSE image of the cross-section of a negative in good condition (i.e. with no delamination).

Table 1 Table of average glass compositions for each sample. Samples with broadly similar compositions are separated by blank lines.

Sample	Delaminated	Date	Na_2O	MgO	Al_2O_3	SiO	SO_4	K_2O	CaO	TiO	Cr	MnO	FeO	As_2O_3	Total
BL 10348/10	**Yes**	**1890**	**11.7**	**<0.1**	**1.5**	**71.9**	**0.4**	**0.7**	**13.4**	**<0.1**	**<0.1**	**<0.1**	**0.2**	**<0.2**	**99.8**
BL 9971	No	1890	11.9	<0.1	1.6	72.7	0.3	0.6	12.5	<0.1	<0.1	<0.1	0.3	<0.2	99.9
BL 11797	**Yes**	**1892**	**11.7**	**<0.1**	**1.6**	**73.1**	**0.4**	**0.6**	**12.3**	**<0.1**	**<0.1**	**<0.1**	**0.1**	**<0.2**	**99.8**
BL 11417	No	1892	11.4	<0.1	1.6	73.7	0.2	0.5	12.3	<0.1	<0.1	<0.1	0.2	<0.2	99.8
BL 12341	**Yes**	**1893**	**13**	**<0.1**	**0.5**	**70**	**0.7**	**<0.1**	**15.4**	**<0.1**	**<0.1**	**<0.1**	**0.2**	**<0.2**	**99.8**
12329_2	**Yes**	**1893**	**12.9**	**0.1**	**0.5**	**70.1**	**0.7**	**<0.1**	**15.4**	**<0.1**	**<0.1**	**<0.1**	**0.2**	**<0.2**	**99.9**
BL11941	No	1893	11.3	<0.1	1.5	73.5	0.3	0.7	12.4	<0.1	<0.1	<0.1	0.1	<0.2	99.7
BL11944	No	1893	11.5	0.1	1.6	72.1	0.2	0.6	13.6	<0.1	<0.1	<0.1	0.1	<0.2	99.9
BL 12894/7	**Yes**	**1894**	**11.6**	**<0.1**	**1.5**	**72.4**	**0.3**	**0.6**	**13.5**	**<0.1**	**<0.1**	**<0.1**	**0.1**	**<0.2**	**100.1**
BL 12575/A	No	1894	11.2	<0.1	1.5	72.7	0.3	0.7	13.3	<0.1	<0.1	<0.1	0.2	<0.2	99.8
BL 13547	No	1896	11.2	0.1	1.8	72.7	0.3	0.7	12.9	<0.1	<0.1	<0.1	0.2	<0.2	99.9
BL 13847	No	1896	11	0.1	1.8	73.1	0.2	0.7	12.7	<0.1	<0.1	<0.1	0.3	<0.2	99.9
BL 13717	No	1896	11.4	0.1	1.6	72.4	0.4	0.7	13.1	<0.1	<0.1	<0.1	0.3	<0.2	99.9
BL 15580	No	1899	11	<0.1	1.8	72.9	0.3	0.7	13	<0.1	<0.1	<0.1	0.2	<0.2	99.8
BL 17405	No	1902	11.5	<0.1	1.7	72.2	0.3	0.8	13.2	<0.1	<0.1	<0.1	0.2	<0.2	99.8
BL 18761	No	1905	11.9	1.5	0.7	70.1	0.5	0.2	14.8	<0.1	<0.1	<0.1	0.2	<0.2	99.9
BL 20160/07	No	1908	11.4	1	0.8	71.3	0.3	0.4	14.5	<0.1	<0.1	<0.1	0.2	<0.2	99.9
BL 19115	No	1905	14	0.4	0.6	72.3	0.6	0.1	11.8	<0.1	<0.1	<0.1	0.1	<0.2	99.9

SEM-EDX analysis

Bulk glass composition

Small samples, a few millimetres in diameter, from the plates were examined using an FEI-Inspect SEM with attached EDX. The surface characteristics of the fragments were imaged using SE mode. To investigate compositional changes in the glass, the fragments were mounted in epoxy resin and polished to a 1µm finish. In BSE imaging mode, areas with different compositions appear as different grey-scale shades. The mounted polished glass samples were also analysed using EDX attached to the SEM. The conditions were an electron beam current of approximately 1 nA and a voltage of 25 kV. The EDX data were quantified using Oxford Instruments INCA software and calculated stoichiometrically as oxides. The accuracy and precision of the results were checked by analysing Corning glass standard A, which is a soda-lime-silica glass. At least three areas were analysed for each sample. The totals in each case were approximately 100 wt% but the results shown have also been normalised for ease of comparison.

The plates contained mainly sodium, calcium and silicon oxides making the glass a soda-lime-silica type typical of nineteenth-century sheet glass [4–7]. Although the composition of the glass plates varied slightly throughout this time

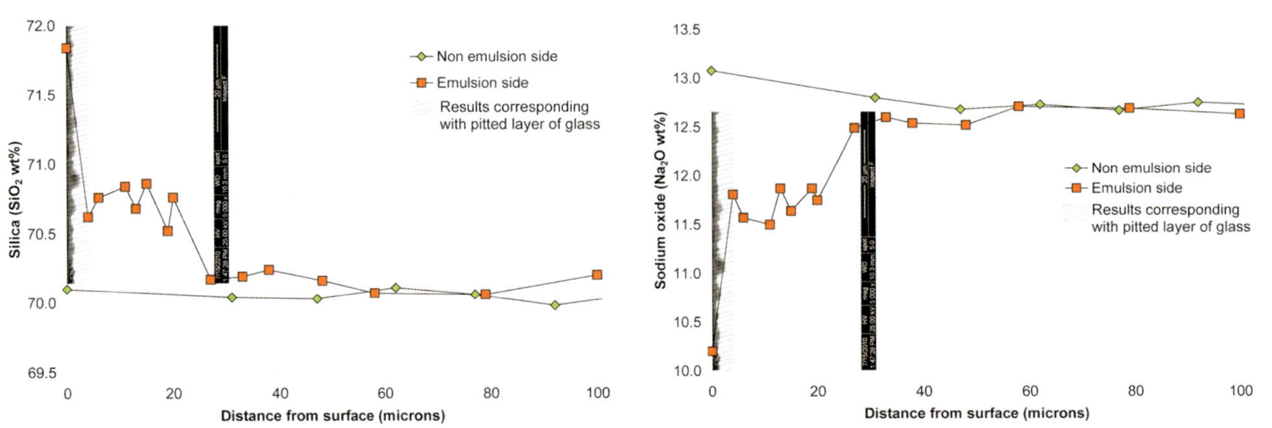

Figure 5 Graphs showing the increased concentrations of silica (left) and reduced concentrations of sodium (right) in the pitted surface of a deteriorating glass plate compared to the rest of the glass. The underlying backscattered SEM images are of the deteriorating glass plate and correspond spatially to the results on the graphs.

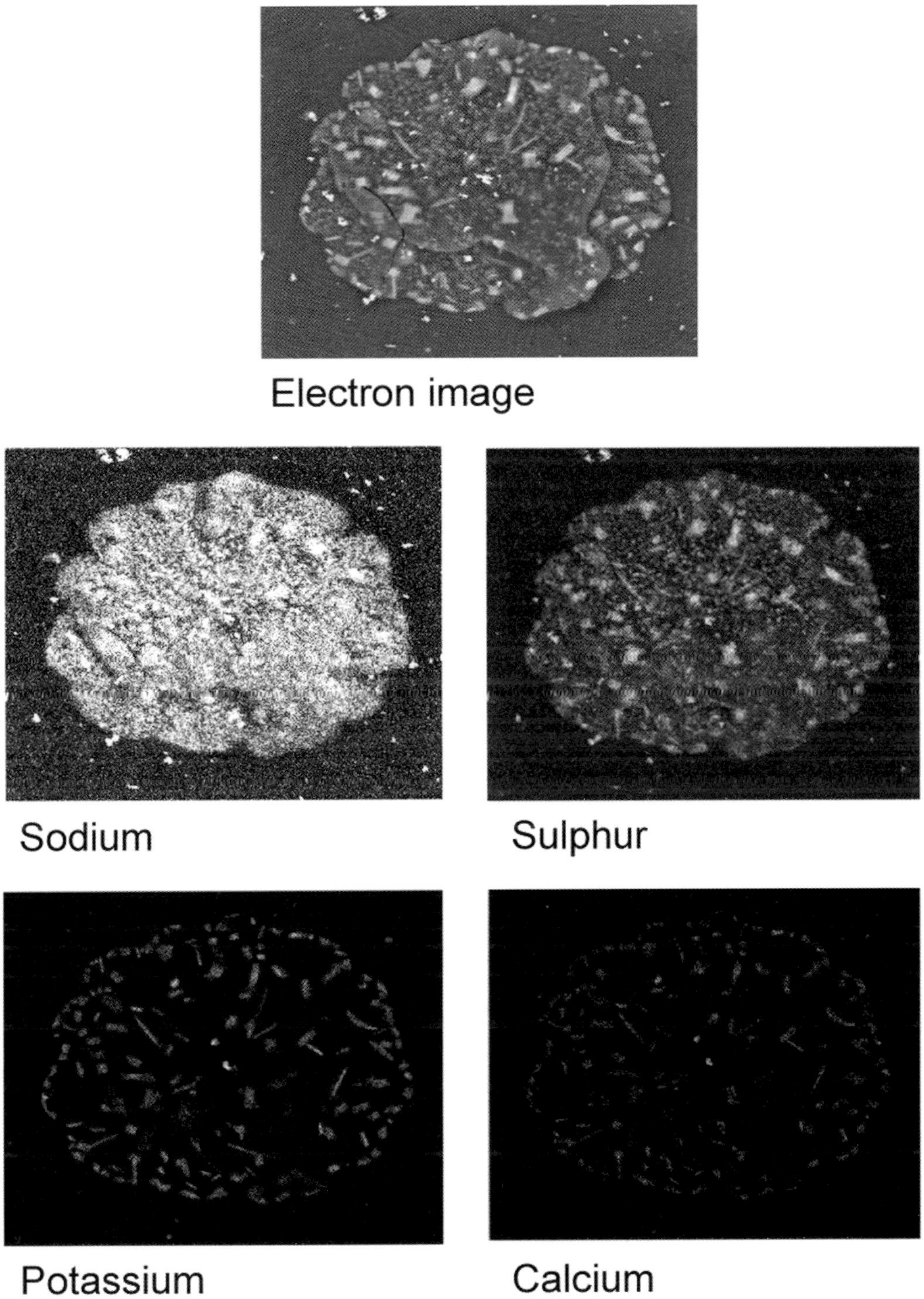

Figure 6 An SEM image (top) of a blister, made up of numerous small crystals, on the surface of the emulsion. The element maps (below) show the increased concentrations of sodium, sulphur and to a lesser extent potassium and calcium in the blister area.

period, the delamination of the emulsion was not restricted to any one particular glass type, as can be seen in Table 1.

The composition of all the analysed glass negatives suggests that they should be stable under normal conditions, and there is little in the glass composition that would explain the corrosion, as has been the case in other studies [3].

Line scans

Cross-sections were also analysed using EDX. To gain a wider range of results, samples could not be limited to previously broken negatives. Therefore small glass samples, again between 1 and 2 mm^2, were removed from the non-image border at the edge of the negatives. Gelatine dry plates have a distinctive blank border caused by the rebate of the camera slide. This allowed very small samples to be unobtrusively removed using a diamond glass cutter with no damage to the main body or image area of the plates. These samples were mounted in resin and polished to expose the glass surface.

X-ray line scans were obtained from the centre of the plate, through to the glass/emulsion boundary and into the resin layer. The results showed that in delaminated negatives some sodium and calcium had been leached out of the glass surface, corresponding with the micro-porous layer seen during the visual analysis of the cross-sections. As a result, the pitted area contained a higher percentage of silica than the rest of the glass. In contrast, the composition of the glass in negatives that were in good condition remained constant right through the plate, even into the glass/emulsion boundary. This is demonstrated by Figure 5.

Identification of crystals

A small quantity of crystalline material was removed from the surface of negatives and analysed using X-ray diffraction (XRD) in order to identify the crystalline phases present. The analyses were performed using a Bruker D8 Advance diffractometer, with a Lynxeye detector, at 40 kV and 40 mA. The crystalline salts were concentrated between the glass and emulsion layer but sometimes erupted through the emulsion surface as small round blisters with a high density of crystals. These crystals may contribute to the translucent appearance of the glass.

Figure 6 shows an SEM image of one of these blisters (top), which is made up of many small indistinct crystals plus occasional larger angular ones. The accompanying four images in Figure 6 are element concentration maps that show the increased amounts of sodium, sulphur and to a lesser extent potassium and calcium at the site of the blister. The results indicate that the majority of the crystals are sodium sulphate whilst the larger angular ones are potassium calcium sulphate.

Conclusions

The Bedford Lemere & Co. collection at the EH Archive contains a significant number of negatives that are affected by a particular form of delamination. Analyses show that the glass surface that was originally in contact with the photographic emulsion layer in delaminated negatives is pitted and chemically altered. This alteration does not occur in negatives in good condition.

The results indicate that some components of the glass, in particular sodium and potassium, are reacting with sulphur-rich compounds at the interface between the emulsion and the glass. Alkali sulphate salts form and the process leaves the glass surface physically and chemically changed, weakening the bond between the emulsion and the glass plate. Thus this silica-rich micro-porous layer may be a contributory factor in the resulting delamination of the emulsion layer.

It may also be the cause of the iridescent appearance visible in all the delaminated negatives examined for this research. However, because this iridescence is also visible in some negatives that are not delaminated, this suggests that other factors are significant in causing negatives to eventually peel, for example fluctuating environmental conditions.

The bulk composition of the glass appears not to be a significant factor in the case of the delaminating negatives in the Bedford Lemere & Co. collection. The plates are a typical nineteenth-century sheet glass soda-lime-silica type and although the composition of the glass changed very slightly over time, the deterioration is not restricted to any particular glass type, suggesting that this is not the cause of the problem.

Other possible causes have been suggested for future investigation. One theory is that occasional poor processing, particularly insufficient washing, results in the retention of residual sulphur-rich processing chemicals by the gelatine layer, eventually degrading the glass. Inconsistencies in the manufacturing process have also been mooted as possible causes.

What is clear is that the causes of the delamination of photographic emulsion in the case of the Bedford Lemere & Co. collection are complex and that further research is needed to provide a definitive explanation.

Acknowledgements

We would like to thank the following individuals for their assistance and contributions to this research: Susie Clark, Angela Karsten, Ian Leith, Mark McCormick Goodhart, Karla Graham, Ian and Angela Moor, Tony Rumsey, David Thickett and English Heritage Archive Photographic Services.

Note

1. This was identified using Fourier transform infrared spectroscopy (FTIR). The method and results are not included in the paper as the varnish layer was ruled out as a contributory factor as it was present on all negatives regardless of their condition.

References

1. McCabe, C., 'Preservation of 19th-century negatives in the National Archives', *Journal of the American Institute for Conservation* **30** (1991) 41–73.
2. Clark, S., 'The conservation of wet collodion positives', *Studies in Conservation* **43**(4) (1998) 231–241.
3. McCormick-Goodhart, M.H., 'An analysis of image deterioration in wet-plate negatives from the Mathew Brady Studios', *Journal of Imaging Science and Technology* **36**(3) (1992) 297–305.
4. Muspratt, S., *Chemistry: Theoretical, Practical and Analytical*, Mackenzie, Glasgow (1860).
5. Powell, H.J., Chance, H. and Harris, H.G., *The Principles of Glass-Making*, George Bell, London (1883).
6. Dungworth, D., *Welch Road, Southsea, Portsmouth: An Investigation of Some Window Glass*, Research Department Report 17/2010, English Heritage, Portsmouth (2010).
7. Dungworth, D. and Wilkes, R., *Wentworth Conservatory, Wentworth Castle, Stainborough, South Yorkshire: Chemical Analysis of the Flat Glass*, Research Department Report 18/2010, English Heritage, Portsmouth (2010).

Authors' addresses

- Sarah Allen, English Heritage Archives Conservation Laboratory, Swindon, UK (sarah.allen@english-heritage.org.uk)
- Jenny Hodgson, English Heritage Archives Conservation Laboratory, Swindon, UK
- David Dungworth, Fort Cumberland, Portsmouth, UK
- Sarah Paynter, Fort Cumberland, Portsmouth, UK

Understanding Viking filigree and granulation with the aid of scanning electron microscopy

Barbara Regine Armbruster

ABSTRACT This paper deals with manufacturing processes of Viking period goldwork, more precisely with the use of scanning electron microscopy (SEM) in the understanding of technological aspects of joining and decorating techniques. Technological investigations of Viking jewellery have profited substantially from SEM applications since the mid-1980s, when this approach was introduced to early medieval archaeology through the study of ornaments from Sweden. The case studies presented focus on Viking gold jewellery from northern Germany, which is richly decorated with filigree and granulation. One objective is to clarify the wire production in Viking goldsmith workshops, in particular the identification and documentation of tool marks and surface textures by SEM. It also aims to present the advances made by SEM with energy dispersive X-ray (EDX) microanalysis and SEM imaging in the interpretation of the use of metallic hard solder as a joining technique in the production of Viking precious metal jewellery.

KEYWORDS scanning electron microscopy, gold technology, Viking period, filigree, granulation

Introduction to Viking jewellery studies

Gold and silver ornaments from the Viking period (from the eighth to the beginning of the eleventh century AD) are well known for their technical and aesthetic qualities, as well as for the designs typical of this period in Scandinavian history. Interlaced figural and floral patterns are realised in filigree and granulation, the most prominent techniques employed in Viking jewellery manufacture [1]. These purely ornamental techniques are characterised by small decorative elements fixed to a sheet base by soldering. During the Viking period, filigree and granulation were used mainly to embellish hollow, three-dimensional pieces of jewellery. The preliminary, undecorated workpiece was constructed of two thin sheets, one raised into relief using a die, and the other, a plain sheet soldered to the back. This type of hollow ornament gives a remarkable three-dimensional optical effect with little precious metal having been used, and it is therefore quite light in weight.

Early investigations of granulation and filigree technologies focused mainly on ancient jewellery. Chemical analyses in the 1980s revealed that copper salt solder was used for Etruscan goldwork in particular [2–4]. Filigree studies showed that both copper salt hard solder, as well as metallic hard solder, appear to have been applied in different geographical areas and periods [5–6]. Since the 1960s, Nordic jewellery studies have included archaeometallurgical aspects, such as technology and chemical material analyses [7]. Duczko (Archaeological Research Laboratory, Stockholm) was the first to use scanning electron microscopy (SEM) for the examination of filigree on Viking precious metal objects [5]. His publication on jewellery from Birka, an important Viking period settlement site in Sweden during the eighth and ninth centuries AD [5], was a breakthrough in the understanding of filigree and granulation techniques. Almost all items discussed in that publication, which were mostly silver filigree ornaments [5], were studied with SEM, providing analytical data as well as images. Recently, SEM has been used in the technological study of gold filigree objects from the Norwegian Hoen hoard [8]. This ensemble from the ninth century AD is the most important gold treasure found in Viking Scandinavia, and SEM imaging and non-destructive, quantitative chemical SEM with energy dispersive X-ray (EDX) analyses were carried out on selected pieces [8].

The purpose of this paper is to review a number of recent case studies of SEM applications on Viking artefacts. It takes into account the manufacturing processes of the early medieval goldsmith, in particular the wire production and the joining technique of soldering applied in the assemblage of sheet elements of hollow items as well as the decoration techniques of filigree and granulation. The following case studies concern tenth-century Viking goldwork from two different locations in northern Germany that were investigated over the last 10 years with an interdisciplinary approach including SEM.

Viking jewellery of the tenth century AD and its manufacturing techniques

Two types of jewellery with filigree and granulation predominate in Scandinavia during the tenth and the beginning of the eleventh century AD: the 'Terslev' and 'Hiddensee' types. The first is named after the Danish silver treasure from Terslev, Zealand, in the collections of the Danish National Museum, Copenhagen [9]. Disc-shaped brooches and pendants with quadripartite interlaced ornament are characteristic of the Terslev type (Figure 1). The second type is named after a significant tenth-century gold treasure from the north German isle

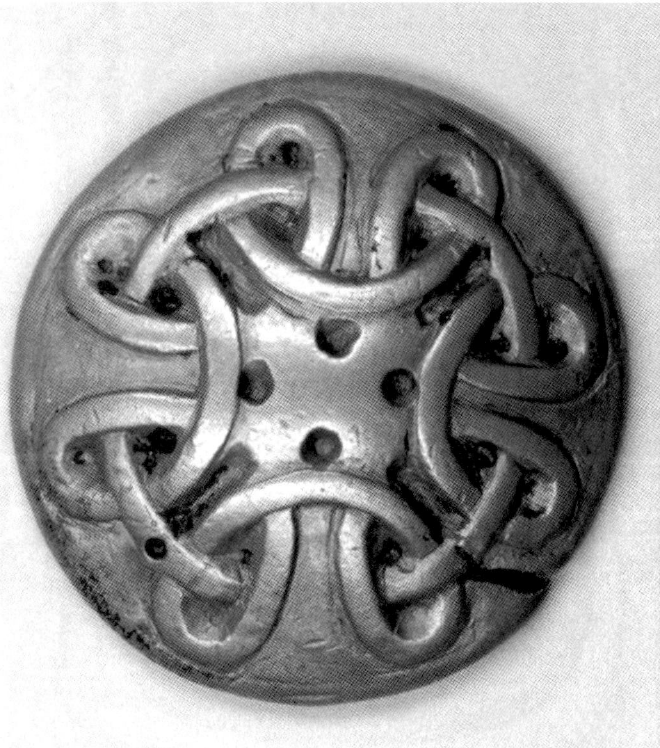

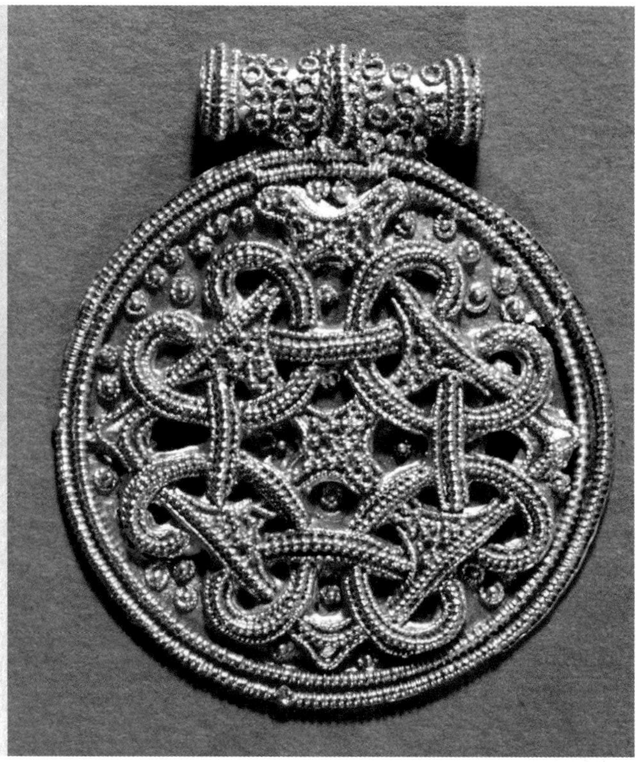

Figure 1 Terslev-type copper alloy die (left, diam. 4.1 mm) and gold pendant (right, 38.2 × 32 mm) from Hedeby. (Image: B. Armbruster.)

of Hiddensee, now in the collections of the Kulturhistorisches Museum, Stralsund [10]. The Hiddensee treasure is composed of 16 gold objects, a neck-ring of four twisted rods, a disc-shaped brooch and 14 pendants with a total weight of 598 g, dating to c. AD 1000. The Hiddensee type is distinguished by disc-shaped brooches and cross- or leaf-shaped pendants designed with a mixture of Nordic animal style and western European interlace ornaments (Figures 2 and 5) [11]. The brooches are characterised by a stylised pattern of three or four interlaced animal bodies, whilst the pendants have suspension loops in the form of animal heads and interlaced tendrils on the pendant bodies.

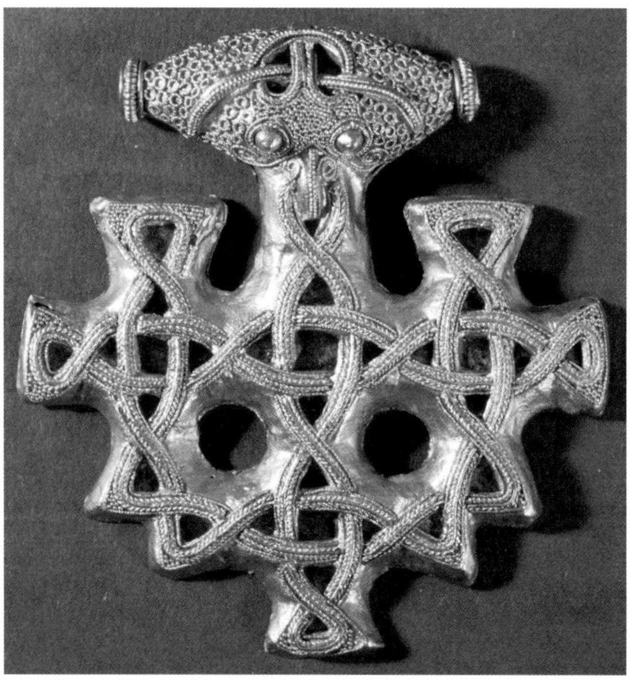

Figure 2 Cross-shaped pendant with suspension loops in the form of an animal head and interlaced tendrils, Hiddensee treasure (64.4 × 64.6 mm). (Image: B. Armbruster.)

Both jewellery styles – Terslev and Hiddensee – are characterised by composite objects comprising numerous individually made elements such as sheet, wires and granules, joined by soldering. Their manufacture is based mainly on plastic shaping with soldering as the joining technique. Gold sheet and rods were worked manually by hammering a cast ingot. Thin wires were made using a draw-plate and tongs. Decorative patterns were formed with wires and granules, applied to thin relief sheets. The pressed relief sheet was formed by working sheet metal into the concave depressions of a die, dies being solid tools cast in copper alloys with the motif in negative relief on the surface.

A large number of dies are known from the Viking world, attesting to the serial production of gold and silver filigree jewellery [10, 11]. Directly related to both jewellery styles discussed is an important toolset of a tenth-century goldsmith comprising 41 copper alloy dies and an iron draw-plate, which was found during excavations at the harbour of the Viking settlement of Hedeby, north Germany [12, 13]. Large numbers of circular and cruciform dies from the Hedeby tool hoard are morphologically very similar to the disc-shaped brooch and to cruciform pendants from the Hiddensee treasure. One die also corresponds to the Terslev style (Figure 1). In addition to the rich toolset, more dies and other remains of fine metalworking, such as crucibles, casting moulds, hammers, anvils and tongs, were found on this important Viking settlement site. At Hedeby, considerable numbers of gold and silver jewellery items have been discovered in tombs, in the settlement area and also as stray finds. The gold objects from Hedeby, as well as the jewellery from the Hiddensee treasure, were examined by stereo optical microscopy (OM) and SEM. X-ray fluorescence (XRF) and SEM-EDX analyses were carried out on both jewellery assemblages.

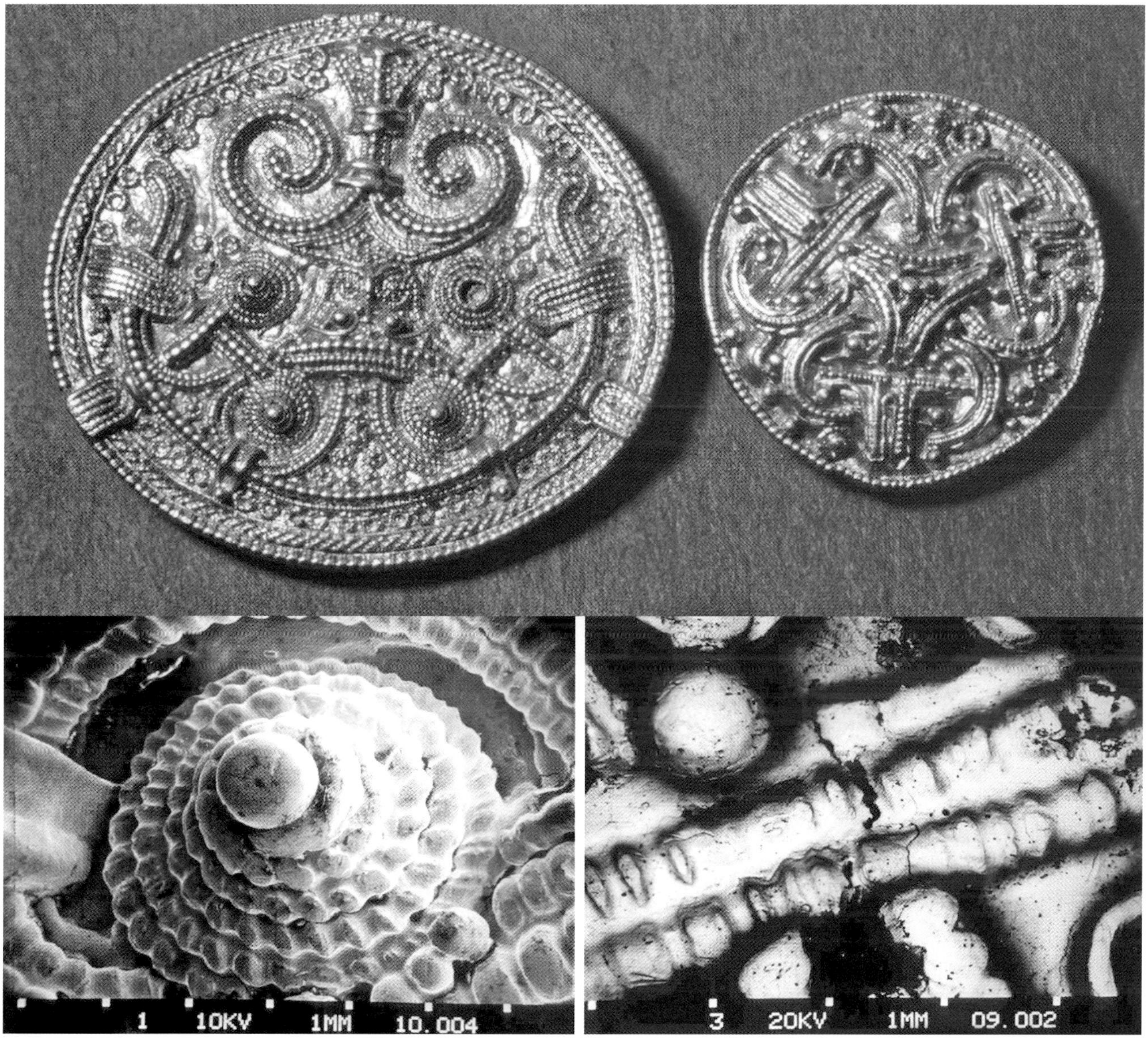

Figure 3 Two filigree brooches from Hedeby (left, 39.3 × 35.9 mm; right, diam. 25.8 mm) (above) and corresponding SEM images of filigree wires: beaded wire cone (below left), abundant solder material and crack (below right) (Max Planck Institute, Heidelberg). (Image: B. Armbruster.)

SEM examination of Viking goldwork from northern Germany

Like OM, SEM allows a close and precise look at tool marks, surface structures and details of the manufacturing processes. But due to the depth of focus, SEM is able to give particularly accurate images of excellent quality. SEM examination reveals an astonishing variety of beaded, twisted or obliquely incised filigree wires used by Viking goldsmiths (Figures 3–6). Higher magnification images show uneven granulation with unsorted granules of different sizes, as well as (in many cases) a large amount of solder material flooding the granulation. The highly detailed images also provide information about the surface structure of the metallic solder in contrast to the smooth surfaces of the base sheet. Damage, such as cracks, missing elements or repairs, is also revealed by SEM (Figure 3). Furthermore, SEM facilitates the study of wear marks caused by the use of the gold ornaments. Finally, the optical examination of surface structures shows the topography of sheet reliefs created by dies.

SEM-EDX microanalyses of 50 gold items and the imaging of a selection of gold objects from Hedeby were undertaken by Pernicka and colleagues (Max-Planck-Institut, Heidelberg) [14, 15]. The settlement finds show a high variability in alloy compositions. Apart from a pair of Terslev-type gold pendants (Figure 1) of identical design, manufacture and

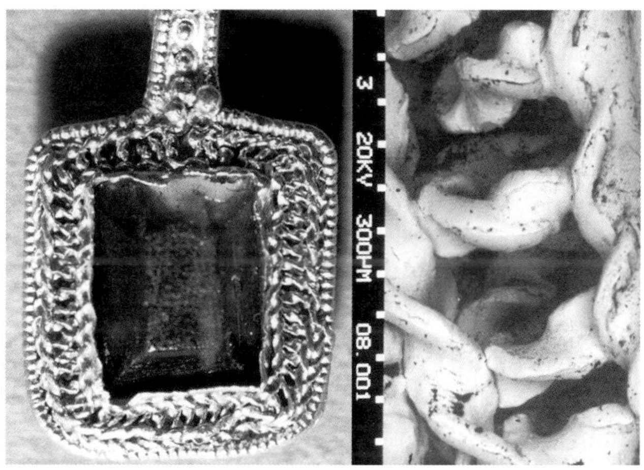

Figure 4 Pendant with glass setting from Hedeby (24.3 × 15.1 mm) (left), and SEM image of the helix of a twisted ribbon. (Image: B. Armbruster.)

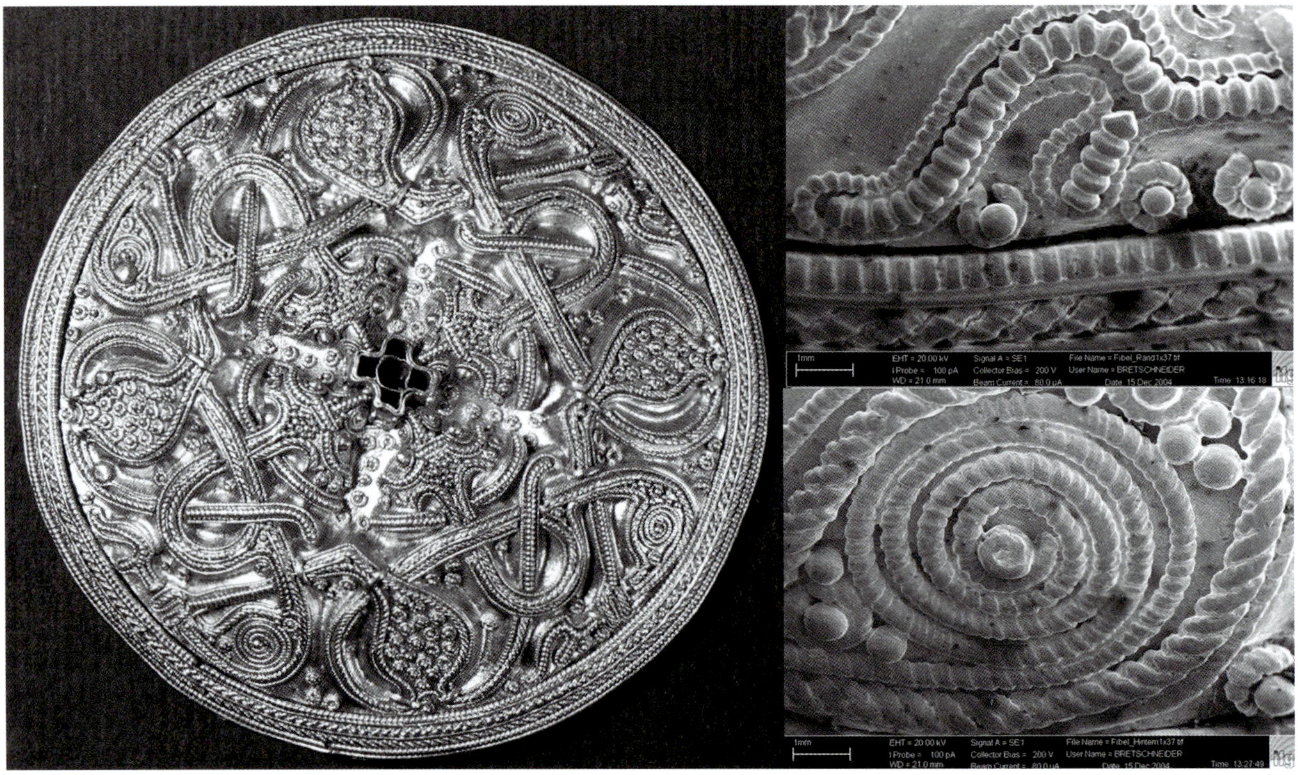

Figure 5 Brooch (diam. 80 mm), Hiddensee treasure with animal-style ornament and cross-shaped setting in the centre (left), and SEM image details of filigree and granulation (Institute for Scientific Instruments, Berlin, Hiddensee-project Armbruster/Eilbracht.) (Image: B. Armbruster.)

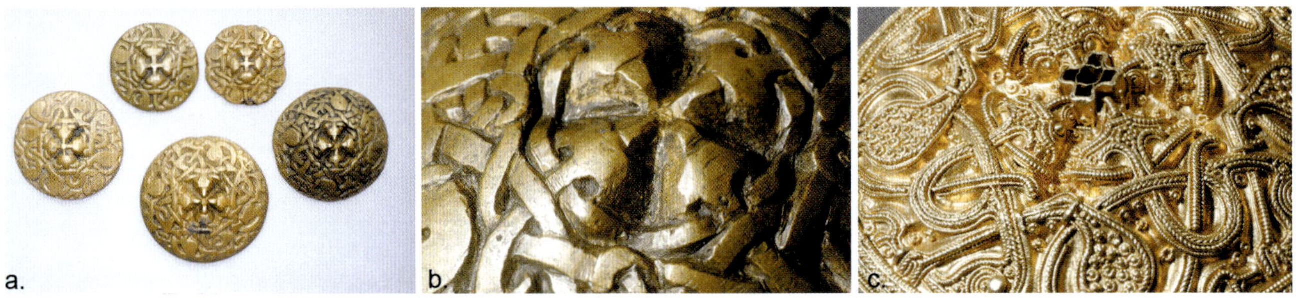

Figure 6 (a) Five bronze dies from the Hedeby toolset, the largest in the centre (diam. 68.5 mm); (b) detail of the centre of the largest die from Hedeby with four animal heads and space for a setting; (c) detail of the centre of the Hiddensee brooch with cross-shaped glass setting between four animal heads. (Images: B. Armbruster.)

composition, nearly every item from Hedeby is made of a different gold alloy. This heterogeneity in the chemical compositions reflects the assorted find circumstances, such as funerary contexts, settlement area and stray finds. In contrast, the alloy composition of the 16 gold items from the Hiddensee treasure is very homogeneous [10]. The results mirror the concept of a jewellery set that was conceived and manufactured as a uniform group.

One particular example, which shows the advantages of material analyses for the interpretation of archaeological artefacts, concerns the so-called 'amethyst pendant' from Hedeby [16]. Due to the lack of any archaeometrical approach in the 1978 detailed study, incorrect conclusions were drawn about the filigree decoration and the identification of a so-called 'amethyst'. Recent SEM material analyses, undertaken for 50 gold items from Hedeby, were able to show that it is not amethyst in the gold setting but coloured glass. Moreover SEM imaging made it possible to interpret and understand the filigree decoration consisting of a long helix of one twisted gold ribbon (Figure 4) [14]. The previous study had misinterpreted the filigree pattern surrounding the glass setting supposing it to have been assembled from a large number of small wire elements.

The disc-shaped brooch from the Hiddensee treasure (114 g; diam. 80 mm) is a particularly complex piece of filigree and granulation on pressed sheet (Figures 5 and 8). A cross-shaped glass setting made of a doubled strip is situated in the centre, embedded between four animal heads. The largest bronze die from the Hedeby harbour toolset is too small in diameter (68.5 mm) to have served for the manufacture of this brooch, but it corresponds to the relief outline of the interlaced animal-style ornament, and also has a space between the animal heads for the insertion of a cross-shaped glass setting (Figure 6). Many varieties of filigree wire types – round-sectioned, twisted wire rope, beaded wire, obliquely beaded wires, or wires of other profiles – were chosen for the decoration, and were often juxtaposed (Figure 5 right). The interlaced aspect of the animal-style patterns, as well as the intertwined ribbons, is an optical illusion achieved by using bent segments of different types of decorative wires for the motif layout. A spiral of obliquely beaded wire is used on this brooch to depict the hip of a mystical beast. Collared granules on small wire rings are applied in addition to the filigree patterns of the animal shoulder, and granulation clusters mark the top of the animal heads (Figure 7). A large band, composed of multiple wires of different sizes and types, outlines the contour of the brooch.

During the Viking period, the fabrication of thin round-sectioned wire differed from the previous periods, as draw-plates and tongs were used rather than the earlier traditions of hammered and strip-twisted wires. Draw-plates with a series of conical holes have been found in several Viking tool assemblages, thus attesting to that method [17].

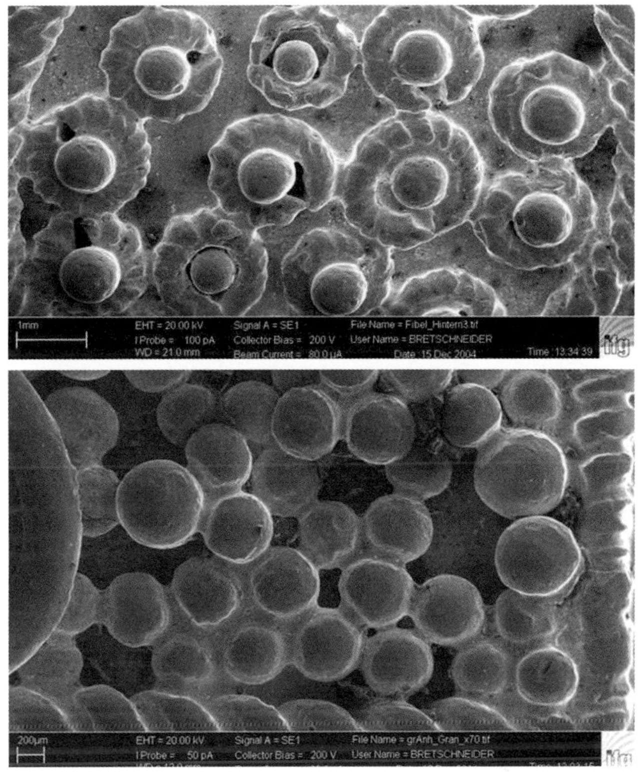

Figure 7 SEM image details of filigree and granulation: collared granules from the brooch (above), granules and abundant metallic solder material from a cross-shaped pendant from Hiddensee (below) (Institute for Scientific Instruments, Berlin, Hiddensee-project Armbruster/Eilbracht). (Image: B. Armbruster.)

Other approaches to the study of goldworking

Technological investigations on Viking goldwork have combined several approaches from material sciences, humanities and crafts. Apart from SEM applications, radiography and non-destructive and invasive analyses have been used. Archaeometric investigations assisted in the identification of an iron draw-plate with conical perforations, found in the set of goldsmithing tools at the harbour of Hedeby [18]. Radiography showed a dense material in the conical holes, and XRF analyses identified this dense material as flakes of silver, thus corroborating the function of that iron object as a draw-plate for the manufacture of silver wire [19].

Our understanding of ancient wire production can benefit from medieval images. Representations of artisans, such as in the *Hausbuch der Mendelschen Zwölfbrüderstiftung* dating to the sixteenth century, give an impression of fine-metal workers, with details of tool equipment, procedures and techniques [20]. Wire drawing by means of a draw-plate fixed to a wooden block is shown in combination with the use of a swing on which the goldsmith sits to improve his drawing force [10, 17]. Ethnoarchaeological fieldwork allows comparative study using actual traditional gold and silver working processes as analogies for the understanding of early medieval wire drawing. For instance, today in western Africa filigree wire is still produced using handmade iron draw-plates in combination with simple tools and equipment. Ancient literary sources give explanations of materials, tools, chemical recipes and technical processes. For example, Presbyter, writing about the manufacture of beaded wire in the twelfth century AD, described the use of a *lima inferius fossa* (beading file) and an *organarium*, the latter used to create beaded wires of thicker section [21]. Foltz (former coordinator of workshops, Römisch-Germanisches Zentralmuseum, Mainz) reconstructed both a beading file and an *organarium* according to Presbyter's descriptions [22] and his practical experiments contributed to the understanding of the manufacture of beaded wire. Experimental wire modelling with triangular, quadrangular, V-shaped or curved sections and grooved blades with one or several teeth, and with a beading file has been carried out by several scholars to give an idea of the different kinds of tools used and the tool marks produced in the manufacture of decorative wires [23, 24]. Duczko used SEM imaging of experimental pieces to illustrate the manufacture of beaded wire; a particular aspect of this beaded wire is the medial seam resulting from the use of a beading file (Figure 8) [5]. The only archaeological evidence of an actual beading file was found at Illerup-Ådal, Denmark, a third-century AD bog site with Germanic sacrifices [25].

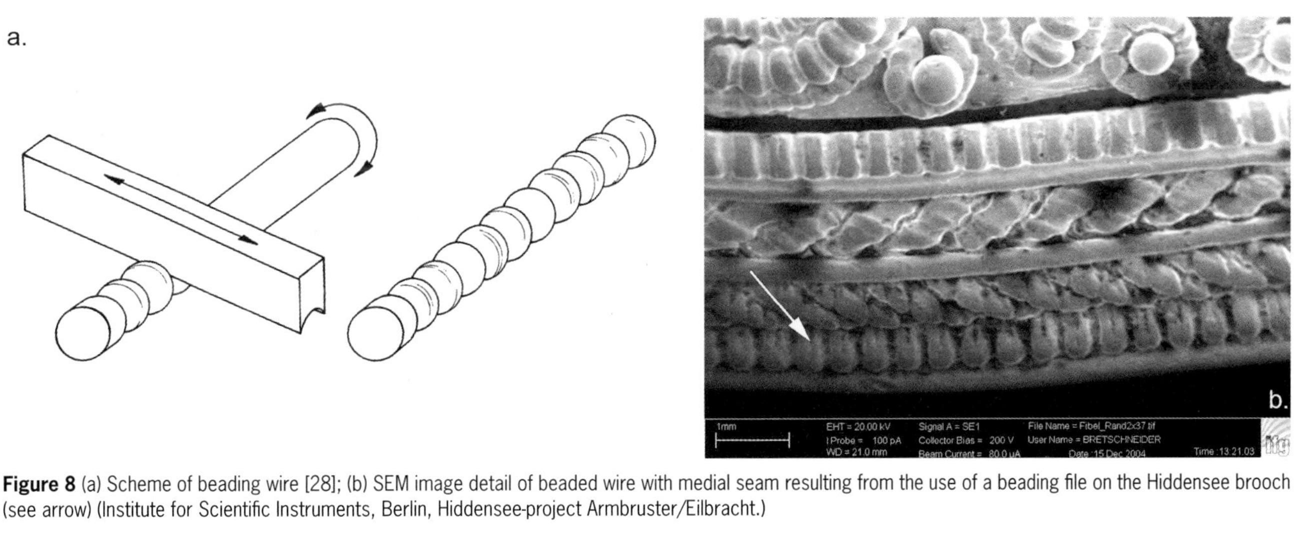

Figure 8 (a) Scheme of beading wire [28]; (b) SEM image detail of beaded wire with medial seam resulting from the use of a beading file on the Hiddensee brooch (see arrow) (Institute for Scientific Instruments, Berlin, Hiddensee-project Armbruster/Eilbracht.)

Case study: the Hiddensee project

Within the scope of an interdisciplinary research project, the gold ornaments from the Hiddensee treasure were studied and documented by material scientists in Berlin. Analytical data on the chemical composition of the alloy were obtained by XRF analysis with synchrotron radiation (BESSY synchrotron centre); SEM images were provided by the Institute for Scientific Instruments, Berlin [26, 10]. The analyses provided broad information about the metal composition; approximately 10 point analyses were also made on each of the items. Particularly interesting results were obtained on the use of metallic solder applied in the filigree and granulation work. The gold alloy solder contains more copper and silver than the base metal and consequently has a lower melting point than the base metal, which is the essential characteristic of a solder. A large amount of solder material is also clearly visible between granules in granulation clusters and between filigree wires (Figure 6). These results match the SEM microanalyses of several filigree and granulation ornaments from the Hedeby material, where the presence of metallic solder is evident [14]. It seems most likely that the solder alloy was prepared by filing or grinding to make small pieces, which were then spread on the artefact as a powder before being melted in the furnace or with a blowpipe, presumably with a flux to protect from oxidation and to facilitate the flow of the solder. Formerly it was believed that Viking goldsmiths used copper salt hard solder in a similar way to Etruscan gold granulation, instead of metallic solder [27].

Conclusions

The present study aims to emphasise the importance of SEM in the study of Viking filigree and granulation work. The benefit of SEM as part of the technological investigation of precious metal production is twofold. First, SEM images provide information about surface structures, with an excellent view at a microscopic level of morphological features such as details on solder, sheet and wire as well as granule production. Secondly, SEM allows an in-depth investigation of tool marks, which can be a guide to the identification of specialised tools, for example, the use of a beading file. The study of wear marks at a microscopic scale also provides a detailed understanding of the purpose of the artefact when in use. Alloy composition analyses supply data on both the metal of the artefact and the solder. Combined with other approaches in an interdisciplinary methodology for the study of precious metal objects, SEM serves as a valuable and sophisticated tool for the appreciation of the operational procedures in ancient metal technology and goldsmithing techniques.

Acknowledgements

I would like to thank colleagues and institutions who were involved in the Hedeby and Hiddensee projects: Claus von Carnap-Bornheim, Kurt Schietzel and Roland Aniol (Archäologisches Landesmuseum Schleswig), Ernst Pernicka and his team (Max-Planck-Institute Heidelberg), Andreas Grüger (Kulturhistorisches Museum Stralsund), Heide Eilbracht (Freie Universität Berlin), Ina Reiche (Centre de Recherche et de Restaurarion des Musées de France, Paris), Martin Radtke (Bundesanstalt für Materialprüfung, Berlin and the Institute for Scientific Instruments, Berlin). I am also grateful to Nigel Meeks and Niamh Whitfield for polishing my English text.

References

1. Eilbracht, H., *Filigran- und Granulationskunst im wikingischen Norden. Untersuchungen zum Transfer frühmittelalterlicher Gold- und Silberschmiedetechniken zwischen dem Kontinent und Nordeuropa*, Zeitschrift für Archäologie des Mittelalters, Beiheft 11, Köln (1999).
2. Parrini, P., Formigli, E. and Mello, E., 'Etruscan granulation: analysis of orientalising jewelry from Marsiliana D'Albegna', *American Journal of Archaeology* **86** (1982) 118–121.
3. Wolters, J., *Die Granulation. Geschichte und Technik einer alten Goldschmiedekunst*, Callwey Verlag, Munich (1986).
4. Nestler, G. and Formigli, E., *Etruskische Granulation. Eine antike Goldschmiedetechnik*, NIE, Siena (1993).
5. Duczko, W., *The Filigree and Granulation Work of the Viking Period: An Analysis of the Material from Björkö*, Birka 5, Kungl. Vitterhets Historie och Antikvitets Akademine, Stockholm (1985).
6. Wolters, J., 'Filigran', in *Reallexikon zur Deutschen Kunstgeschichte*, Band 8, ed. K.-A. Wirth, Munich (1987) 1062–1184.
7. Oldeberg, A., *Metallteknik under vikingatid och medeltid*, Seelig, Stockholm (1966).
8. Astrup, E., 'Techniques, craftsmanship and composition of gold in the Hoen hoard', in *The Hoen Hoard: A Viking Gold Treasure of the Ninth Century*, ed. S.H. Fuglesang and D.M. Wilson, Acta ad archaeologiam et artium historiam pertinentia XIV, Rome (2006) 156 pls 55, 57, 59, 65.
9. Friis Johansen, K., 'Sølvskatten fra Terslev', *Aarbøger for Nordisk Oldkyndighed og Historie* (1912) 189–263.
10. Armbruster, B.R. and Eilbracht, H., *Wikingergold auf Hiddensee*, Archäologie in Mecklenburg-Vorpommern 6, Schwerin (2010).
11. Duczko, W., 'Scandinavian dies of the Viking Age', in *Outils et Ateliers d'Orfèvre des Temps Anciens*, ed. C. Eluère, Antiquités nationales, mémoire 2, Société des amis due Musée des antiquités nationales et du château de Saint-Germain-en-Laye, Saint-Germain-en-Laye (1993) 185–190.
12. Armbruster, B.R., 'Goldsmiths' tools at Hedeby', in *Land, Sea and Home: Proceedings of a Conference on Viking-period Settlement at Cardiff, July 2001*, ed. J. Hines, A. Lane and M. Redknap, Society for Medieval Archaeology Monograph 20, Leeds (2004) 109–123.
13. Kleingärtner, S., *Der Pressmodelfund aus dem Hafen von Haithabu*, Ausgrabungen in Haithabu 12, Neumünster (2007).
14. Armbruster, B.R., 'Goldschmiede in Haithabu – Ein Beitrag zum frühmittelalterlichen Metallhandwerk'; in *Berichte über die Ausgrabungen in Haithabu*, ed. K. Schietzel, Das archäologische Fundmaterial VII(34), Neumünster (2002) 85–205.
15. Pernicka, E., 'Röntgenfluoreszenzanalyse der Goldobjekte von Haithabu', in *Berichte über die Ausgrabungen in Haithabu*, ed. K. Schietzel, Das archäologische Fundmaterial VII(34), Neumünster (2002) 199–200.
16. Arrhenius, B., 'Ein Amethystanhänger aus Haithabu', in *Berichte über die Ausgrabungen in Haithabu*, ed. K. Schietzel, Das archäologische Fundmaterial III(12), Neumünster (1978) 9–23.
17. Armbruster, B.R., 'Craft remains from Viking Age precious metal working', in *Trade and Communication Networks of the First Millennium AD in the Northern Part of Central Europe: Central Places, Beach Markets, Landing Places and Trading Centres*, ed. H. Jöns, S. Kleingärtner, B. Ludowici, J. Scheschkewitz and M. Hardt, Neue Studien zur Sachsenforschung Band 1, Hannover (2010) 200–201 figs 11–14.
18. Armbruster, B.R., 'Notes on wire production during the Viking Age', in *Proceedings of the 33rd International Symposium on Archaeometry, 22–26 April 2002, Amsterdam*, ed. H. Kars and E. Burke, Geoarcheological and Bioarchaeological Studies 3, Amsterdam (2005) 289–292.
19. Aniol, R., 'Ein Zieheisen aus Haithabu', Anhang 2, in *Berichte über*

die Ausgrabungen in Haithabu, ed. K. Schietzel, Das archäologische Fundmaterial VII(34), Neumünster (2002) 201–205.

20. Treue, W., Goldmann, K., Kellermann, R., Klem, F., Schneider, K., Stromer, W.V., Wißner, A. and Zirnbauer, H., *Das Hausbuch der Mendelschen Zwölfbrüderstiftung zu Nürnberg. Deutsche Handwerksbilder des 15. und 16. Jahrhunderts*, Munich (1965).
21. Hawthorne, J.G. and Smith, C.S., *On Diverse Arts: The Treatise of Theophilus*, University of Chicago Press, Chicago (1963).
22. Foltz, E., 'Einige Beobachtungen zu antiken Gold- und Silberschmiedetechniken', *Archäologisches Korrespondenzblatt* **9** (1979) 213–222.
23. Whitfield, N., 'The manufacture of ancient beaded wire: experiments and observations', *Jewellery Studies* **8** (1998) 57–86.
24. Tamla, Ü. and Varkki, H., 'Learning the technologies of making beaded wire', *Estonian Journal of Archaeology* **13**(1) (2009) 36–52.
25. Carnap-Bornheim, C.v., 'The social position of the Germanic goldsmith A.D. 0–500', in *Roman Gold and the Development of the Early Germanic Kingdoms: Aspects of Technical, Socio-Political, Socio-Economic, Artistic and Intellectual Development, A.D. 1500. Symposium held in Stockholm 14–16 November 1997, KVHAA Konferenser 5*, ed. B. Magnus, Stockholm (2001) 264 fig. 1.
26. Armbruster, B.R., Eilbracht, H., Reiche, I., Grüger, A., Radtke, M. and Riesemeier, H., 'The Vikings in Berlin', in *BESSY Highlights*, Berliner Elektronenspeicherring-Gesellschaft für Synchrotronstrahlung, ed. G. Andrén, H. Henneken and M. Sauerborn, Berlin (2005) 32–33.
27. Lønborg, B., *Vikingetidens metalbearbejdning*, Fynske Studier 17, Odense Bys Museers, Odense (1998).
28. Williams, D. and Ogden, J., *Greek Gold: Jewellery of the Classical World*, British Museum, London (1994).

Author's address

Barbara Regine Armbruster, TRACES - UMR 5608 du CNRS, Université de Toulouse II Le Mirail, Toulouse, France (barbara.armbruster@univ-tlse2.fr)

A pilot application of scanning electron microscopy and high-resolution X-radiography for the conservation of paintings

Aviva Burnstock, Alexander D. Ball, Lauren E. Howard and Genevieve Silvester

ABSTRACT This paper introduces the application of a combination of scanning electron microscope (SEM) imaging and *in-situ*, high-resolution X-radiography using the Gatan X-ray ultra microscope (XuM) for the examination of paint samples from a late nineteenth-century painting on canvas before and after treatment with reagents proposed for the removal of semi-opaque surface coatings and dirt. The comparison of high resolution images of the paint surface before and after treatment taken using a conventional high resolution SEM, combined with indication of cracking in the bulk paint layers shown by XuM imaging, provided data that were useful for selecting an appropriate reagent for treatment of the painting. Images generated using XuM could be interpreted in a similar way to conventional X-radiography of paintings, with enhanced resolution of fine features in the paint that contained lead white in an oil medium. It also provided some information about the proportion and relationship of pigment to organic material in the paint. Interpretation of the effects of the reagents on the paint required both the XuM data and SEM surface images. The limitations of the method and interpretation of the images in relation to sample preparation is evaluated.

KEYWORDS paintings, cleaning, X-radiography, SEM, ultra microscope, XuM

Introduction

Scanning electron microscopy (SEM) is a well-established tool for examination of the surface and composition of paintings and painting materials. The application of SEM, coupled with light microscopy (LM) in studies of the cleaning of paintings has provided data about the alteration of surfaces during conservation treatment that has informed conservation practice [1]. The development of variable pressure SEM (VP-SEM), that allows the study of the same sample before and after treatment, has further advanced the quality of data available to conservators [2].

One of the limiting factors of the application of SEM imaging to investigate options for cleaning painted surfaces is that surface imaging methods do not provide detailed topographical information about the paint surface prior to treatment. This is a problem that is familiar to conservators who undertake empirical tests to remove darkened and semi-opaque coatings from the surface of paintings. It is often difficult to see through these obscuring layers to the paint layers below, either by eye or using LM – the tools most used to evaluate possible approaches to cleaning. Using SEM imaging alone, in secondary electron (SE) imaging mode, it is not possible to examine the paint surface beneath the coatings and dirt prior to treatment. While it is possible to image beneath the surface using backscattered electron (BSE) imaging, it is challenging to differentiate the surface from subsurface features accurately enough to detect small changes using this method.

Macroscopic methods for non-invasive technical study, such as infrared reflectography (IRR), technical photography and conventional X-radiography, have proved useful for characterising the condition of paintings. However the resolution of the data derived from these techniques is not sufficient to provide information about small-scale changes to the paint layer that might occur during the removal of surface coatings.

Thus a technique is needed to characterise the condition of the paint beneath the opaque coating, prior to the removal of the coating. This would facilitate a comparison between the paint before treatment with areas treated using different reagents used to remove or partially remove the surface coatings.

Conventional X-radiography of paintings is a non-invasive technique that has been applied to study the condition and techniques used for paintings for almost 100 years. X-radiation penetrates the thickness of paintings and may provide information about the paint, preparation and support (in this case, canvas). The radiographic image is created by the contrast between X-ray absorbent inorganic materials, such as compounds of lead present in the paint layer(s) as pigments, and organic and inorganic compounds of lower X-ray absorption. The X-radiographic image is captured digitally or on film. Many research institutes and museums have accrued significant collections of X-radiographs of paintings (as well as sculptures and other historical objects), and most conservators and art historians are familiar with this standard technique and the interpretation of radiographs.

Submicron resolution micro-CT systems (sometimes described as nano-CT systems) have only recently become

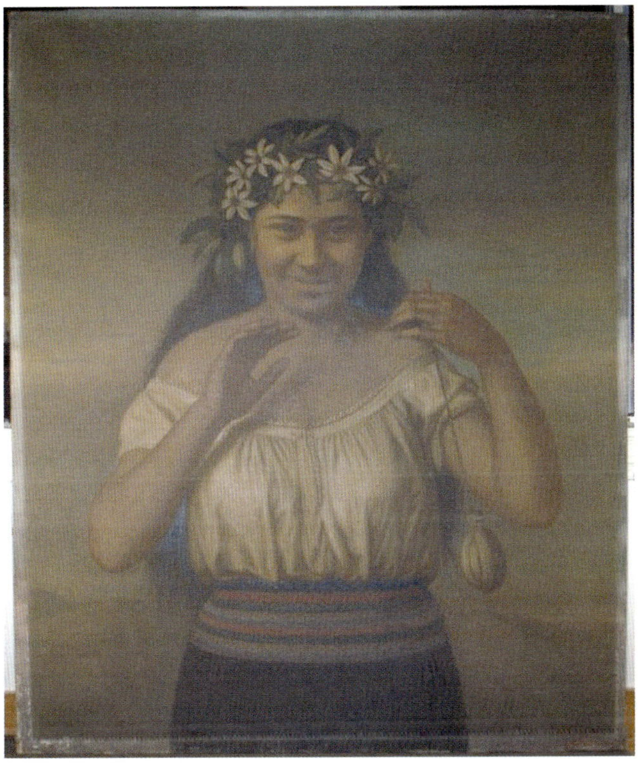

Figure 1 *Maori Girl with Poi* by Godfrey Lindauer, signed and dated 1886, Royal Collection, photograph taken before treatment.

available as an instrument readily accessible to researchers in the scientific laboratory. Comparable results could previously only be achieved using a synchrotron source that requires significant funding and advance booking for limited periods of use. Typically two types of laboratory system are available; cabinet-based instruments, where the X-ray source, sample manipulator and detector are housed in a lead-shielded cabinet, and SEM-hosted systems where the SEM electron beam is directed onto a metallic target and generates the X-ray beam. Since the X-rays generated are relatively low energy, they are entirely contained by the SEM's own shielding. The advantage of SEM-hosted systems is that the manipulator system and target are typically removable or retractable, leaving the SEM capable of functioning normally when the tomography system is not required. The Gatan X-ray ultra microscope (XuM) at the Natural History Museum (NHM) used in this study has previously been used to examine samples as diverse as microscopic diatoms, marine invertebrates, microfossils, polymers, mineral grains and micrometeorites. This technique potentially provides the opportunity to examine the paint at a comparable resolution to a synchrotron-hosted facility and could reveal sub-micron scale features such as cracking in the paint and facilitate the detection of changes in the paint layer after treatment.

The present study investigated the potential application of a combination of the Gatan XuM coupled with high resolution topographical SEM-SE imaging for assessing the effects of selected cleaning reagents on paint from *Maori Girl with Poi* by Godfrey Lindauer dated 1886 (Figure 1).[1] The surface of the painting was covered with a semi-opaque combination of dirt and aged varnish, not untypical of paintings made in the nineteenth century, and examination of the work prior to treatment suggested that the painting had never been previously conserved. To make the image more visible the current conservator sought to remove or reduce the surface material. In keeping with standard conservation practice a series of empirical tests was carried out that varied in the reagent and method used for reduction or removal of the surface coatings. The difficulty in judging the effects of the reagents on the paint, whether through chemical reaction or physically as a result of the cleaning methodology, became apparent during the tests. In particular, the question of whether the fine surface cracking, that was evident on the original paint surface after treatment that removed the coating most effectively, was characteristic of the aged original paint or induced by the cleaning treatment. Other questions of interest to the conservator related to the extent of the removal of the coating in the test areas, and to the physical relationship between the coating and dirt and to the paint surface that was not clearly visible in cross-sections viewed in the light microscope.

Experimental

Instrumentation

X-radiography of the whole painting was carried out at the Courtauld Institute of Art, London, using Fuji 50 X-ray film and wet processing. The exposure was made at 20 kV and 4.3 mA for 30 seconds. The radiograph was made up of a four plate composite.

Samples and sample preparation

Samples of paint and the underlying ground were taken from adjacent sites from the sky to the left of the girl's head. One sample from the sky was prepared as a paint cross-section that allowed characterisation of the layers of ground, paint, coatings and dirt in the sample before treatment. The paint cross-section was prepared using a standard method by mounting in a clear casting acrylic resin with styrene (Cold Setting Clear Casting Resin fromTiranti.co.uk). The sample was examined and photographed using a Leica Aristomet light microscope with a Kodak DCX 620X SLR digital camera at the Courtauld Institute of Art. Elemental analysis of the cross-section was carried out using a JEOL S100 SEM coupled with standard light element INCA energy dispersive X-ray spectrometer (EDX) in the Department of Physics at King's College, London. The sample was carbon coated before analysis.

Separate samples from the same passage of sky were removed before and after treatment for imaging using sub-micron X-radiography and SEM. Areas in the sky, adjacent to the area where the untreated samples were removed, were treated (*in situ*) using two reagents: X and Y. Both reagents removed the surface coatings evenly but incompletely and samples were subsequently taken for comparison with the untreated sample of sky paint. Samples were approximately 2 mm in length and less than 1 mm wide. The samples were mounted, using acrylic adhesive, onto the end of aluminium pins (Figure 2) in an orientation such that the longest dimension of the sample was vertical. Samples were then coated with 20 nm of gold-palladium. XuM and SEM analyses were carried out in the Electron Microscopy and Mineral Analysis (EMMA) laboratories at the NHM London.

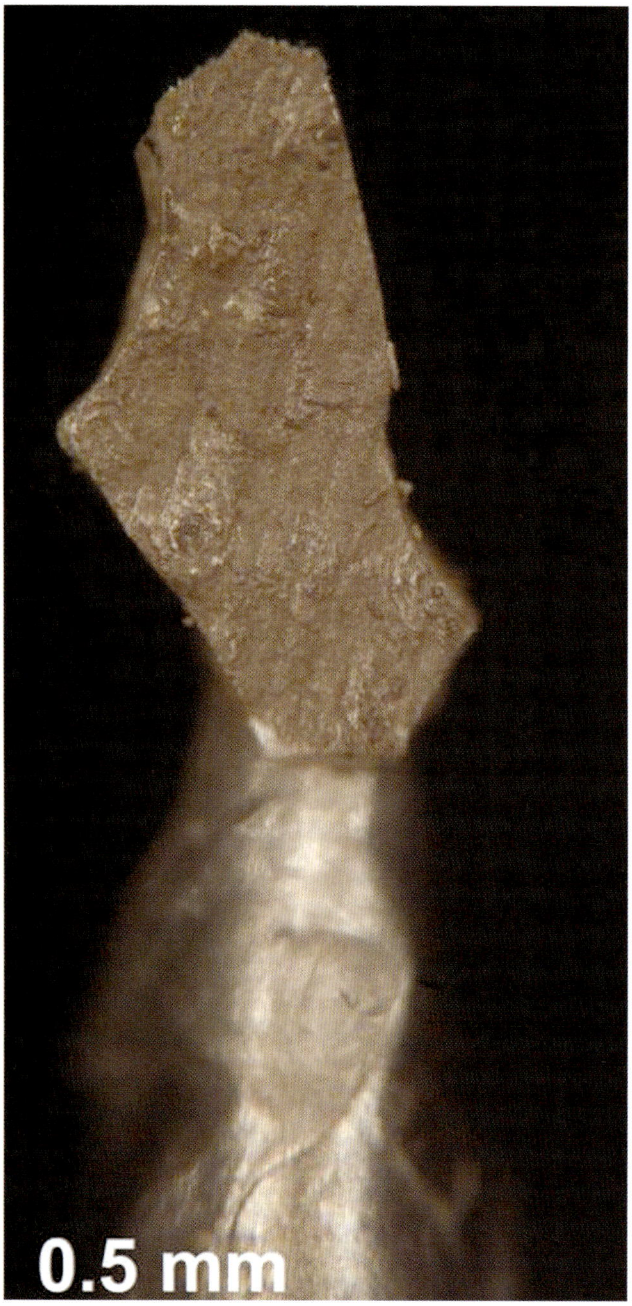

Figure 2 Sample of paint from the sky before treatment, mounted for XμM imaging on an aluminium pin.

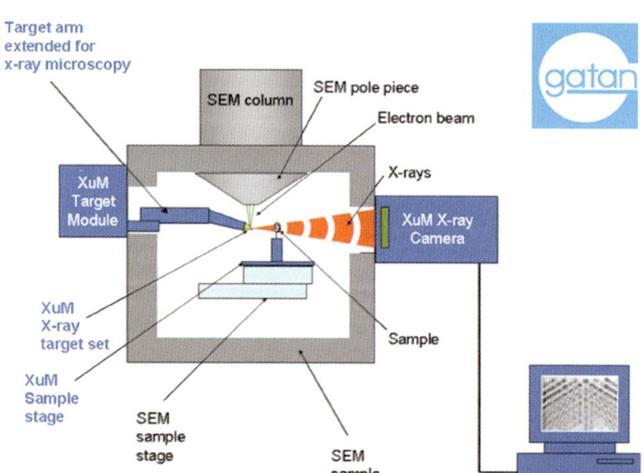

Figure 3 Schematic diagram for the XμM-SEM set-up.

High resolution X-radiography was carried out using a Gatan XuM fitted to a Philips XL30 field emission gun SEM. A diagram of the set-up of the X-ray computed tomography accessory inside the SEM is given in Figure 3. The electron beam of the SEM (set at 30 kV acceleration voltage, largest spot size (200 μm aperture)) produces a cone of X-rays by focusing the beam onto a metal target (gold in this instance). The pin-mounted sample was inserted into a precision rotation stage, which is positioned within the cone of X-rays generated by the impact of the electron beam into the gold target. The resulting X-ray absorption image is projected onto a charge-coupled device (CCD) behind the specimen. This CCD output consists of a 2D X-ray image where differences in greyscales reflect changes in density of the specimen. The system is capable of generating a full 3D computed tomography dataset, by collecting 2D images over 360 degrees at 1° increments of rotation. However, only 2D data were acquired for this study. The XuM is capable of achieving 250 nm

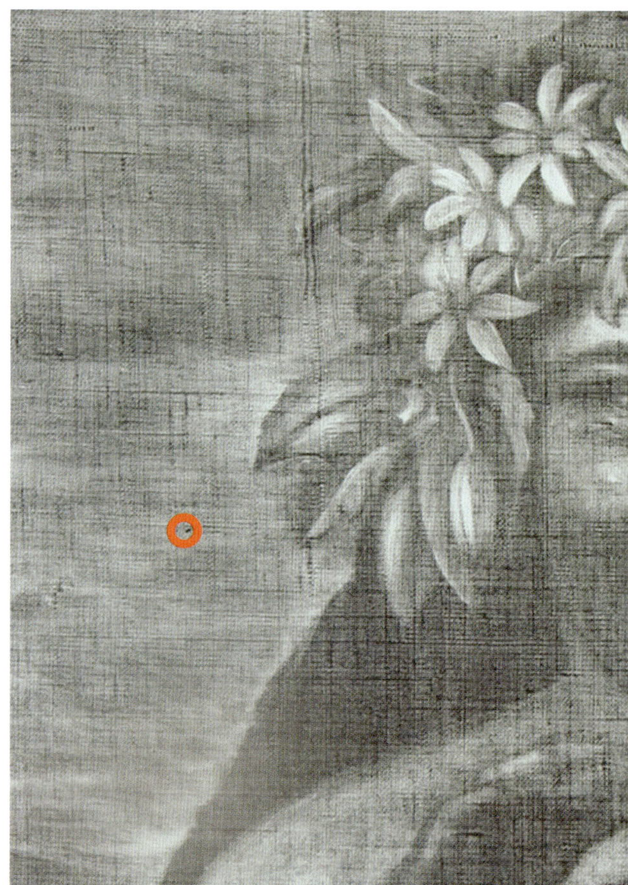

Figure 4 A detail of a conventional X-radiograph of the *Maori Girl with Poi*, showing the area of sky examined in this study. Detail shows the passage of sky paint examined using XμM-SEM, and the red spot indicates sample site.

Figure 5 A paint cross-section from the sky before treatment, showing the layered structure of ground, paint and superficial coating and dirt layers.

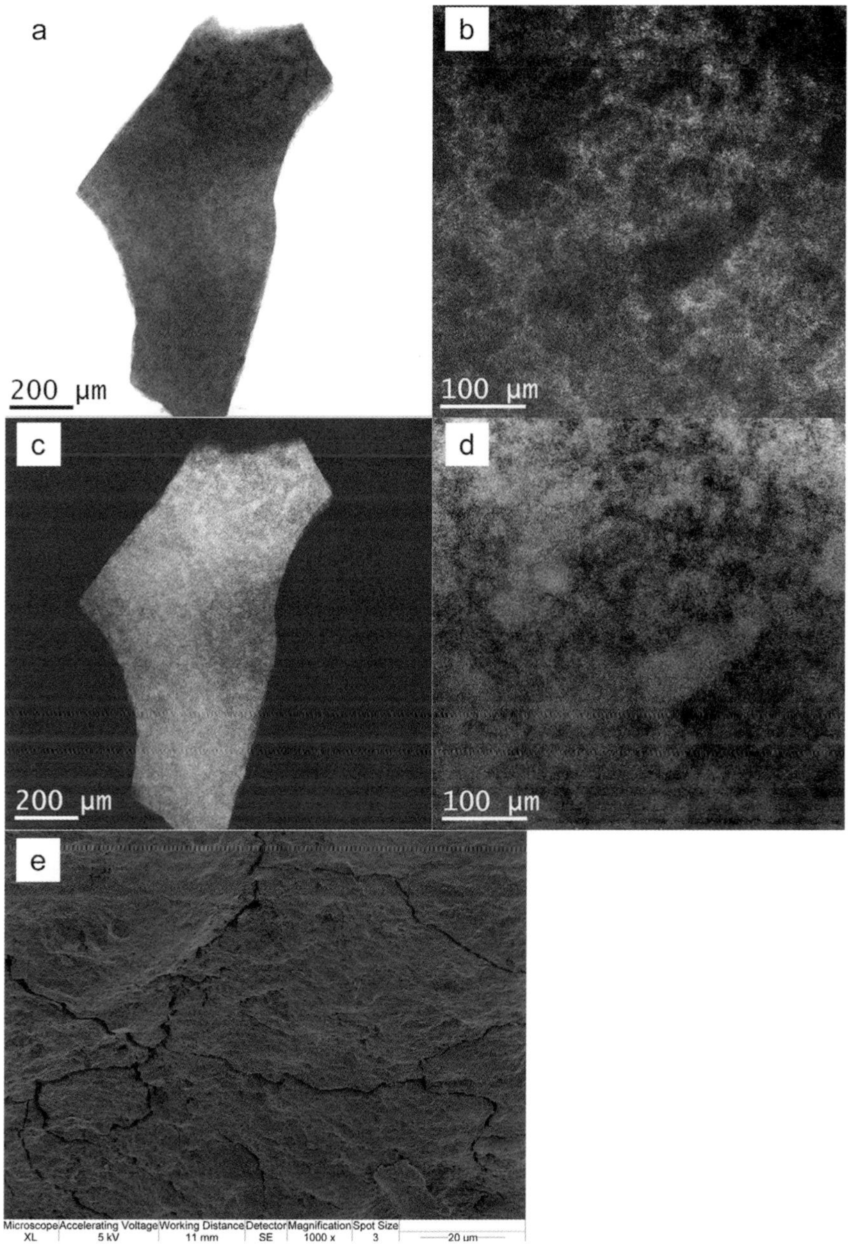

Figure 6 A sample taken from the sky paint before treatment: (a) and (b) XµM images taken at two magnifications in X-ray negative mode and (c) and (d) the same image in X-ray positive mode. (e) SEM-SE micrograph of the surface of the untreated sample from the sky.

resolution for 2D X-radiographs. The energy of the X-rays produced is approximately 10 KeV, which provides data comparable to that of conventional macro-scale X-radiographs (~20 KeV).

The same samples were subsequently imaged in the XL30 FEG SEM using the instrument's SE detector, 5 kV acceleration voltage, spot size 3 and 30 µm aperture. This allows a direct comparison between the 'depth' radiograph image and the 'surface' SEM image.

Results

The traditional macro-scale radiograph of *Maori Girl with Poi* (Figure 4, showing a detail of the sky to the left of the girl's head) shows general features of the painting technique, for example the use of X-ray absorbing lead carbonate white paint for the sky and the flesh that contrasts with passages of paint containing a lower proportion of lead white and other pigments, with relatively low absorption of X-rays, that were used for the girl's hair. The radiograph in this case provides little detailed information evidence about the condition of the paint beneath the semi-opaque coating and dirt, however the radiograph registers a small paint loss in the sky as a dark spot (in an area where the lead white X-ray absorbing paint layer was missing). This was the site from which samples were taken for this study, from the edge of the existing loss.

The structure of the paint and ground layers was confirmed by LM examination of the paint cross-section, which was prepared from a sample taken from the edge of this loss in the sky before treatment (Figure 5). Elemental analysis of the sample, using EDX, provided evidence of a calcium-rich layer of ground, indicative of calcium carbonate that can be seen in the lower layer of the cross-section, and includes a few particles of bone black [3]. On top of the ground is a more substantial layer of sky paint that comprises a few particles of synthetic

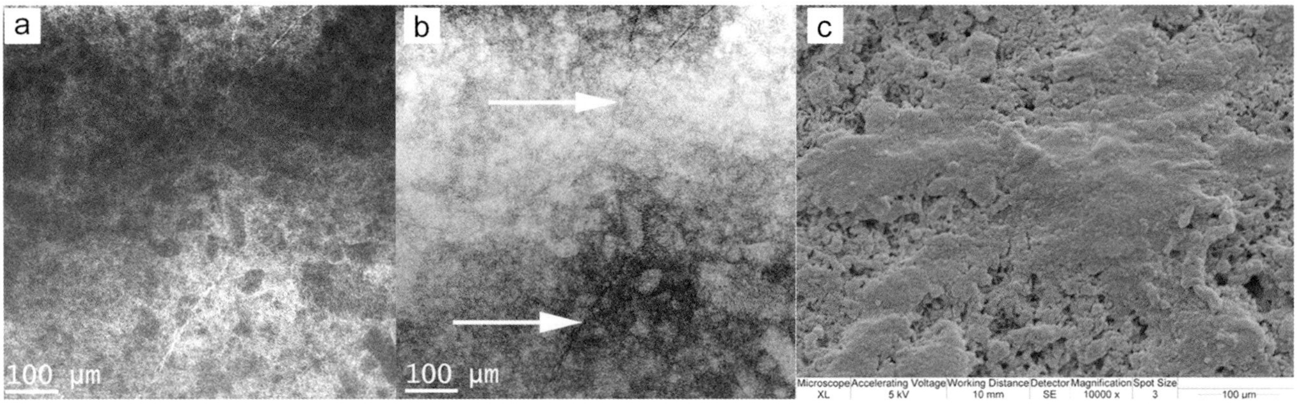

Figure 7 Sample from the sky after treatment using reagent X: (a) XµM X-ray negative image; (b) XµM X-ray positive image (the arrows point to cracks in the paint); (c) SEM-SE micrograph of a sample from the sky.

ultramarine blue pigment and red and yellow earth pigments in a matrix of lead carbonate white.

Staining tests and Fourier transform infrared spectroscopy (FTIR) (carried out in a previous study of the painting) suggested that the binding medium for both paint and ground was a drying oil [3]. The organic coating and dirt layer on the paint film is evident in the cross-section over the upper paint layer. The coating comprises a layer of semi-translucent organic material of about 2 µm in thickness, and a layer of approximately 1 µm in thickness comprising inorganic particles and other organic matter characteristic of surface dirt. The dirt layer is partially embedded in the coating below, that is on the paint surface. In this case, it was not possible to ascertain from studying the cross-section information about the nature of the coating or its relationship to the paint beneath, or to characterise the paint surface, in order to predict its quality and appearance after removal of the coating.

Figure 6 shows a sample taken from the sky paint before treatment: the positive mode images are comparable to traditional macro-scale X-radiographic images of the painting. The images show the relationship of pigment (6b and 6c show dark aggregates in the negative images, 6d and 6e show light aggregates in the positive mode) contrasting with the organic medium. Neither the macro-scale radiograph nor the XuM images show the fine cracks evident in SEM-SE micrographs of the surface of the organic coating and dirt in samples from the painting before treatment (Figure 6e), or the cracking in the chalk ground layer. This is because the dominant signal in the radiograph images comes from the lead white pigment, which masks any fine-scale cracks in the other less X-ray absorbent layers.

While SEM-SE images illustrate the fine cracking in the surface coating, it was not possible to reliably characterise the paint surface using more deeply penetrating BSE imaging due to the lack of specific information about the depth of penetration of these images and location of the paint surface. Nor was it possible, using LM, to penetrate the coating and view the surface. After treatment using reagent X, which removed most but not all of the coating and dirt, fine cracks in the paint were visible in the XuM images. The XuM images in Figure 7 (7a X-ray negative and 7b X-ray positive) show fine cracks, indicated by arrows, in the bulk paint film, beneath the residual coating and dirt evident in the SEM-SE image (Figure 7c) that disguise the fine cracks in the bulk paint.

A sample taken from a neighbouring area where the coating and dirt was removed by an alternative reagent Y examined using XuM (Figure 8a and 8b) showed no fine cracking but two wider cracks in the top edge of the sample. The angle of these cracks suggests that they are stress cracks induced by sample-taking after treatment, and are not induced by the cleaning reagent. The SE image of this sample (Figure 8e) showed an even thin layer of coating remaining on the surface of the paint after treatment with reagent Y.

Discussion

In all the samples, the absence or presence of cracking in the original paint surface was impossible to identify through the initial superficial coatings or the residual layer of organic material on the paint surface by eye, LM or SEM-SE imaging alone.

The use of XuM to characterise subsurface changes on a micrometre scale was useful in the present study for characterising the bulk paint, including cracking in the paint layers beneath layers of organic surface coating and dirt. The interpretation of the XuM images necessitated characterisation of the paint layer structure and composition using other techniques. The localisation of cracking in the lead white containing paint layers was confirmed by data from samples prepared as cross-sections combined with elemental analysis of the ground layer and superficial layers that comprised materials that provided no contrast either in the macro-scale radiograph or the XuM images. Thus, in this case study, it would appear that the integrity of the paint layers was compromised by fine cracking induced by treatment with reagent X, while images taken after cleaning using reagent Y were very similar to those obtained for the sample examined before treatment, indicating that reagent Y is a more suitable choice for removal of the coatings.

The drawbacks of the technique include potential problems with interpretation of the XuM images. The process of taking a sample could induce cracking that could be misinterpreted as a change in the sample induced by cleaning. Secondly, the problem with taking samples from paintings is that on the micrometre scale on which SEM imaging operates, local topographical differences in samples of closely similar visual character may lead to misinterpretation. This problem is implicit in the interpretation of all SEM images of samples

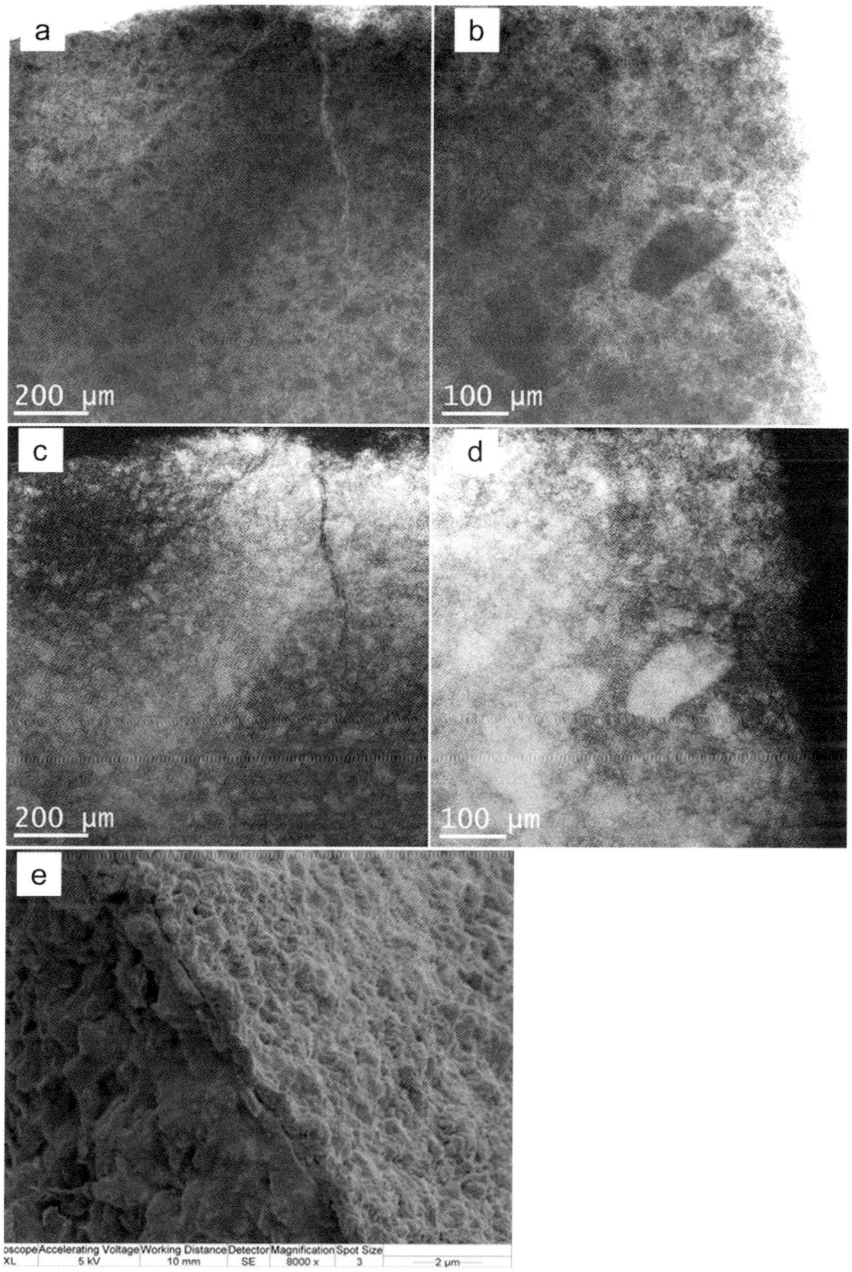

Figure 8 The sky after cleaning with reagent Y: (a) and (b) XuM X-ray negative mode; (c) and (d) XuM X-ray positive mode; (e) SEM-SE micrograph.

from inhomogeneous surfaces and is largely overcome by examination of a larger data set and from increased familiarity with the interpretation of images of paint surfaces. While experience and repetition make inferences more reliable there are significant limitations on the availability of samples from paintings for any purpose, and the predominant trend in methodology is to favour non-invasive analytical approaches to diagnostics for cleaning and technical study.

The characterisation of paint, coatings and underlying layers may be significantly more complex in paintings or painted surfaces that comprise sequences of layers that contain materials that absorb radiation of the same or different range of energies. In these cases, it would be more challenging to interpret changes in the materials that could be in any layer that registers on the XuM image. This could be mitigated by the application of 3D computed tomography of samples that is available but was not carried out in the present study. These data (relatively time-consuming to acquire, necessitating continuous exposure for around 10 hours) might be useful for locating changes in different layers but would be subjected to the same possible sample taking artefacts as the imaging studies presented in 2D in this study.

The results of the present study compare favourably with the resolution that could be obtained using a synchrotron X-ray source, however the equipment used in this study is less expensive and more readily available for conservation diagnostic studies. The application of alternative techniques for characterising the paint surface beneath semi-opaque coatings include white light confocal microscopy that has been successfully applied [4] but is not as yet widely utilised in cleaning studies (largely because imaging is compromised by the opacity and thickness of surface coatings, and resolution in visible light). The use of optical tomography for the study of the structure of paintings (see for example [5, 6]) has so far found limited application.

Conclusions

The combination of high resolution X-radiography and high resolution SEM-SE imaging was useful for the characterisation of micro-scale cracking in samples from a painting that could not readily be visualised through a residual coating of dirt and varnish using any other technique. The interpretation of changes in the surface and underlying paint was informed by the high resolution topographical images that enhanced the information that could be gleaned from examination of the painting surface using the light microscope or visual assessments commonly employed by conservators. Although the procedure for analysis using XuM and SEM imaging was rapid and straightforward (given the availability of the equipment), the interpretation of more complex paint layers might benefit from the acquisition of data in 3D that is available but not undertaken in this study. The drawbacks of the methods include the necessity of taking samples from the paintings. However, both chamber size and a greater understanding of the effects of the SEM chamber environment on paintings and other objects in the future might provide guidelines that facilitate the examination of whole works of art using these methods.

Acknowledgements

We would like to thank William Luckhurst for the use of the SEM-EDX at King's College, London.

Note

1. Gottfried Lindauer, *Maori Girl with Poi* (1886) is from the Royal Collection, and is typically housed in Buckingham Palace, London. The painting was commissioned for the Colonial and Indian Exhibition held in London 1886, and entered the Royal Collection as a gift to the Prince of Wales at this time. There is no further conservation history.

References

1. Burnstock, A. and Kieslich, T., 'A study of the clearance of solvent gels used for varnish removal from paintings', in *Preprints of ICOM-CC 11th Triennial Meeting, Edinburgh, 1–6 September 1996*, vol. 1, ICOM Committee for Conservation (1996) 253–262.
2. Morrison, R., Bagley-Young, A., Burnstock, A., van Keulen, H. and van den Berg, K.J.. 'The effects of triammonium citrate and variable pH for surface cleaning unvarnished oil paintings', *Studies in Conservation* **52**(4) (2007) 255–270.
3. Silvester, G., A technical examination condition and treatment report of *Maori Girl with Poi* d.1886 by Gottfried Lindauer from the Royal Collection (unpublished), Conservation archive of the Courtauld Department of Conservation and Technology, London (2011).
4. Daffara, C., Fontana, R. and Pezzati, L., 'NIR confocal microscopy for painting diagnostics'. *Proceedings of the International Society for Optics and Photonics (SPIE)* **7391**, 73910M (2009), DOI:10.1117/12.827673.
5. Liang, H., Martin-Simpson, S., Podoleanu, A. and Saunders, D., 'Optical coherence tomography: a non-invasive technique applied to painting conservation', *Proceedings of the International Society of Optics and Photonics (SPIE)* **67**(334) (2006)1–3.
6. Targowski, P., Iwanicka, M., Tymińska-Widmer, L., Sylwestrzak, M. and Kwiatkowska, E.A., 'Structural examination of easel paintings with optical coherence tomography', *Accounts of Chemical Research* **43**(6) (2010) 826–836.

Authors' addresses

- Aviva Burnstock, Department of Conservation and Technology, Courtauld Institute of Art, London, UK (aviva.burnstock@courtauld.ac,uk)
- Alexander D. Ball, Electron Microscope Unit, Natural History Museum, London, UK
- Lauren E. Howard, Electron Microscope Unit, Natural History Museum, London, UK
- Genevieve Silvester, Courtauld Institute of Art, London, UK

Organic cores from the Iron Age Snettisham torc hoards: technological insights revealed by scanning electron microscopy

Caroline Cartwright, Nigel Meeks, Duncan Hook, Aude Mongiatti and Jody Joy

ABSTRACT Scanning electron microscope examination of four small twisted-wire bronze torcs from Snettisham (Norfolk) in the British Museum and four examples from Norwich Castle Museum revealed that each had an organic core around which the torc wires had been twisted. Although these cores are now in the form of charred wood, originally they were coppiced wood from alder, hazel, dogwood, field maple, willow and elder. They were sufficiently flexible to provide the appropriate support for the torcs during construction. Relevant to this study is the fact that wood retains most of its qualitative features when charred or converted to charcoal, but temperature-dependent anatomical changes occur. These changes vary from one species to another, but experimental charring under laboratory conditions gives accurate information on temperature-related anatomical alterations. This paper describes a case study using modern coppiced alder wood to compare with the charred alder cores of torcs 251 and C.17. From the state of the anatomical features present, it was possible to establish the temperature peak at which this material became charred, i.e. no higher than 360°C, thus providing interesting new information about the construction of these artefacts.

KEYWORDS charred wood, twisted wire, bronze torcs, cores, Iron Age

Introduction

Many hoards of Iron Age metalwork have been uncovered from Snettisham (Norfolk); these include gold, silver and copper-alloy torcs (neck-rings), coins and ingots. The first discoveries of torcs were made in 1948 during deep ploughing. Excavations followed and three hoards (A, B and C) were found, which included torc fragments and coins [1]; these are now in the Norwich Castle Museum (NCM). Two further hoards (D and E) were uncovered in 1950, including the 'Great Torc'; these are in the British Museum (BM). Isolated finds were made intermittently until 1990, when another large hoard (F) was found through metal detecting. The site was excavated for the BM by Ian Stead, yielding seven more hoards containing torcs [2]. Full details of the discovery and contents of the Snettisham treasure will be published in a BM monograph [3].

The treasure comprised many separate deposits, carefully placed in the ground. Some contained a large number of complete gold, silver and bronze artefacts, whilst others comprised numerous pieces of scrap and partly fused mixed precious metals. In addition to its sheer quantity, the metalwork is rightly famous for the quality of artistry and workmanship of the torcs. Many different kinds of torc are represented, including tubular torcs and twisted-wire torcs with an array of terminal types. Associated coins were used to date the hoards to the late Iron Age, c.125–60 BC although some of the torcs are thought to be approximately 100–200 years older [4].

This paper focuses on the organic cores that were observed in several of the twisted multiple-wire bronze torcs through optical microscopy (OM) and scanning electron microscopy (SEM), and evaluates what technological advantages were provided by the use of such material during manufacture. Central to this evaluation was the identification of the wood taxa and consideration of their physical characteristics for their role as core material. Making comparisons with experimentally fired examples allowed the charring temperatures of the archaeological material to be determined. Consideration was given to when the charring had occurred and whether the metalworking processes had resulted in the charring of all the core wood at the same temperature and to the same degree, or whether the charring reflected localised heating.

Scientific research has been undertaken on the technology and materials of the Iron Age gold, silver and bronze torcs and fragmentary material from Snettisham at the BM and NCM (for example [3, 5]). The study of these torcs by OM and especially using SEM with energy dispersive X-ray (EDX) analysis indicated a high level of craftsmanship and the use of bronze and gold-silver-copper alloys for manufacture of the torcs.

The small twisted-wire bronze torcs appear to be much smaller and simpler versions of the multiple twisted-wire torcs such as the famous 'Great Torc'. For example, torc 242 is approximately 6 mm in diameter and comprises five 2 mm-diameter twisted square-section wires that are all twisted around a (now charred) 2 mm-diameter wood core. These particular bronze torcs are a group of corroded, fragmentary objects that have four or more tightly twisted wires and a 'hollow' core [3].

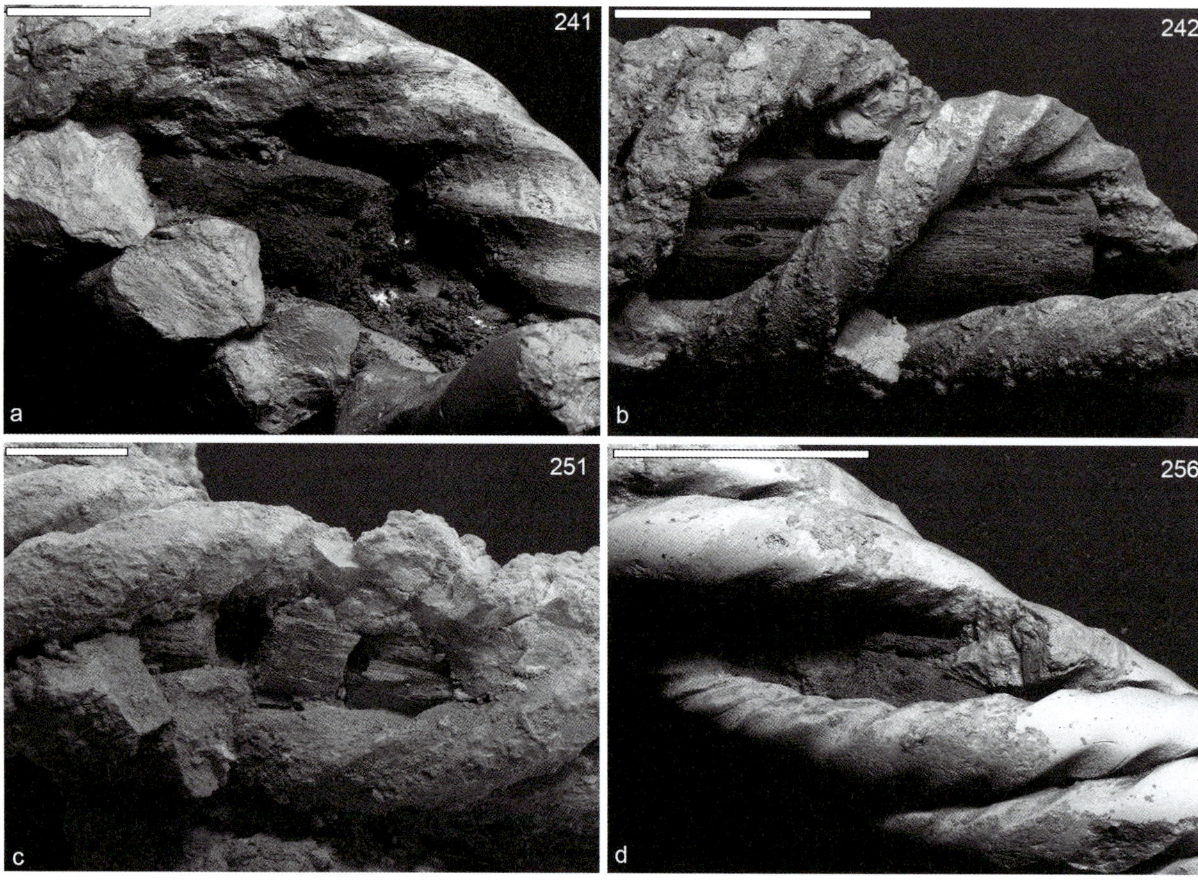

Figure 1 (a–d) SEM backscattered electron (BSE) images (JEOL JSM840) of the charred wood cores revealed at broken and fractured regions of the corroded bronze wires of torcs 241, 242, 251 and 256. (Scale bars: 2 mm, 5 mm, 2 mm and 5 mm respectively.)

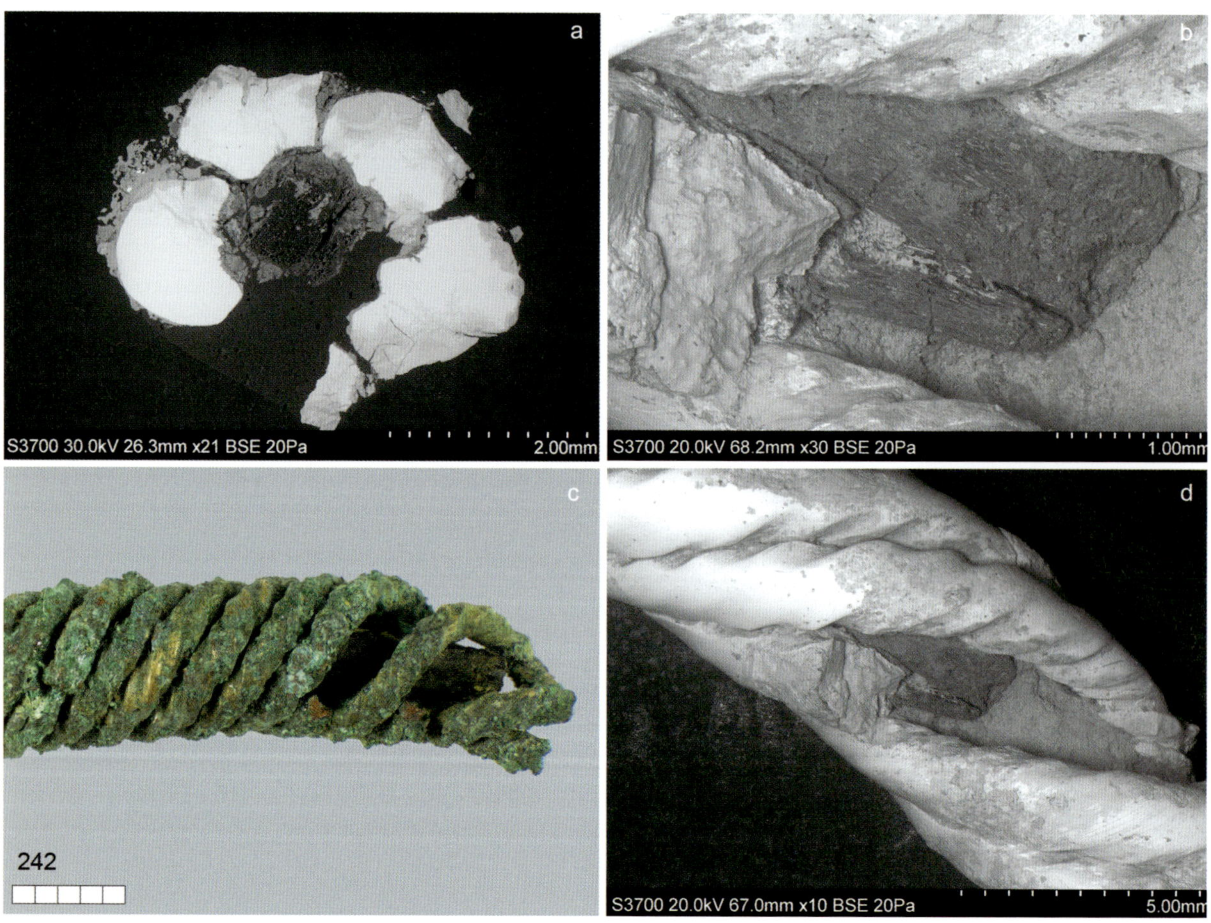

Figure 2 (a) VP-SEM BSE image (Hitachi S-3700N) of polished TS through the five corroded wires and core of torc 242; (b) VP-SEM BSE image of the charred hazel (*Corylus avellana*) core of torc 242; (c) OM image of torc 242 (scale bar: 5 mm); (d) VP-SEM BSE image of the charred hazel (*Corylus avellana*) core of torc 242.

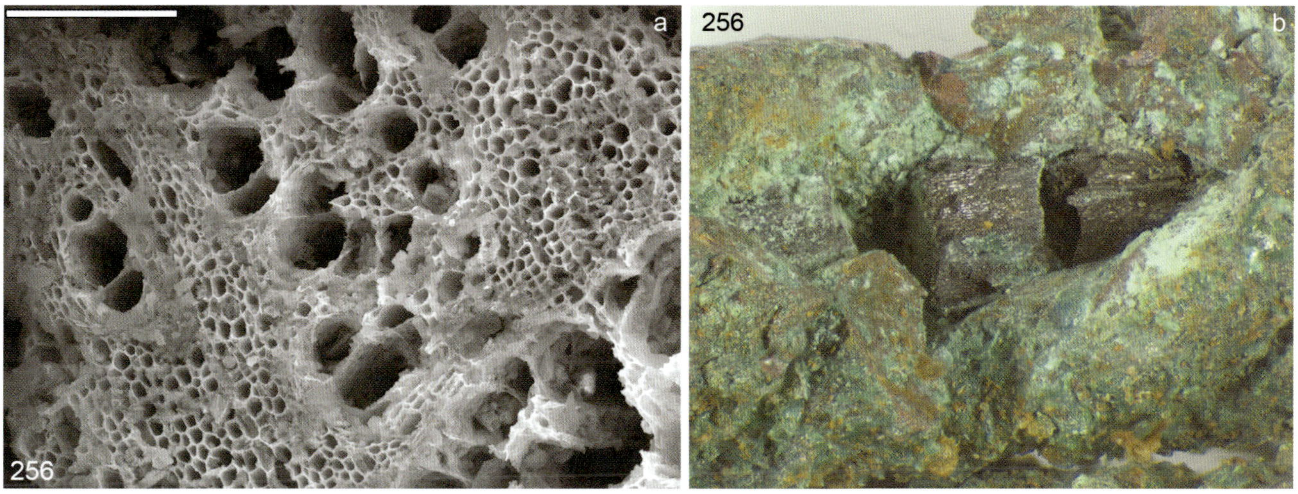

Figure 3 (a) SEM image (JEOL JSM840) of the TS of the charred dogwood (*Cornus sanguinea*) core of torc 256 (scale bar: 100 μm); (b) OM image of the fragmented section of torc 256 revealing the charred core.

Table 1 Identification of woody taxa used in organic cores of some Snettisham twisted-wire bronze torcs.

Accession number	Description	Identification of wood
Snettisham 241 (BM)	Miniature twisted-wire bronze torc or bracelet	*Acer campestre*, field maple
Snettisham 242 (BM)	Twisted- wire bronze miniature torc	*Corylus avellana*, hazel
Snettisham 251 (BM)	Twisted-wire bronze torc	*Alnus glutinosa*, alder
Snettisham 256 (BM)	Heavily worn twisted bronze torc/bracelet wires	*Cornus sanguinea*, dogwood
C.20 a (NCM)	Twisted-wire bronze torc terminal fragment	*Salix* sp., willow
C.20 b (NCM)	Twisted-wire bronze torc terminal fragment	*Sambucus nigra*, elder
C.17 (NCM)	Twisted-wire bronze torc fragment	*Alnus glutinosa*, alder
B.27 (NCM)	Twisted-wire bronze torc fragment	*Corylus avellana*, hazel

Wood identification and experimental charring

For the initial examination, a JEOL JSM840 SEM was used in high vacuum (HV) mode; full details of the methods will be included in the monograph [3]. In this SEM, imaging of the composite objects comprising bronze and charred wood was achieved using a scintillator backscattered electron (BSE) detector (K.E. Centaurus), offset from the electron beam axis to give topographic images with no charging effects. Selected areas were also documented by secondary electron (SE) imaging to show details of cell structures by carefully avoiding any charged areas. Subsequently, a variable pressure (VP) Hitachi S-3700N SEM was used for additional imaging of the twisted-wire bronze torcs and also for the examination of the experimentally charred alder (see below), operating typically at 15 kV and a pressure of 30 Pa.

Standard techniques of wood identification and terminology determined by the International Association of Wood Anatomists (IAWA) are recommended for the characterisation of modern wood [6, 7]. For each sample, the key features are compared with reference collection specimens and published descriptions [8]. This protocol can often be applied to archaeological or historical wood, providing it is modified to accommodate the effects of the conditions of preservation such as charring [9]. Each sample needs to be prepared to expose transverse, radial longitudinal and tangential longitudinal sections/surfaces (TS, RLS and TLS) for identification. For modern and some archaeological or historical woods, thin-sections of approximately 12–14 μm are cut on a microtome, mounted on microscope slides and examined by OM. Charcoal is fractured by hand to expose fresh TS, RLS and TLS for examination using the SEM for fine resolution and high magnification of crucial diagnostic features. Fracturing rather than cutting the charcoal fragment exposes the clearest view of the cellular structure, as cutting with a scalpel or microtome blade creates fine debris that fills the cells. Variations of these standard techniques applicable to charcoal were used for both the charred cores and experimental samples (see below).

The charred wood cores of twisted-wire bronze torcs

The use of organic cores appears to have been an integral part of the manufacturing process during the twisting of some of the Snettisham bronze torcs (Figure 1). The charred wood cores of four small twisted-wire bronze torcs or bracelets from the BM were examined in the SEM (see above), initially where the charred wood cores were exposed at fractured ends or by corroded breaks in the metal (Figures 1–3). Although standard procedures of examining anatomical features were observed, certain restrictions were presented by the charred wood sections of the BM cores available for examination. However, because of good preservation, sufficient diagnostic features could be distinguished in each case and identifications of the woody taxa were achieved (Tables 1 and 2). Two of the charred wood cores (241 and 242) were sampled for radiocarbon dating [4], thus permitting access to TS, RLS and TLS. In addition, tiny samples were permitted from the cores of four twisted-wire bronze torc fragments from NCM, again allowing routine TS, RLS and TLS examination (Table 1).

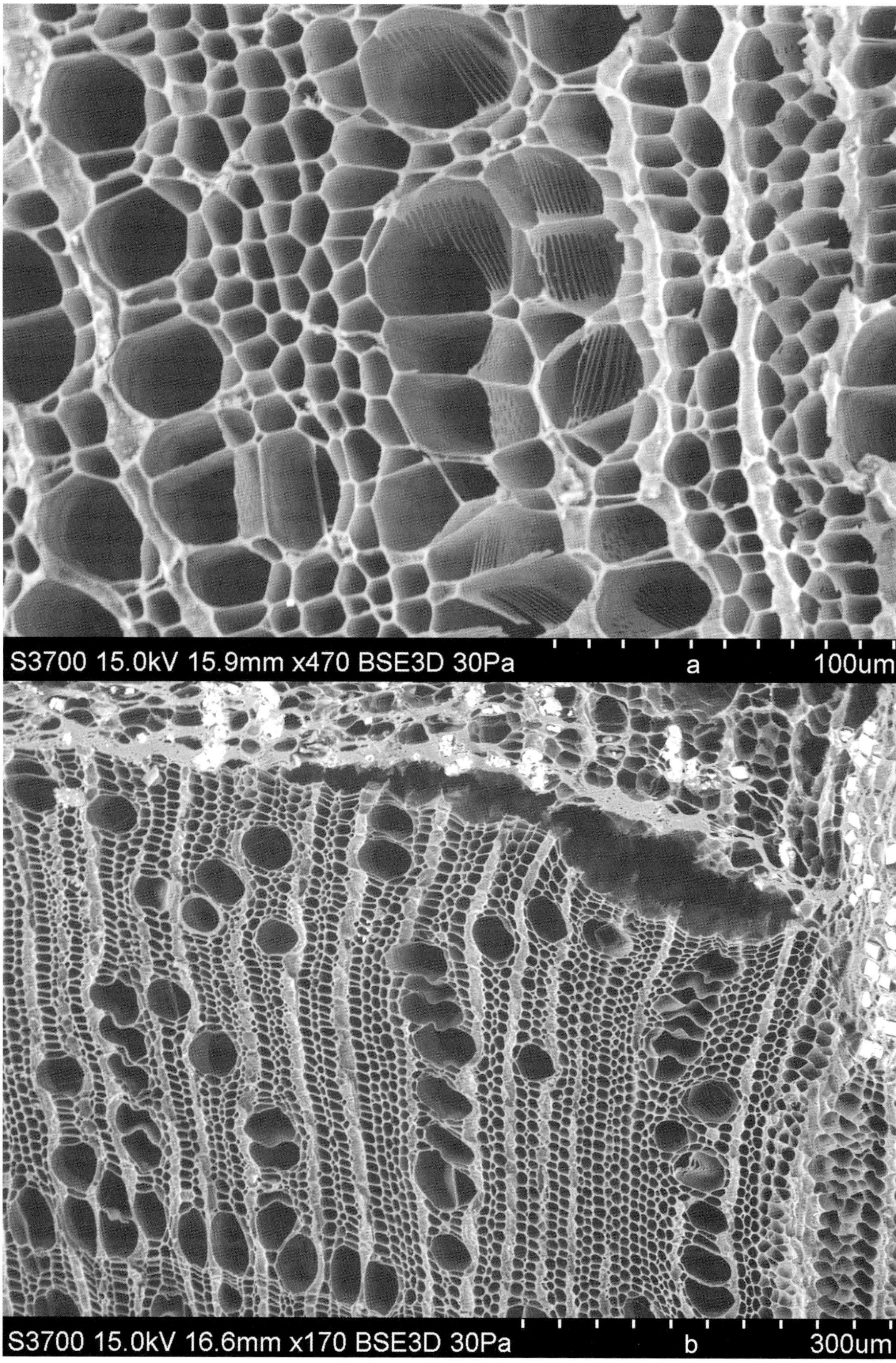

Figure 4 (a) VP-SEM (Hitachi S-3700N) image of the TS of the experimentally charred (360°C) alder (*Alnus glutinosa*) coppiced wood, with the angular vessels still unchanged and intact, but the scalariform perforation plates starting to buckle and fracture; (b) another area of the same sample displaying a tangential bark split and heat-crinkled vessels (especially those in vessel chains).

Table 2 Key anatomical features of the six woody taxa identified in the organic cores of some Snettisham twisted-wire bronze torcs.

Key diagnostic features	*Acer campestre*, field maple	*Alnus glutinosa*, alder	*Cornus sanguinea*, dogwood	*Corylus avellana*, hazel	*Salix* sp., willow	*Sambucus nigra*, elder
growth ring boundaries	distinct; a few rows of radially flattened fibres occur at growth ring boundaries	distinct; growth ring boundaries undulate near aggregrate rays	sometimes distinct; growth ring boundaries may undulate	distinct; growth ring boundaries may undulate	distinct	usually distinct because of a clear difference in size of earlywood and latewood vessels or of marginal bands of thin-walled vascular tracheids in latewood
diffuse-porous / ring-porous / semi ring-porous	diffuse-porous	diffuse-porous to semi ring-porous	diffuse-porous to semi ring-porous	semi ring-porous to diffuse-porous	semi ring-porous to diffuse-porous	semi ring-porous to diffuse-porous
vessel arrangement	widely spaced, solitary, often in short (2–4) radial multiples	radial multiples of 4 or more vessels common; clusters in earlywood	generally solitary; mostly similar in size; relatively widely distributed across growth ring	in diagonal (dendritic) and/or radial arrangement, in multiples, radial rows or clusters	solitary or in short radial files with 2–3 vessels, occasionally in clusters	thin-walled vessels grouped in oblique / radial / tangential bands or clusters; sometimes dendritic
vessel shape (other than round or oval)		angular solitary vessel outline				angular solitary vessel outline
perforation plates	simple	scalariform	scalariform	scalariform	simple	simple and scalariform
intervessel pit arrangement	alternate	opposite or alternate	scalariform or opposite	alternate; shape of pits polygonal	alternate; shape of pits polygonal	alternate; shape of pits polygonal
vessel-ray pits	with distinct borders, similar to intervessel pits	with distinct borders, similar to intervessel pits	with distinct borders, similar to intervessel pits	with distinct borders, similar to intervessel pits	large, rounded or angular pits with much reduced borders or apparently simple	with much reduced borders to apparently simple: pits rounded or angular
helical (spiral) thickenings in vessels	present			present		
tyloses in vessels					occasional thin-walled tyloses present	occasional thin-walled tyloses present
fibres	non-septate; thin- to thick-walled	non-septate; thin-walled; fibre-tracheids rare	non-septate; thin- to thick-walled	non-septate; thin- to thick-walled	non-septate; fibres very thin-walled but also some that are thin- to thick-walled	septate and non-septate; thin- to thick-walled;
fibre pitting		with simple to minutely bordered pits or distinctly bordered	with distinctly bordered pits; common in both radial and tangential walls	simple to minutely bordered pits or distinctly bordered; mainly restricted to radial walls	simple to minutely bordered pits	simple to minutely bordered pits
axial parenchyma	absent or extremely rare apotracheal diffuse or marginal; scanty paratracheal;	apotracheal axial parenchyma diffuse, or diffuse-in-aggregates	apotracheal axial parenchyma diffuse; scanty paratracheal parenchyma	diffuse apotracheal axial parenchyma	sparse apotracheal parenchyma, occasionally terminal / marginal parenchyma in discontinuous uniseriate bands	axial parenchyma absent or extremely rare; scanty paratracheal, diffuse or in marginal bands
rays	mostly 2–4 seriate; homocellular; all ray cells procumbent	exclusively uniseriate; homocellular; all ray cells procumbent	heterocellular; multiseriate rays; mostly 3–5 cells wide; larger rays commonly 4–10 seriate; with 1 to over 4 rows of upright /square marginal cells	1–3 cells wide; heterocellular rays; square and upright cells restricted to marginal rows	uniseriate; heterocellular with square and upright marginal cells	heterocellular; larger rays commonly 4–10 seriate, mostly with 1–4 rows of square and upright marginal cells
sheath cells			present			present
aggregate rays		present, with bi- to triseriate rays common		present		
calcium oxalate crystals	prismatic crystals in chambered axial parenchyma cells	not observed	not observed	present	not observed	not observed

Table 3 Experimental kiln charring of alder; anatomical changes as a result of increasing temperatures.

Temperature	Bark	Vessels	Scalariform perforation plates within vessels	Fibres	Ray parenchyma	Axial parenchyma
300°C 1 hour and 2 hours	no change	no change	no change	no change	no change	no change
360°C 1 hour and 2 hours	split in some places but not in others	unchanged and intact in some areas; vessel chains crinkle and deform in other areas	start to deform and fracture	no change	no change	in some areas slight deformation of cell walls
400°C 1 hour and 2 hours	splits widen	vessel outlines appear crinkled and deformed	noticeably deformed and fractured	some buckling of fibre walls	some distortion and displacement of cell walls	some fracturing and deformation of cell walls
600°C 1 hour and 2 hours	splits widen and bark becomes completely detached from xylem	many vessels are markedly deformed and some have collapsed in on themselves where cell walls have failed	only vestiges survive, deformed and fractured	buckling of fibre walls	distortion and displacement of cell walls	marked deformation of cell walls

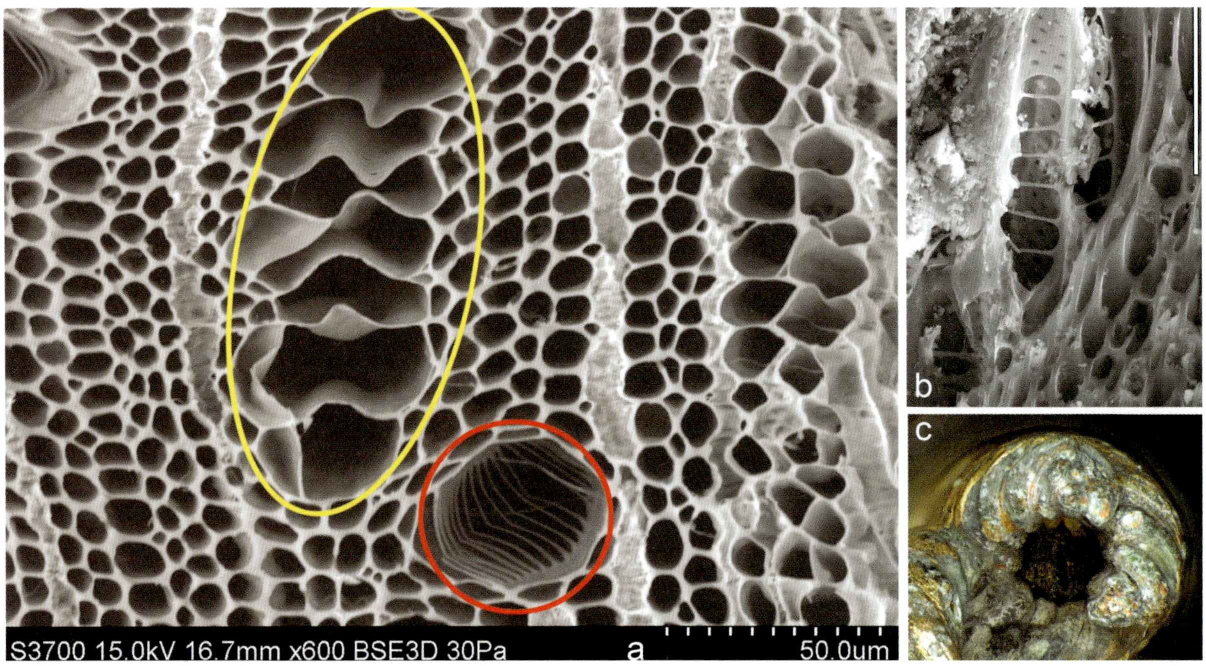

Figure 5 (a) VP-SEM (Hitachi S-3700N) image of the TS of the experimentally charred (400°C) alder with its vessel multiples crinkled and deformed (outlined by the yellow oval), and the scalariform perforation plates (outlined in red) starting to buckle and distort; (b) SEM image (JEOL JSM840) of an unaltered alder scalariform perforation plate in TLS in the charred core of torc 251 (scale bar: 50 μm); (c) OM image of torc 251 revealing the charred core.

Field maple (*Acer campestre*), a wood with good bending properties, was used for the core of torc 241 (Figure 1). The cores of torc 242 (Figures 1 and 2) and B.27 were of hazel (*Corylus avellana*), whose high density wood resists transverse compression relatively well [10], and, consequently, has useful flexibility for many purposes. Alder (*Alnus glutinosa*) wood, whose properties include good elasticity, was selected for the cores of torc 251 (Figure 1) and torc fragment C.17. The core of torc/bracelet 256 (Figures 1 and 3) was of dogwood (*Cornus sanguinea*), another highly flexible and springy raw material. The two cores of C.20 are samples from torc terminals rather than from twisted-wire bronze bodies (C.20 a and C.20 b in Table 1). Their cores have been identified as willow (*Salix* sp.) and elder (*Sambucus nigra*) wood. Willow wood is light and whilst it can be bent easily, it will ultimately buckle because of its very low transverse compressive strength [10]. Like willow, elder wood is noted for its elasticity.

SEM identification of the key characteristic features of the charred cores revealed that six woodland trees are represented (Tables 1 and 2), probably selected because they coppice well. Evidence for the woodland management practice of coppicing has been recorded from as early as the Neolithic period, notably from the Somerset Levels and Cambridgeshire [11]. Coppicing, accomplished using simple hand tools, encourages growth of new, slender flexible stems that were ideally suited for the construction of the tightly twisted torcs.

Given that it is possible, on a species-by-species basis, to link changes in wood anatomical features to charring temperatures (see below), charring experiments using coppiced wood formed the next logical investigative phase.

Experimental charring of coppiced alder wood

Wood retains most of its qualitative features when converted to charcoal, but temperature-dependent anatomical changes (including the dimensions and appearance of the cells) occur during the process [12]. These changes vary from one species to another, but experimental replication under laboratory

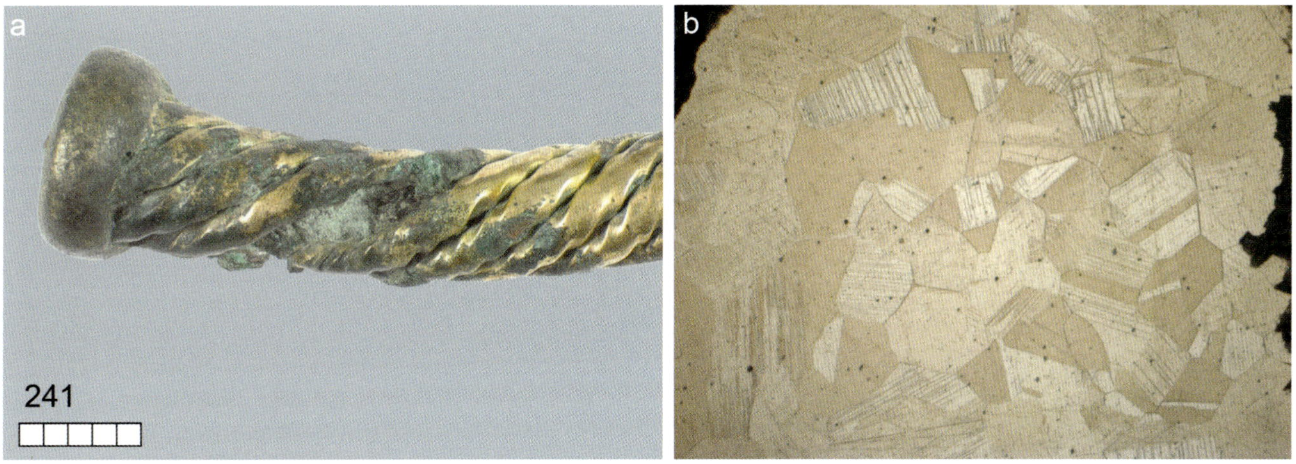

Figure 6 (a) OM image of the cast-on terminal of torc 241 (scale bar in mm); (b) OM image of bronze wire microstructure in TS; it is fine grained and annealed with very little eutectoid, showing homogenised structure from hammering and annealing. Field of view c. 0.3 mm.

kiln-charring conditions gives accurate information on temperature-related anatomical alteration. Although all six woody taxa identified in the twisted-wire bronze torcs were experimentally charred, due to space constraints this paper describes only one case study using alder wood, which was used to compare directly with the alder cores of torc 251 (BM) and torc fragment C.17 (NCM). The results from the alder experiments were representative of those obtained for the other taxa.

A coppiced alder wand from Suffolk (collected in 2009) provided samples for the laboratory-based charring experiments in a Carbolite® ashing furnace. Following the protocol of previous studies [12], alder wood samples were charred for one hour and two hours at 300°C, 360°C (the temperature at which charring commences [9]), 400°C and 600°C to establish at which temperatures the cells show anatomical changes. Table 3 summarises the results and should be read alongside Table 2, which provides the key anatomical features of fresh *A. glutinosa* wood. At 300°C, both for one and two hours of kiln time, there were no observed changes to the anatomical structure of bark cells, vessels, scalariform perforation plates within the vessels, fibres, ray parenchyma cells or axial parenchyma cells. At 360°C (charring times as above), changes started to occur within the anatomical structure: vessels were unchanged and intact in some areas (Figure 4a) but vessel multiples (or chains) crinkled and deformed in other areas (Figure 4b), scalariform perforation plates within the vessels started to deform and fracture (Figure 4a), bark split in some places but not in others (Figure 4b), fibres and ray parenchyma cells were largely unchanged but in some areas there was slight deformation of cell walls of axial parenchyma. At 400°C and 600°C (charring times as above), there was progressive bark splitting and eventual detachment from the xylem, vessel deformation increased (sometimes to a point at which vessels collapsed inwards), scalariform perforation plate deformation (Figure 5a) and fracturing ultimately resulted in only vestigial survival, fibre walls buckled, ray and axial parenchyma cells became distorted and displaced.

These results were compared with the charred alder wood cores from torcs 251 and C.17 (Figure 5). From the state of all the anatomical features present, it was possible to establish the temperature peak at which this material became charred. The clearest indicator can be seen in Figure 5b, which shows that there is no buckling or deformation of the bars of the scalariform perforation plate (and only one possible fracture); a temperature peak of not more than 360°C is suggested (see Table 2). As far as it is possible to ascertain, the cores appear to have been charred to the same degree throughout, suggesting exposure to even heating to the same temperature during the metalworking processes rather than localised heating.

Discussion

Annealed pairs of, or multiple, bronze wires were twisted together to produce the composite 'hollow' multiple-wire body of the torc; the use of a thin flexible wood internal core would prevent the wires from collapsing in on themselves during twisting. Once the composite multiple-wire body was formed, the torc/bracelet would then have been bent into its required final curved shape. The core would also have been important in avoiding wire collapse or 'kinking' during bending, and in creating a uniform twisted wire tube. The cores remain *in situ*, either because the metalsmiths did not, or could not, remove them, and in some cases the ends of the wires were shaped or cut to their final length to prepare for the cast-on terminals. Wax models of the terminals would have been added onto the torc wire ends and then encased and moulded in clay to prepare for the lost-wax casting of the bronze terminals (Figure 6a). Charred wood would have been too brittle to function as a core during the twisting and bending processes described above and thus the charring of the cores must have occurred as a result of heating processes during the subsequent finishing of the torcs, or possibly a post-manufacturing charring event was involved. Some of the possible processes are discussed below, and while firm conclusions cannot be drawn, the similarities between the cores examined suggest that the heating process was regular and controlled, and thus part of some deliberate manufacturing process.

For those twisted-wire bronze torcs with cast-on terminals, the clay moulds prepared *in situ* on the ends of the torc would have had to be carefully dried and heated to drive off the water and harden the mould without cracking. This was probably done by placing the prepared torcs at the edge of a hearth, and, as they dried, moving them into hotter regions to melt and burn out the wax of the model terminals and fully harden the mould [13]. Additionally, immediately prior to casting, the terminal moulds on the curved torc would have been heated to several hundred degrees Celsius to minimise thermal shock and prevent the clay mould from shattering when the molten bronze was poured in. This heating would also prevent the molten bronze from solidifying prematurely, allowing it to flow fully into the moulds. It is therefore possible that the wood cores of the torcs were charred during these pre-heating stages or affected when the terminals were cast with molten bronze at approximately 1000°C.

However, if the clay moulds were baked above their firing temperature (minimum 600°C) or pre-heated to red heat before the molten bronze was poured in, rather than simply dried, this would have resulted in the cores at the torc ends being exposed to temperatures of well over 400°C. The apparent even charring of the cores suggests that localised heating to such temperatures did not occur, although the good heat conductivity of bronze should be borne in mind. Furthermore, the presence of the wood (or indeed charred wood) during the casting-on process might have been problematic, producing combustion gases [14]. It is possible that, instead, separately cast terminals were soldered onto the torcs, although evidence for this has not been observed thus far [3].

The evidence from the cores indicates exposure to temperatures below the full annealing temperature for bronze (>600°C), suggesting that although the wires would have been fully annealed prior to twisting (to prevent cracking and facilitate tight twisting around the wood core), a further full annealing stage was not carried out for shaping and completing the torc. The bronze wire microstructure in TS (Figure 6b) is fine-grained and annealed with very little eutectoid, showing homogenised structure from hammering and annealing to make the wire, plus a little work hardening, resulting in a relatively springy torc [3]. Although clearly not fully annealed, it is possible, however, that the torcs were gently heated after the twisting and/or bending processes.

The questions of whether the use of organic cores was a widespread technological innovation or the product of a single workshop, and whether the wood was locally sourced, may not be straightforward to answer. Firstly, some other artefacts, such as Bronze Age gold lock-rings, also display wood cores [15] although, interestingly, no wooden cores were observed in the studied gold alloy wire torcs from Snettisham [3, 5]. Secondly, the six taxa represented in the cores (hazel, field maple, alder, dogwood, willow and elder) have an extremely wide distribution in the British Isles and across Europe. Therefore the wood used for the cores cannot be assumed to be local, although a local source would have been possible. While six different taxa have been used, all show good elasticity and coppice well and it is not clear, therefore, from the samples examined, whether there was any deliberate taxon selection for particular purposes or torc forms.

Torc terminals C.20 a and C.20 b have provided an interesting example of how taxon identification can be used to reinforce or refute their original attribution. Rainbird Clarke suggested in 1954 [1; 52, pl. XI] that these terminals were very similar and belonged to the same torc, although one terminal differed slightly from the other due to the presence of an incised line around its end. If these are indeed part of the same torc, it is extremely interesting as coupling two different types of thin osiers together (elder and willow) would create a rigid framework (similar to basket bases) onto which the bronze could then be twisted into the requisite torc shape. It is perhaps more likely, however, that the terminals are not from the same torc.

Conclusions

There are still many unanswered questions about Snettisham and its metalwork hoards. The archaeological excavations have shown no signs of a metalworking or a habitation site. There are no local metal sources of copper and tin, it is by no means clear who commissioned the work, and whether the objects were made locally or came from further afield. Archaeometric studies may not explain the reasons for the existence of this hoard but they reinforce the quality of craftsmanship and enhance the corpus of knowledge regarding the large variety of methods used to produce the objects.

The results have provided new information about the extent to which the use of organic cores was an integral part of the manufacturing process during the twisting of some of the bronze wire torcs. Although these cores are now charred, originally they were slender wands of coppiced wood, with sufficient flexibility to provide the appropriate support to prevent collapse during construction of these torcs. It is possible that uncoppiced branches were available, but they would have required more preparation for use, been less flexible and more prone to breaking. Through the experimental charring of coppiced alder wood, and documentation of the resultant anatomical changes, comparisons with the alder cores of torcs 251 and C.17 have highlighted the significance of the 360°C temperature peak. These comparisons suggest that the temperatures at which charring occurred was around 360°C and is similar in both the terminals and body of all of the torcs examined. This suggests that the twisted-wire bronze torcs were substantially complete before charring took place and that the heating appears regular and controlled, and thus part of some deliberate manufacturing process.

Acknowledgements

We are grateful to Ian Stead for access to his site archive and valuable discussion. We thank Norwich Castle Museum, especially John Davies, for permission to study and sample selected torcs and Tim Pestel for access to the material. We also thank Ian Jones (School of Metallurgy and Materials, University of Birmingham), Paul Craddock, Susan La Niece and Catherine Higgitt for useful feedback and Antony Simpson for assistance with the figures.

References

1. Rainbird Clarke, R., 'The Early Iron Age treasure from Snettisham, Norfolk', *Proceedings of the Prehistoric Society* **20** (1954) 27–86.
2. Stead, I.M., 'The Snettisham Treasure: excavations in 1990', *Antiquity* **65** (1991) 447–465.
3. Joy, J. (ed.), *The Snettisham Treasure: Excavation Report and Catalogue*, British Museum Press, London [in preparation].
4. Garrow, D., Gosden, C., Hill, J.D. and Bronk Ramsey, C., 'Dating Celtic art: a major radiocarbon dating programme of Iron Age and Early Roman metalwork in Britain', *Archaeological Journal* **166** (2009) 79–123.
5. Meeks, N., Mongiatti, A. and Joy, J., 'Metallurgy and analysis of precious metal torcs from the Iron Age Snettisham Treasure', in *Under the Volcano: Proceedings of the International Symposium on the Metallurgy of the European Iron Age held in Mannheim, Germany, 20–22 April 2010*, ed. E. Pernicka and R. Schwab, Forschungen zur Archäometrie und Altertumswissenschaft 5, Verlag Marie Leidorf, Rahden/Westf. [in preparation].

6. Wheeler, E.A., Pearson, R.G., LaPasha, C.A., Zack, T. and Hatley, W., *Computer-aided Wood Identification*, North Carolina Agriculture Research Service Bulletin 474, North Carolina (1986).
7. Wheeler, E.A., Baas, P. and Gasson, P.E., 'IAWA list of microscopic features for hardwood identification', *IAWA Bulletin* **10** (1989) 219–332.
8. Schweingruber, F.H., *Anatomy of European Woods*, Verlag Paul Haupt, Bern (1990).
9. Cartwright, C.R., 'The Bronze Age wooden tomb furniture from Jericho: the microscopical reconstruction of a distinctive carpentry tradition', *Palestine Exploration Quarterly* **137** (2005) 99–138.
10. Van Casteren, A., Sellers, W.I., Thorpe, S.K.S., Coward, S., Crompton, R.H. and Ennos, A.R., 'Why don't branches snap? The mechanics of bending failure in three temperate angiosperm trees', *Trees – Structure and Function* Online First (2011) 1–9. DOI: 10.1007/s00468-011-0650-y.
11. Pryor, F., French, C. and Taylor, M., 'An interim report on excavations at Etton, Maxey, Cambridgeshire, 1982–84', *Antiquaries Journal* **65** (1985) 275–311.
12. Leme, C.L., Cartwright, C.R. and Gasson, P.G., 'Anatomical changes to the wood of *Mimosa ophthalmocentra* and *Mimosa tenuiflora* when charred at different temperatures', *IAWA Journal* **31** (2010) 333–351.
13. Meeks, N.D., Tulp, C. and Söderberg, A., 'Precision lost-wax casting', in *Proceedings of the 1st International Workshop on Experimental and Educational Aspects of Bronze Metallurgy, Wilhelminaoord, October 1999*, ed. C. Tulp, N.D. Meeks and R. Paardekooper, VAEE, The Netherlands (2001) 56–78 (www.journal.exarc.net, in preparation).
14. Paul Craddock, personal communication, May 2012.
15. La Niece, S. and Cartwright, C.R., 'Bronze Age gold lock-rings with cores of wax and wood', in *Metals and Societies: Studies in Honour of Barbara S. Ottaway*, ed. T.L. Kienlin and B.W. Roberts, Universitätsforchungen zur Prähistorischen Archäologie Aus dem Institut für Archäologische Wissenschaften der Universität Bochum Fach Ur- und Frügeschichte, Band 169, Verlag Dr. Rudolf Habelt GMBH, Bonn (2009) 307–312.

Authors' addresses

- Caroline Cartwright, Department of Conservation and Scientific Research, The British Museum, London, UK (ccartwright@thebritishmuseum.ac.uk)
- Nigel Meeks, Department of Conservation and Scientific Research, The British Museum, London, UK
- Duncan Hook, Department of Conservation and Scientific Research, The British Museum, London, UK
- Aude Mongiatti, Department of Conservation and Scientific Research, The British Museum, London, UK
- Jody Joy, Department of Prehistory and Europe, The British Museum, London, UK

Iron Age glass beads from Carthage

Katherine Eremin, Patrick Degryse, Nathaniel Erb-Satullo, Monica Ganio, Joseph Greene, Andrew Shortland, Marc Walton and Lawrence Stager

ABSTRACT A large number of glass beads were found within urns containing the cremated remains of children and occasionally animals from the Carthage Tophet, dating to the eighth to fourth century BC. The glass beads were analysed to determine their composition and microstructure to identify the alkali source used and assist with determining the likely provenance of the beads. The main analytical technique used was scanning electron microscopy with energy dispersive microanalysis with additional techniques including X-ray diffraction and Raman spectroscopy to more closely characterise the individual phases and matrix glass. This paper concentrates on the most common bead types, which were characterised by high iron contents and high levels of crystalline phases, including magnetite, hematite, wollastonite, wuestite and barium sulphate. The glass was extremely vessicular and contained remnant quartz and (rarely) feldspar. Many samples showed extensive alteration, evident from both the microstructure and composition. These beads represent some of the earliest natron glass known from a firm context, although absolute dates are lacking. The most likely scenario for the manufacture of these beads is local production from imported glass, which was probably coloured locally with iron-rich metallurgical waste.

KEYWORDS Carthage, glass, analysis, Iron Age

Introduction

Excavations in the 'Precinct of Tanit', the so-called 'Tophet' of Carthage, Tunisia, by the American Schools of Oriental Research (ASOR) Punic Project between 1976 and 1979 under the direction of L.E.Stager yielded cremated remains of human infants, and in some cases young ovicaprids (sheep/goat) and birds, buried in cinerary urns often surmounted by carved stone markers, some of which bear Phoenician inscriptions [1–4]. The excavated materials and records are currently housed at the Semitic Museum, Harvard University. The excavation is now being prepared for publication and the first comprehensive scientific analysis of the finds is being undertaken in conjunction with this.

The Carthage Tophet is a complex burial site consisting of a large number of cremation urns. The nature of the burials remains hotly debated – some authors favour child sacrifice [3] whilst others reject this in favour of normal infant mortality [5]. Use of the Tophet is believed to have spanned the period from the very earliest days of Carthage in the eighth century BC through to 146 BC, when ancient Carthage was destroyed by the Romans.

This paper concentrates on the glass artefacts found in the urns, which include rare monochrome blue glass beads, small numbers of blue and white eye beads, some red beads with white and yellow spots and large numbers of black and brownish beads. These later assemblages can be divided into simple black ring beads, eye beads with a triangular black core and black and white eyes, and small beads with a mottled, brownish or dark brown-black colour. Chronologically, the urns, and hence the glass beads enclosed in these, span several phases of use of the Tophet, although most glass beads date to the sixth to fourth century BC (Table 1). Since the alkali source used for glass production in the ancient Mediterranean changed from plant ash to natron during the first millennium BC [6], analysis was undertaken to determine whether these are plant ash or natron glass. It was not possible to sample the monochrome blue beads or the blue and white eye beads due to their rarity and completeness but samples were obtained from the black, mottled and dark beads, as these included large numbers of broken fragments and identical multiples.

Table 1 Details of urns and glass beads within the Carthage Tophet.

Phase	Dating BC	No. of urns	Glass beads
1–3	8th to 7th	69	Monochrome blue (rare)
4–6	6th to 5th	238	Monochrome blue (rare), black ring, black eye bead, small mottled, small dark, blue and white eye beads (rare)
6+	4th	136	Blue and white eye beads (rare), red spotted, black ring, small mottled, small dark

Table 2 SEM-EDX data for mounted beads: in order to identify each bead, the sample number consists of the urn number with a unique letter as in some cases several beads share a registration number. Typology of beads: S – small, BR – black ring, BEB – black core for eye bead, RS – red spotty, BLK? – fragments, SRO – small rounded.

Sample no.	Field no.	Date BC	Glass	Typology	Na_2O	MgO	Al_2O_3	SiO_2	P_2O_5	SO_3	Cl	K_2O	CaO	TiO_2	Fe_2O_3	PbO	CuO	BaO
5924A	x777T211	6th to 5th	A	BR	14.8	0.4	0.4	66.7	nd	0.2	0.7	0.3	8.5	0.3	7.8	nd	nd	nd
5924B	x777T211	6th to 5th	A	BEB	15.2	0.5	0.4	68.2	nd	0.1	0.9	0.2	8.3	0.3	5.9	nd	nd	nd
5588A	x609	6th to 5th	A	BR	18.3	0.4	0.8	62.0	nd	nd	1.1	0.5	8.5	0.1	6.9	1.3	nd	nd
6393A	x1827T889	6th to 5th	C	BEB	0.6	1.0	5.3	40.1	1.0	0.3	0.0	1.8	2.4	0.5	46.3	1.0	nd	nd
5849A	x985T	6th to 5th	A	BR	19.4	0.2	0.3	68.3	nd	0.4	1.1	0.3	4.9	nd	5.2	nd	nd	nd
5849B	x967T251	6th to 5th	A	BRB	19.3	0.2	0.3	68.3	nd	0.3	1.2	0.3	5.0	0.1	5.2	nd	nd	nd
5849C	x982T258	6th to 5th	A	BR	19.9	0.2	0.3	67.7	nd	0.3	1.1	0.3	5.0	0.1	5.1	nd	nd	nd
5929A	x867T239	6th to 5th	A	BR	15.5	0.4	0.3	68.1	nd	0.1	1.1	0.1	10.3	0.1	4.0	nd	nd	nd
5929B	x867T239	6th to 5th	A	BR	15.7	0.4	0.4	68.4	nd	nd	1.0	0.3	10.0	0.1	3.9	nd	nd	nd
5929C	x867T239	6th to 5th	A	BR	15.5	0.4	0.4	68.5	nd	0.1	1.1	0.2	10.1	0.1	3.7	nd	nd	nd
5596A	x1453T361	6th to 5th	A	BR	17.8	0.3	0.4	64.1	nd	0.3	1.0	0.2	7.5	0.1	8.3	nd	nd	nd
3185A	x85T	4th	A	BR	19.0	0.2	0.4	66.2	nd	0.5	1.0	0.2	5.6	0.1	6.8	nd	nd	nd
5904A	x2015T	4th	A	BR	17.7	0.5	0.5	64.6	nd	0.2	1.1	0.1	7.9	0.1	7.1	nd	nd	nd
5933A	x1224T294	4th	A	BR	17.9	0.3	0.4	64.7	nd	0.3	1.0	0.1	7.9	nd	7.3	nd	nd	nd
5933B	x1224T294	4th	A	BR	17.7	0.4	0.5	63.5	nd	0.3	0.9	0.3	7.8	0.1	8.6	nd	nd	nd
5933C	x1224T294	4th	A	BR (matrix)	17.0	0.5	0.5	63.9	nd	0.3	0.9	0.3	9.5	nd	7.1	nd	nd	nd
5933C	x1224T294	4th	B	BR (inclusion)	16.5	0.3	3.3	49.7	0.3	0.3	1.0	0.2	8.2	0.3	19.7	nd	nd	nd
5826A	x798T226	6th to 5th	C	S1	0.8	0.9	3.1	39.1	1.5	0.2	0.3	1.1	4.2	nd	47.3	nd	1.2	0.1
5826A	x798T226	6th to 5th	B	S1 (inclusion)	15.0	0.3	1.4	51.4	0.2	0.4	0.5	0.3	7.1	nd	20.6	nd	0.1	2.5
5826B	x799T227	6th to 5th	C	S1	0.7	0.9	4.3	41.0	1.4	0.2	0.2	1.4	4.1	nd	45.3	nd	0.5	0.2
5598A	x1442T357	6th to 5th	C	S1	0.4	0.7	3.7	30.4	3.6	0.3	0.0	1.6	3.5	0.2	53.5	1.1	0.4	nd
5525A	x352Ta	4th	C	S1	0.5	0.8	1.7	32.2	3.9	0.2	0.6	0.6	5.0	nd	54.1	nd	0.5	0.2
3187A	x71T	4th	B	S2	14.5	0.3	2.0	61.4	nd	nd	1.1	0.9	6.3	nd	11.0	2.6	nd	nd
3187B	x71T	4th	B	S2	14.4	0.3	2.0	61.9	nd	nd	1.1	0.9	6.1	nd	10.7	2.6	nd	nd
3187C	x71T	4th	B	S2 (matrix)	14.4	0.4	1.9	61.3	0.1	nd	1.0	0.9	6.2	nd	11.1	2.7	nd	nd
3187C	x71T	4th	B2	S2 (inclusion)	11.9	0.1	4.0	55.2	0.7	nd	1.2	0.8	4.7	nd	18.8	2.7	nd	nd
5976A	x1663T393	8th to 7th	E	SRO	19.7	0.3	0.5	68.0	nd	0.5	0.9	0.4	4.3	0.1	4.6	nd	0.5	nd
5965A	x1838T897	4th	F	RS	16.8	0.3	0.8	67.4	nd	0.3	1.0	0.2	5.8	nd	4.9	nd	2.4	nd

Methodology

Selected beads were sampled and mounted in resin for examination by scanning electron microscopy with energy dispersive microanalysis (SEM-EDX), X-ray diffraction (XRD) and Raman spectroscopy. The samples were polished with successively finer grades of abrasive papers (240, 320, 400 and 600 grade) and finished with polishing using 6 μm and 1 μm monocrystalline diamond suspension. The mounted, polished samples were carbon coated for SEM-EDX analysis. In all cases, the samples contained a full cross-section from the surface to interior of the beads selected. This ensured that analysis was undertaken on the least altered interior glass. The SEM analysis was undertaken with a JEOL JSM-640LV scanning electron microscope operated at 20 kV in high vacuum mode. Samples were imaged using a four quadrant backscattered electron detector (BSE) to examine altered and pristine glass and individual phases, and the phases and glass matrix were analysed using an Inca X-sight Oxford Instruments energy dispersive detector. Raman spectroscopy used a Bruker Senterra dispersive Raman microscope with an Olympus BX51M microscope and 785 nm, 633 nm and 532 nm lasers. The microscope has ×20, ×50 and ×100 objectives, with laser spot sizes approximately 5, 2 and 1 μm respectively and the instrument is controlled via OPUS software version 5.5. XRD was performed with a Bruker D8 Multipurpose Diffractometer. Samples were collected over a 2theta range of 30° to 90° using Cu K α radiation at 40 kV and 40 mA. The primary X-ray beam was collimated to approximately 0.5 mm with a monocapillary and the diffracted X-rays detected with a GADDS detector. Polished mounted samples were examined with a polarising microscope under incident light.

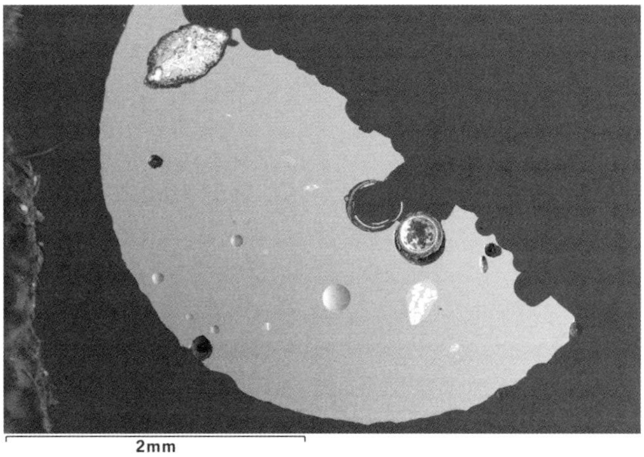

Figure 1 Black glass ring bead 5933C showing large iron oxide inclusions.

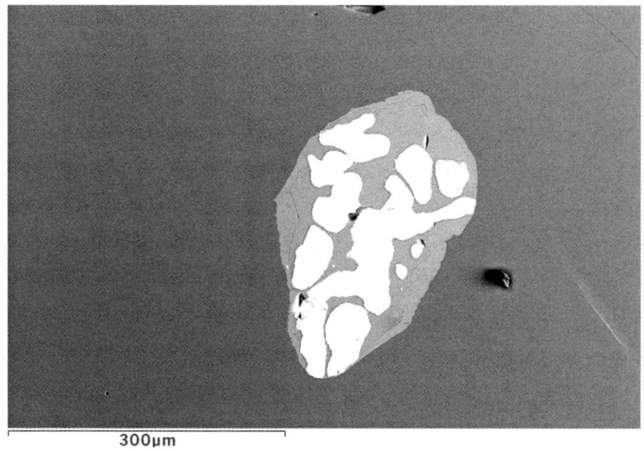

Figure 2 Detail of inclusion in black glass ring bead 5933C. The inclusion contains iron oxide, wollastonite (mainly at the edge of the inclusion) and a compositionally distinct glass.

Results

SEM examination indicated a variety of microstructures in the beads, which can be divided into original features and those that can be attributed largely to alteration, as discussed below. The compositional information from SEM-EDX is summarised in Table 2. A number of beads share the same field number, hence in the table individual beads are denoted by the urn number and a sequential alphabetic identifier with the field number given in the adjacent column. Beads with the same urn and field number were found together and are visually extremely similar; these are distinguished only by A, B, C etc.

Black glass: ring and eye beads

Visual examination shows that the black glass beads were formed by wrapping thin strands of glass around a core; these strands are clearly defined by trails of bubbles. The core is still present in many beads and has a yellow appearance. SEM-EDX analysis and imaging shows it is calcite-rich clay with many angular quartz particles and occasional feldspar. The black glass is characterised by varying amounts of crystalline phases, bubbles and an inhomogeneous glass matrix. In all instances, the black colour results from high to extremely high levels of iron in the glass matrix, which itself is characterised by high soda and lime and low magnesia and potash (almost always less than 0.5 wt% oxide). Several samples have an extremely streaky appearance in the BSE images due to variations in the iron content of the matrix glass.

The main phases present are iron oxides (magnetite, hematite and probably wuestite) and wollastonite. The iron oxides occur in a range of morphologies and sizes, from fine equant crystals to coarse dendritic segregations. Remarkably, the dendrites of wuestite as seen in the glass are typical for the remains (slag) of iron metallurgy. Wollastonite occurs as small elongated crystals, often in clusters, and frequently nucleated around fine crystals of iron oxide. Again, wollastonite is a phase not uncommon in metal slag. The glass matrix contains some rounded quartz (100–200 μm in size) with rare feldspar (both fairly pure sodium and plagioclase feldspar occurring). In some instances, the rounded quartz has an irregular outline due to dissolution into the glass matrix and is surrounded by shrinkage cracks. Many beads contain aggregates of iron oxides and wollastonite both within the glass matrix and as discrete inclusions surrounded by a separate glass phase, as seen in bead 5933C, shown in Figures 1 and 2. Analysis of the distinct glass in the inclusion in Figure 2 revealed higher levels of alumina and iron and lower levels of silica than the main glass (see below).

The black glass beads analysed spanned the full chronological range, from phase 4 to phase 7 (i.e. the sixth to the fourth century BC) and include both black ring beads and the triangular black cores of black and white eye beads. However, ring beads dominate both the urn assemblage and the samples analysed. The composition of the unaltered black glasses was denoted glass type A and is remarkably consistent over this period, as shown in Table 2, with the exception of a heavily altered eye bead and the glass in the inclusion shown in Figure 2. The glass from the inclusion is very similar to that in some mottled beads, glass type B, discussed below. The altered glass in the eye bead is similar to altered glass, termed glass C, in some mottled beads and small dark beads as discussed below. The high level of alteration of the eye-bead glass is apparent in the microstructure seen in the SEM, varying from coherent, uniform altered surfaces depleted in soda extending 10s to 100 μm into the glass to thicker, irregular alteration zones comprised of multiple lamellae depleted in soda and potash and with concentrations of heavier elements.

Although more data are required, it appears that beads within a particular urn are closer in composition than those from different urns of the same date. This may indicate that beads were produced in small batches and each urn contains a distinct batch. This does not necessarily mean that the beads were produced specifically for the burials, but does imply that they were at least acquired together. There was no significant compositional difference between different typologies within an urn, i.e. between black ring beads and the black triangular cores to black and white eye beads. Although the white layers in the eyes were too altered for quantitative analysis, SEM revealed abundant particles of calcium antimonate indicating an original white glass.

Black glass coloured by high levels of iron has been discussed by Moretti and Gratuze [7], Gratuze [8] and Van der Linden et al. [9]. In all cases, the authors discuss the compositions but not the microstructure of the glass. Moretti and Gratuze [7] and Gratuze [8] discussed black glass from

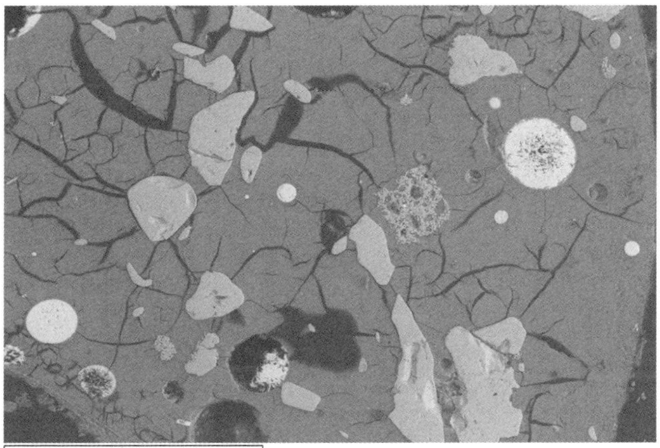

Figure 3 Detail of small mottled bead 5826B showing the equant and lacy iron oxides and spherical baryte. Note the pronounced cracking of the altered glass.

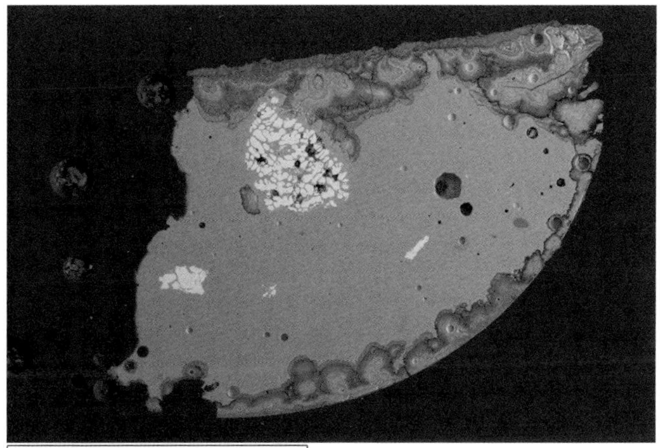

Figure 4 Small mottled bead 3187C showing the iron oxide inclusions and alteration of the glass surface.

a number of Late Bronze Age/Early Iron Age to Roman sites, attributing the colour to iron in all examples. In contrast, Van der Linden et al. [9] divided Roman period black glass into dark green, dark purple and dark brown and attributed the colour to high levels of iron, manganese and the ferri-sulphide (Fe^{3+}–S^{2-}) chromophore respectively. In the samples studied here, the colour can be attributed solely to the presence of high levels of iron. For the Roman examples, Moretti and Gratuze [7] suggested that iron oxide was added in European workshops to raw glass imported from the Eastern Mediterranean, since the black Roman glass was a typical Roman natron-based glass apart from a higher percentage of Fe_2O_3 (6.3–8.3 wt%) and MnO (0.5–1 wt%). The Iron Age black glass presented by Gratuze [8] was also a natron glass but with much higher Fe_2O_3 contents (10–25 wt%) and lower Al_2O_3 (<2 wt%) and CaO than the Roman glass. Some Bronze Age examples had higher levels of potash and magnesium and were probably produced from plant ash but were otherwise similar to the Iron Age examples [8]. In this study, black glass type A had much lower levels of alumina and lower levels of magnesia and potash than the Roman glass presented by Moretti and Gratuze [7] or Van der Linden et al. [9]. However the iron levels of this black glass type A are lower than in the Iron Age or Bronze Age examples studied by Gratuze [8]. This seems to be a distinct compositional variant of the black iron-rich glass.

Small dark beads

The small dark beads can be subdivided into dark black-brown and mottled (brown-green-red) varieties. All are significantly smaller and more rounded in shape than the more open ring beads. They were again formed by wrapping a strand of glass around a core, which is similar in composition to that used for the black glass, but examination suggests use of a single thicker glass strand.

This glass exhibits two different microstructures: S1 has a high percentage of crystalline magnetite, hematite, wollastonite and barium sulphate. A typical microstructure is shown in Figure 3 for bead 5826B and occurs in both dark and mottled examples. Magnetite occurs mainly as blocky, equant grains whilst hematite occurs as 'lacy' networks of fine grains. The surrounding glass matrix is extremely deteriorated, exhibiting fine lamellar and/or feathery alteration and extensive cracking. Some beads also contain abundant barium sulphate, which occurs both in large spherical voids and as fine spherical inclusions in the glass. The barium sulphate contains significant strontium in all instances and occurs in both large voids and as micro-spheres throughout the glass, suggesting it could be primary rather than a secondary alteration product.

A set of mottled beads from urn 3187 exhibits the alternative structure: S2, with fewer crystalline phases and little or no barium sulphate, as shown in Figure 4 for bead 3187C. These are generally better preserved, although with thick, irregular surface alteration zones with multiple compositional lamellae. Magnetite dominates and occurs as large dendritic masses and fine equant grains. Although the morphology of the dendritic iron oxide in SEM and polished section light microscopy (shown in detail in Figure 5 for bead 3187C) is suggestive of wuestite, only magnetite was identified in both the Raman and XRD spectra.

In the highly crystalline beads, alteration of the glass matrix is reflected in the composition, termed glass C in Table 2, which is highly variable and characterised by very low levels of alkali and silica and high levels of iron and alumina. In contrast, the less altered beads from urn 3187 retain high levels of soda and the glass composition is distinguished from black glass A mainly by higher iron and alumina with resultant lower silica. This glass composition was termed glass B. A compositionally similar glass occurs as rare islands in the heavily altered matrix of bead 5826A. One of these islands of unaltered glass is shown in Figure 6 and the glass composition given in Table 2 (denoted S1 inclusion).

Other coloured beads

Two other samples were obtained for SEM-EDX analysis: a fragment of a dark blue to black glass bead from phase 2 urn, bead 5967A, eighth to seventh century BC, and a red bead decorated with white and yellow spots from a phase 6/7 urn, bead 5965A, fourth century BC. The compositions are shown in Table 2. Both are again natron glasses, with very low levels of magnesium and potassium and fairly high levels of iron. The red spotted beads have higher levels of alumina than either black glass type A or the early phase 2 bead and the colour can be attributed to

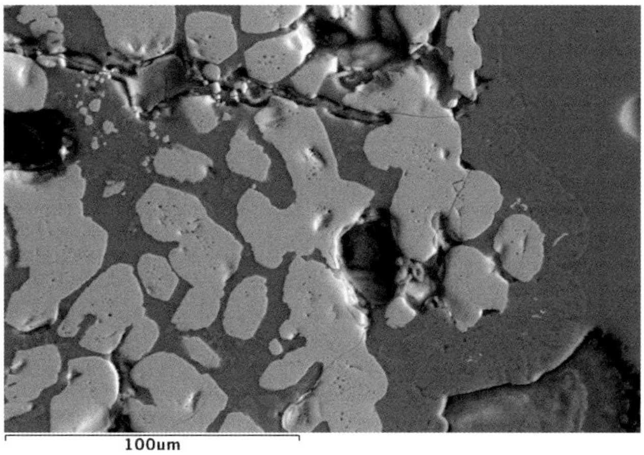

Figure 5 Detail of inclusion in bead 3187C, showing dendritic iron oxide with wollastonite within a distinct glass from that forming the main matrix.

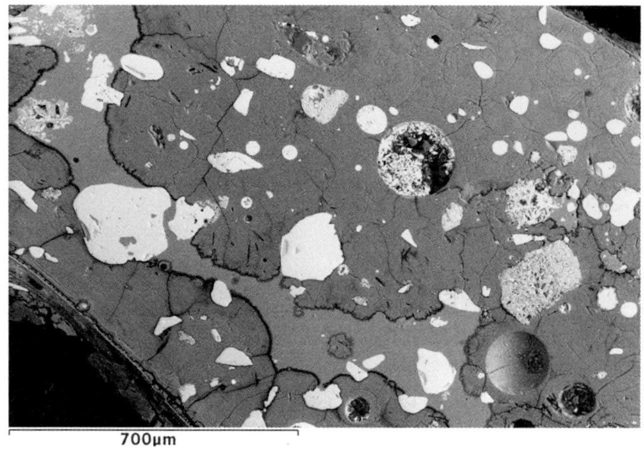

Figure 6 Remnant glass containing iron oxides and wollastonite within altered glassy matrix in a small mottled glass bead 5826A. The matrix glass contains iron oxides and barium sulphate.

the presence of significant copper, presumably present in the reduced form. The decoration on the spotted beads is extremely deteriorated but analysis indicates lead antimonate in the yellow areas and calcium antimonate in the white areas.

Discussion

The high levels of iron and the dendritic nature of the iron oxide minerals in the glass beads, the latter typical for iron metallurgy waste, suggest that iron slag was used to deliberately colour and/or form the black and mottled glass beads. In the absence of primary glass-working evidence from Iron Age Carthage, it seems likely that the beads were either imported or produced locally from scrap glass and/or imported raw glass. This imported material is unlikely to have been iron-rich (since such high iron glasses are unusual), suggesting the possibility of local addition of iron as a colorant during local bead production. Evidence of iron working has been discovered at a number of sites within Carthage, most notably at the Byrsa [10] and Bir Massouda [11, 12] and significant amounts of iron-working debris occur within the current port [13]. Industrial scale iron working appears to have been ongoing by the late seventh century and continued into the fifth century BC [11, 12]. There is some debate as to whether actual iron smelting occurred at Carthage rather than simply iron smithing of raw metal imported and shaped into usable objects [10, 11, 12, 14, 15]. Regardless, the current evidence indicates that metal slag was abundant and metalworking was an important industry in Punic Carthage.

Use of iron ores, iron salts or metallurgical slags to colour Iron Age black glass has been proposed by Gratuze [8]. If metal slag was used, the composition would contain contributions from the fuel ash and furnace or crucible lining as well as the ore itself. Use of an iron slag seems most likely. Iron ores can have appreciable silica, phosphorus and lime depending on geological setting but are unlikely to have high levels of soda. Although some alkali will be inherited from the fuel ash and/or clays used in the furnace or crucible, it is again unlikely that very high levels of soda would be inherited. The presence of a soda-rich glass in all types of beads therefore suggests that the slag was not used directly to make a glass but was instead added to an existing natron glass to achieve a very dark green to black glass.

There is little evidence of glass working from ancient Carthage compared to that for metalworking. The most abundant evidence for glass working at a single site comes from the Roman cemetery which dates to the second–seventh century AD [16, 17]. Glass in general is somewhat lacking in the excavation reports compared to other materials, as discussed by Docter and Sonneveld [18]. They attributed the fairly small amount of glass found at the settlement sites to deliberate recycling and garbage collection. The extant Punic glass finds from settlement sites included vessel fragments, beads and wasters, and covered a range of dates from the late eighth to the fifth century BC. None of the glass described was black in colour. The wasters come from the Middle Punic levels, around 480 to 425 BC, and provide the first definitive evidence for the production of glass on site in Carthage [18]. The beads are dominated by ring beads with some eye beads and rare face beads. This is reminiscent of the assemblage from the Tophet where the majority of the beads are flattened; more spherical ring types of eye beads are rare. The colour of the beads and in particular of the wasters is very different from that of the beads analysed here and unfortunately no analytical data are yet available. The one glass find with published analysis is from the Bir Massouda and is deteriorated soda-lime glass or glaze coloured with antimony [12].

The importance of the glass assemblage from the Tophet increases when one considers the paucity of glass finds of similar date from other sites. A number of features combines to suggest the possibility of local manufacture, the most convincing being the use of high levels of iron as a colorant in conjunction with the presence of industrial scale iron working in Carthage at a similar period to the likely date of the glass. It is probably no coincidence that the black and mottled glass beads span the sixth to the fourth century BC and the iron-working installations are attributed to similar dates. Other factors suggestive of local manufacture, possibly specifically for burials, are the fairly low quality of the glass, the very large numbers of beads, the limited typologies and the apparent lack of similar beads from settlement sites. Local production does not necessarily imply primary glass production at Carthage. It is possible, and perhaps more likely, that iron waste was added as a deliberate colorant to imported and/or recycled cullet. Ongoing trace element and isotopic analysis may differentiate between these two options. The current data suggest local production of a small range of dark beads using metallurgical waste as the colorant. The groups differ mainly in the density of phases and levels of iron in the matrix glass and are hence believed to have a common origin.

Conclusions

Study of the common varieties of glass beads from the Carthage Tophet showed that these were all natron glass coloured with high levels of iron. The glass composition and presence of abundant iron oxides within the glass suggest the use of metallurgical waste from iron working. Excavations have demonstrated the presence of iron-working installations and large amounts of iron-working waste at a number of sites in Punic Carthage. It is proposed that the beads may have been produced locally using iron slags from the nearby iron-working installation as a colorant.

Acknowledgements

We would like to thank Adam Aja, Semitic Museum, Harvard University, Richard Newman, Museum of Fine Arts, Boston, and Scott Speakman, Massachusetts Institute of Technology for assistance with this project.

References

1. Stager, L.E. 'The rite of child sacrifice at Carthage', in *New Light on Ancient Carthage*, ed. J. G. Pedley, University of Michigan Press, Ann Arbor (1980) 1–11.
2. Stager, L.E. 'Carthage: a view from the Tophet', in *Phonizier im Western*, ed. H.G. Niemeyer, *Madrider Beitrage* **8** (1982) 155–166.
3. Stager, L.E. and Wolff, S.R. 'Child sacrifice at Carthage: religious rite or population control? Archaeological evidence for a new analysis', *Biblical Archaeology Review* **10** (1984) 31–51.
4. Stager, L.E., Greene, J.A., Garnand, B.K. and Birney, K., 'Carthage: "The Precinct of Tanit (Tophet)". The American Schools of Oriental Research Punic Project. The Stager Excavations (1976–1979)', in *Studies in the Archaeology and History of the Levant* 6, Semitic Museum Publications [forthcoming].
5. Lancel, S., *Carthage: A History*, tr. A. Neville, John Wiley and Sons, Oxford and Cambridge, MA (1995).
6. Shortland, A.J., Schachner, L., Freestone, I. and Tite, M., 'Natron as a flux in the early vitreous materials industry: sources, beginnings and reasons for decline', *Journal of Archaeological Sciences* **33** (2006) 521–530.
7. Moretti, C. and Gratuze, B., 'Vetri romani di Aquileia e di altri siti europei: analisi chimiche e studio comparative', *Rivista della Stazione Sperimentale del Vetro* **32** (2002) 19–28.
8. Gratuze, B., 'Les premier verres au natron retrouve's en Europe occidentale: composition chimique et chrono-typologie', in *Annales du 17eme Congres de l'Association Internationale pour l'Histoire du Verre, Antwerp*, ed. K. Janssens, P. Degryse, P. Cosyns, J. Caen and L. Van't dack (2009) 8–14.
9. Van der Linden, V., Cosyns, P., Schalm, O., Cagno, S., Nys, K., Janssens, K., Nowak, A., Wagner, B. and Bulska, E., 'Deeply coloured and black glass in the northern provinces of the Roman Empire: differences and similarities in chemical composition before and after AD 150', *Archaeometry* **51**(5) (2009) 822–844.
10. Lancel, S., *La colline de Byrsa à l'époque* punique, Editions Recherche sur les civilizations, Paris (1983).
11. Docter, R.F., Chelbi, F. and Telmini, B.M., 'Carthage Bir Massouda: preliminary report on the first bilateral excavations of Ghent University and the Institut National du Patrimoine (2002–2003)', *BA Besch* **78** (2003) 43–70.
12. Docter, R.F., Chelbi, F., Telmini, B.M. et al., 'Carthage Bir Massouda: preliminary report on the first bilateral excavations of Ghent University and the Institut National du Patrimoine (2003–2004)', *BA Besch* **81** (2006) 37–89.
13. Hurst, H. and Stager, L.E., 'A metropolitan landscape: the late Punic port of Carthage', *World Archaeology* **9**(3) (1973) 334–346.
14. Keesmann, I., 'Untersuchungen zur Metallurgie im archaischen und punischen Karthago', *Mediterranean Archaeology* **14** (2001) 95–108.
15. Niemeyer, H.G., 'Archaeological evidence of early iron technology at Carthage and other Phoenician settlements', *Mediterranean Archaeology* **14** (2001) 83–94.
16. Sterrett-Kreuse, A.E., *Evidence for Glass Production from the Yasmina Necropolis at Carthage*, unpublished MSc dissertation thesis, University of Cincinnati (2006).
17. Sterrett-Kreuse, A.E., 'Evidence for glass working from the Yasmina Necropolis at Carthage', in *Annales du 17eme Congres de l Association Internationale pour l'Histoire du Verre, Antwerp*, ed. K. Janssens, P. Degryse, P. Cosyns, J. Caen and L. Van't dack (2009) 240–246.
18. Docter, R.F. and Sonneveld, J., 'Punic glass from Carthaginian settlement excavations', *Carthage Studies* **3** (2009) 125–145.

Authors' addresses

- Katherine Eremin, Straus Center for Conservation, Harvard Art Museums, Cambridge, MA, USA (Katherine_eremin@harvard.edu)
- Patrick Degryse, Geology, Earth and Environmental Sciences, Centre for Archaeological Sciences, KU Leuven, Leuven, Belgium
- Nathaniel Erb-Satullo, Department of Anthropology, Harvard University, Cambridge, MA, USA
- Monica Ganio, Geology, Earth and Environmental Sciences, Centre for Archaeological Sciences, KU Leuven, Leuven, Belgium
- Joseph Greene, Semitic Museum, Harvard University, Cambridge, MA, USA
- Andrew Shortland, Centre for Archaeological and Forensic Analysis, DEAS/CDS, Cranfield University, Shrivenham, UK
- Marc Walton, The Getty Conservation Institute, Los Angeles, CA, USA
- Lawrence Stager, Semitic Museum, Harvard University, Cambridge, MA, USA

Reconstructing firing practices of Middle Minoan polychrome ware: the role of bloating pores in slips

Edward W. Faber

ABSTRACT Anthropological studies of pottery production have demonstrated that the firing process is more complicated to interpret than suggested by some archaeological studies. The problem facing the archaeologist is that many of the anthropologically defined characteristics of a firing are archaeologically invisible. As an attempt to further understand the firing process from archaeological pottery, this paper addresses the use of scanning electron microscope (SEM) microstructural analysis of black slip decoration as a means of reconstructing the heating rate during firing. By comparing the development of bloating pores in the slip with the equivalent estimated firing temperature for the vessel, it is possible to infer variations in the heating rate used by different production groups. Using a case study of Middle Minoan polychrome ware, the potential for using differences in the firing process to discriminate between the regional production groups making similar pottery is further explored.

KEYWORDS firing processes, Middle Minoan pottery, slip, bloating pores, scanning electron microscopy

Introduction

Reconstructing the firing processes used for pottery production, and thus the potters' technological behaviour and choices central to these processes, is crucial in understanding the *chaîne opératoire* of archaeological ceramics. The limitations in recovering this information archaeologically can be illustrated by the dichotomy between the complexities of the firing process, as shown by anthropological and experimental studies, compared to the difficulties of reconstructing the same processes from the archaeological record that often results in characterising the firing process as just a firing temperature. The rationale behind the use or rejection of specific technological behaviour, including different approaches to firing practices, may be determined by cultural symbolism, traditions, social meanings and constraints, or individual choice [1–5] rather than purely in terms of optimal technological choice [6], therefore reducing the firing process to a single or a few attributes limits the potential understanding of the potters' actions and choices. This paper suggests a method of expanding our understanding of the firing process with an investigation of the heating rate used for firing archaeological pottery. Scanning electron microscope (SEM) microstructural analysis is used to explore the potential for reconstructing the rate of heating during firing, by comparing the microstructure of slip decoration with the equivalent firing temperature estimates based on the vitrification microstructure of the body. By applying this approach to a case study of the different regional groups of Middle Minoan polychrome ware from Crete, it intends to further our understanding of the regional technological styles and behaviour of the potters.

Approaches to understanding firing processes

The disparities between the anthropological and archaeological understandings of pottery firing processes are derived from the difference between the observational and interrogative nature of anthropological studies, as opposed to the interpretational nature of archaeological studies, where inferences have to be made from the physical remains of the pottery or the production sites. Thus, many of the properties anthropologists suggest as indispensible for characterising the complex processes involved in firing are archaeologically invisible. Instead, when examining archaeological ceramics, it is necessary to use criteria that can be consistently recovered from the material being studied, such as changes in microstructure or mineralogy for firing temperature determination. The drawbacks of archaeometric study on reconstructing firing processes is that much of the work concentrates on how to determine the maximum firing temperature rather than on its archaeological significance and interpretations of the firing procedures [7, 8]. This has led to considerable criticism of such archaeometric approaches, even so far as to question whether the determination of firing temperatures is justifiable and meaningful in archaeological terms [7] .

Anthropologists have argued that firing temperatures are a meaningless way to characterise technical variation in firings because the temperatures reached do not allow for differentiation between the different firing techniques and procedures [7–9]. Instead Gosselain suggests [7] that the heating rate is a more diagnostic parameter for the different firing techniques as, on average, the duration of the firing becomes longer as the firing structure becomes more enclosed. However, he does not suggest how the heating rate can be identified from the pottery

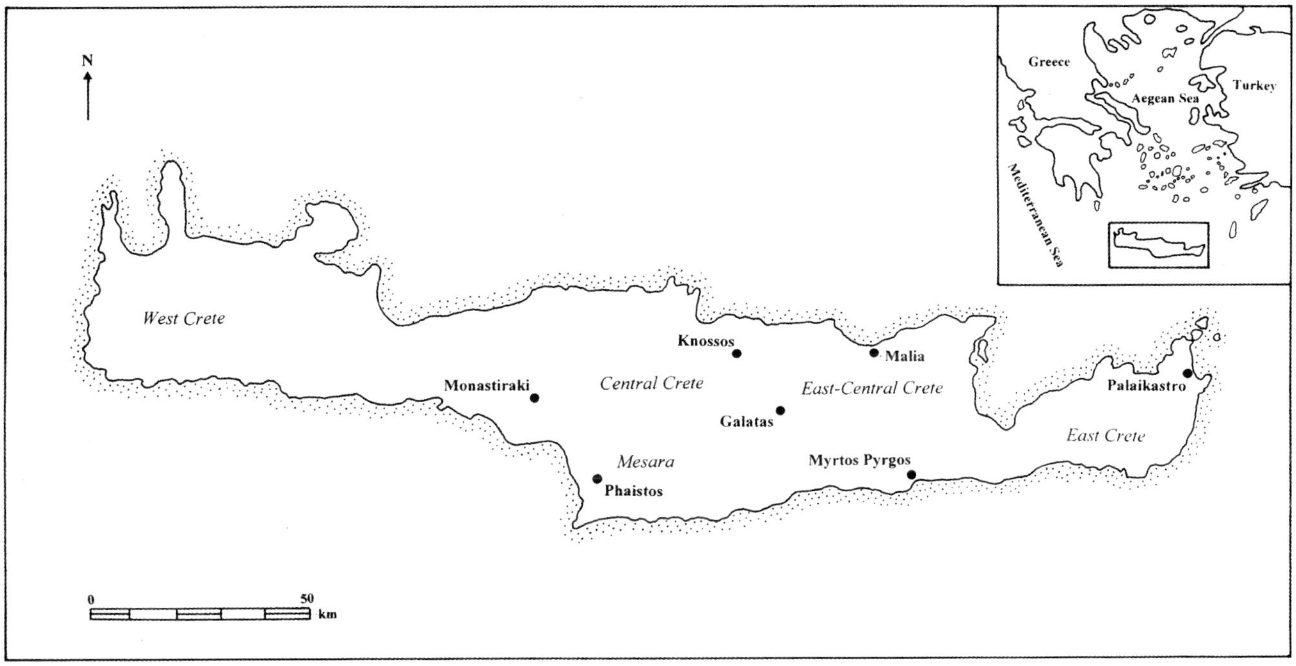

Figure 1 The sites from which ceramics were sampled for this project.

so such an approach may be difficult to apply archaeologically. Expanding upon this research, Livingstone Smith [9] has suggested that the firing process should be described in terms of a range of procedures and conditions in order that they can be properly characterised rather than merely equating firing temperatures with actual technologies. This research astutely identifies that there is no correlation between a specific set of thermal characteristics (the duration, maximum temperature, heating rate and soaking time) and a specific firing procedure, and thus Livingstone Smith argues that firing temperature estimates cannot be used to compare and differentiate between different firing technologies [9].

Such criticisms have been addressed within an archaeological context by Tite [8]. Rather than estimating the actual maximum temperature reached during a firing, he suggests it would be more appropriate to estimate an 'equivalent firing temperature', which is the temperature that, if held for a duration of one hour, would produce the observed change in mineralogy or microstructure and reflects the heat input of the process. While certainly a much better approach than characterising firing processes with a simple firing temperature determination, the concept of 'heat input' relies on using a set combination of (laboratory) conditions that pre-empts the possibility of identifying specific choices and actions of the individual potter.

Polychrome ware on Crete

Middle Minoan polychrome ware, including 'Kamares Ware', is high quality pottery that has complex decorative motifs and a distribution often restricted to the palace complexes or sites of special significance [10, 11]. The polychrome decorative effect is found on a multitude of vessel shapes, including hemispherical cups, bridge-spouted jars, pithoi and fruitstands [10–12] that may have been used for the storage, display, pouring and drinking of liquids and foods, and appears to indicate the performance of ritual consumption or celebration [13, 14]. The first examples of polychrome ware occur at the end of the Prepalatial period (c.2050–2000 BC to c.1925–1900 BC [15]), but the majority of examples of this ware derive from Protopalatial period deposits (c.1925–1900 BC to c.1750–1720 BC [15]). Arguments have been made for the identification of regional production of polychrome ware purely on stylistic grounds, particularly within the broad divisions of Central Crete, East-Central Crete (Malia-Lasithi) and East Crete, although recent analytical work has provided clearer evidence of the regional production [13, 16–19]. Polychrome ware usually has a black-slipped surface on which abstract and naturalistic decorative motifs were applied before firing using two or more over-slip pigments, normally in white, red, orange and occasionally purple. The occurrence of both a reduced iron slip and oxidised iron decoration indicates the use of a three-stage oxidising-reducing-oxidising (O-R-O) firing for polychrome ware [20–23]. There are rare examples of deliberately red-slipped polychrome ware, which are found at the site of Malia [24]. These appear to indicate the use of a different firing process, probably with an oxidising atmosphere throughout.

Sampling and analytical methodology

Within the context of identifying and characterising regional production, 208 examples of Middle Minoan polychrome ware were sampled from seven sites across Central, East-Central and East Crete: Knossos, Phaistos, Malia, Palaikastro, Myrtos Pyrgos, Monastiraki and Galatas (Figure 1), to investigate their provenance and the technology used in their production. The samples were analysed using an integrated approach of petrographic (thin-sections), geochemical (neutron activation analysis) and microstructural (SEM) techniques. The results discussed in this paper are based on the SEM analyses undertaken on the slip decoration and the clay fabric of the body; fuller discussions of the analytical work are reported elsewhere [17–19].

Table 1 The equivalent firing temperature estimates (°C) for the vitrification stages of noncalcareous and calcareous clays, following Maniatis and Tite [26] and Kilikoglou [29]. 'O' = oxidising firing atmosphere. 'R' = reducing firing atmosphere.

Vitrification stage	Noncalcareous (O)	Noncalcareous (R)	Calcareous (O)	Calcareous (R)
No vitrification		<750		<750
Initial vitrification		750–800		750–800
Incomplete vitrification			800–850	800–850
Extensive vitrification	850–950	800–900	850–1050	850–1050
Advanced vitrification			1050–1080	1050–1080
Continuous vitrification	>950	>900	>1080	>1080

Fresh fracture specimens of 184 polychrome ware samples were examined using a Philips 515 SEM with an EDAX 9900 energy dispersive system (EDX), at N.C.S.R. 'Demokritos', using a 25 kV accelerating voltage. This analysis assessed both the microstructure of the ceramic body and the nature of the surface decoration, using the work of Tite and Maniatis [25–28] as a basis. With this information it is possible to estimate the equivalent firing temperature of the vessel and to infer information about raw material selection, decorative techniques and firing procedures [28, 29]. The equivalent firing temperature estimates for these samples were based on comparisons with re-fired samples that were fired under laboratory conditions, using a heating rate of 180°C per hour with a soaking time of 60 minutes at the maximum temperature. The equivalent firing temperature estimates for the different vitrification microstructures are shown in Table 1.

Results

Most of the Middle Minoan polychrome ware samples analysed were made from a calcareous clay paste (n=173), although some were made from a noncalcareous clay paste (n=11). For the sake of clarity any discussion of firing temperature estimates has incorporated the noncalcareous ceramics into the same firing temperature divisions (not microstructural divisions) used for the calcareous ceramics. Although a wide range of vitrification microstructures was observed in the body clay of Middle Minoan polychrome ware, most samples were fired between 850°C and 1080°C. This may appear to be a wide range for firing temperature estimates, but it does encompass the products of several regional production centres that have individual approaches to firing regimes [18, 19].

There was one unusual vitrification microstructure observed in the fabric of two samples of polychrome ware made from fine-grained noncalcareous clays (one of which is red-slipped polychrome ware). The body clay displays a predominantly non-vitrified microstructure close to both edges (Figure 2a) while the core of the ceramic exhibits a heavily vitrified microstructure (Figure 2b). This structure represents the result of a fast firing, with a faster heating rate and shorter firing duration than normal. Due to the fast heating rate the organic component in the core does not fully burn away during the initial stages of firing. This results in an intense, localised reducing atmosphere in the core that causes vitrification to occur at a lower temperature. The clay on the edges of the ceramic is not exposed to the intense reducing atmosphere and therefore does not develop the same degree of vitrification due to the short duration of the firing. Such a microstructure is more likely to be observed in noncalcareous pastes because the heavily vitrified microstructure forms in noncalcareous clays at a lower temperature than in calcareous clays and thus the effects of the reducing atmosphere on vitrification become more pronounced. These two samples are the clearest evidence in Middle Minoan polychrome ware for interpreting the heating rate used during firing.

The high quality slips produced on polychrome ware were the result of a complex technique requiring skilled manipulation of raw materials, expertise in decorative techniques and an accomplished control of firing conditions. Whilst regional differences in the quality of black slips produced from the Early Minoan onwards are probably due to variability in raw materials and in the competence of potters [17–29], Middle Minoan polychrome ware is considered amongst the highest quality ceramic products of the period, and thus it would appear less likely that variations in the slips are due to poor production

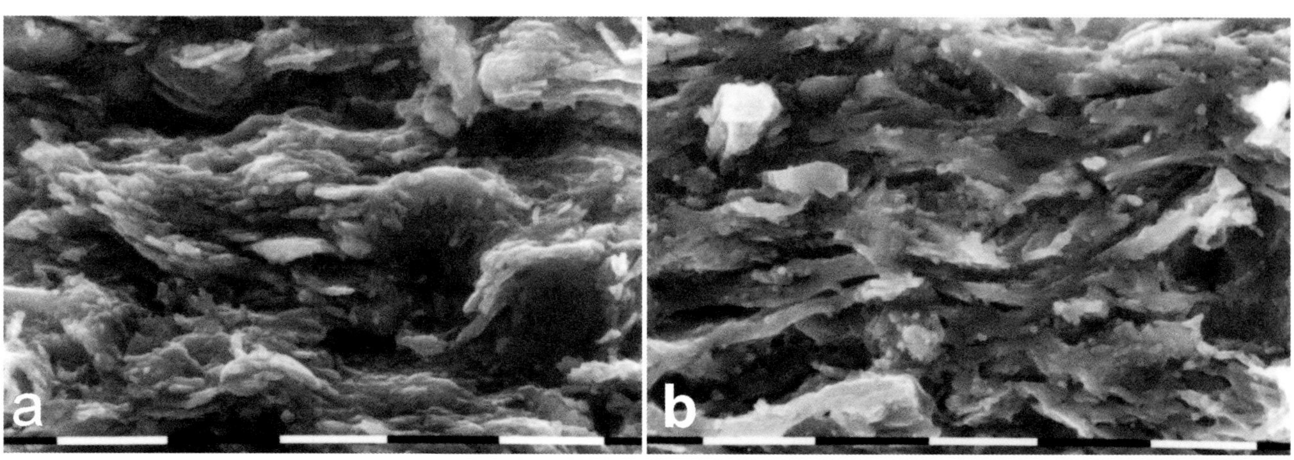

Figure 2 SEM secondary electron images of the fast-firing vitrification microstructure in the noncalcareous fabric of a Middle Minoan red-slipped polychrome ware (MAL0028): (a) edge of the sample showing signs of initial vitrification and (b) core of the sample showing a more heavily vitrified microstructure. Scale bar = 10 μm.

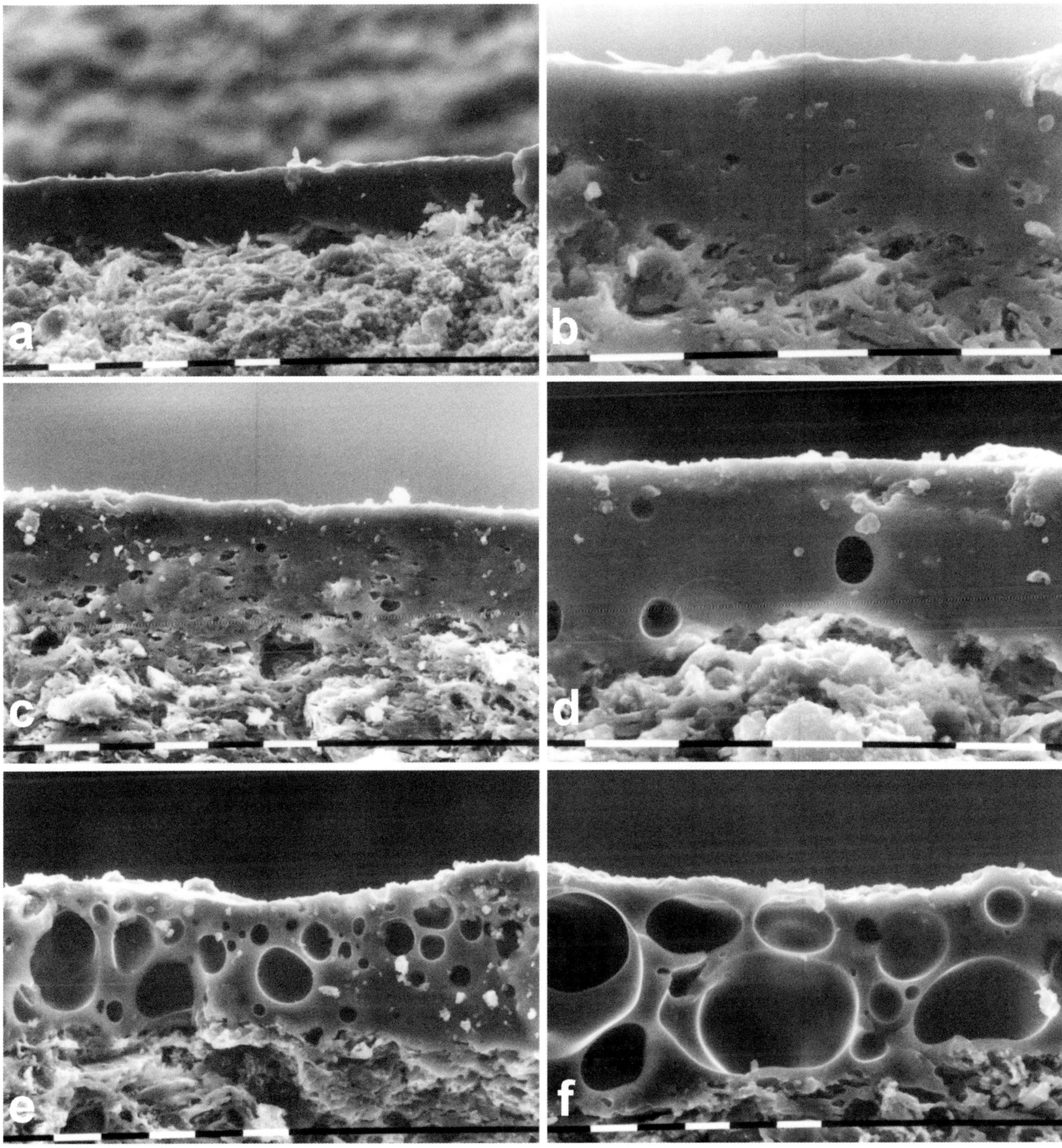

Figure 3 SEM secondary electron images of the different microstructures present in the black slips of a Middle Minoan polychrome ware: (a) vitrified, no bloating (PK0017); (b) vitrified, few FB (MAL0015); (c) vitrified, frequent FB (KNO95262); (d) vitrified, few CB (KNO95294); (e) vitrified, frequent CB and few FB (KNO95286); (f) vitrified, frequent vCB, frequent CB and few FB (KNO95345). Scale bar = 10 μm.

Table 2 The size of bloating pores in the slip.

Description	Abbreviation	Size
Fine bloating pores	FB	<1–c.5 μm
Coarse bloating pores	CB	>c.5–c.25 μm
Very coarse bloating pores	vCB	>c.25 μm

techniques. Many of the chemical or microstructural attributes of the black slips on polychrome ware show little variation, with no apparent correlation to differences in date or typology. The slips were made from a fine-grained suspension of a noncalcareous illitic clay [17, 18]. The thickness of the different slip layers varies between 5 and 90 μm, with most samples falling in the range of 5–35 μm [18].There appears to be little relationship between these attributes and the proposed provenance of the ceramics.

Perhaps the most striking of the microstructural traits observed in the slip of polychrome ware are the complexities in the development of bloating pores in the glassy phase. Bloating pores in slips have been referred to previously in terms of size (fine, medium and coarse with an indication of the size range in microns) and by their relative concentration [25, 26, 29]. However, this descriptive structure does not fully convey the range of variation observed in the Middle Minoan polychrome ware. There is no clear division between samples with fine bloating pores and those with coarse bloating pores; instead there is a continuum of varying concentrations and sizes (Figure 3). In addition, within these samples the term 'coarse

Table 3 The relative concentrations of bloating pores in the slip.

Description	Quantity seen in c.100 µm length
None	Essentially no bloating pores, but allowing for the occasional single pore
Few	c.3–c15/20
Frequent	>15/20 up to and including a sponge-like microstructure

Table 4 Cross-tabulation between the equivalent firing temperature estimate and the maximum size and quantity of bloating pores in the slip. Incongruous bloating patterns are highlighted in bold.

	<750°C	750–850°C	850–950°C	950–1050°C	1050–1080°C	>1080°C
None	1	3	11	0	**9**	**2**
Few FB	0	6	5	17	**30**	**5**
Frequent FB	0	**4**	**3**	8	6	**1**
Few CB	0	**3**	**1**	6	22	4
Frequent CB	0	0	0	4	16	1
Few vCB	0	0	0	0	2	1
Frequent vCB	0	0	0	0	8	4

bloating pore' did not appear sufficient to describe the range of sizes usually thus labelled. There was an upper size measurement that most bloating pores did not exceed, roughly about 25 µm (Table 2), but there were some exceptions. Bloating pores were encountered with a maximum diameter of up to about 100 µm, described here as 'very coarse bloating', and these gave a very different impression to samples with bloating pores up to about 25 µm (compare Figure 3e with Figure 3f). Using relative terms of the concentration of the bloating pores without quantification also did not express the full range of variation. As an attempt to address this, the categories used to describe the bloating were defined by the approximate number of bloating pores observed in a 100 µm length of the black slip to ensure comparability in the concentration levels observed between samples, although the entire visible cross-section of the slip was examined for each sample (Table 3).

The black slips on polychrome ware exhibit a smooth vitrification microstructure with a continuum of varying amounts and sizes of bloating pores from samples with no bloating pores, through various concentrations of fine bloating pores, varying combinations and quantities of both fine and coarse bloating pores, to samples with only coarse or very coarse bloating pores. Since previous research on bloating in slips has suggested that there is a progressive relationship between firing temperature and the size and quantity of the bloating pores it might be a reasonable assumption that this variation reflects differences in the firing temperature [25–29]. However, at least in the case of these polychrome ware samples, this relationship would appear to be more complex. Instead of a progressive development in the size and quantity of the bloating pores with increasing temperature, significant numbers of polychrome ware samples were identified where the size and relative concentrations of the bloating did not appear to correspond to the equivalent firing temperature estimates (Table 4). Such samples either have no or low levels of bloating at higher firing temperatures, or have relatively frequent and/or coarse bloating at lower firing temperatures, which may suggest that more complex development processes are occurring here than just a correlation between maximum firing temperature and the size and quantity of bloating pores. Thus it would appear that while bloating structures may be affected by firing temperature, an increase in the firing temperature is not the only determinant of the size and quantity of bloating pores.

Discussion

The density and size of the bloating pores in a slip have been used to refine the firing temperature estimates for ceramics on the basis that there is a progressive development in the size and quantity of the bloating pores with increasing temperature [25, 29]. For instance, in calcareous clays that exhibit the stable, cellular microstructure in the body (characteristic of a firing temperature of 850–1050°C), the temperature range can be narrowed to 850–950°C, if the black slip exhibits no or very few fine bloating pores, and will be above 950°C if there are abundant coarse bloating pores [29].

This research demonstrates that bloating pores in slips can broaden our understanding of the firing process, but to address this further it is important to consider the factors that cause bloating to occur and which determine how the bloating develops. There are two conditions that must be present for bloating to occur in ceramics. First, they must be made from clay that will produce a high temperature glassy phase with a viscosity sufficiently high to trap gas and, secondly, they must contain a substance that will disassociate and liberate gas when this glassy phase is present [25, 30, 32]. Either the clay meets these conditions and thus bloating pores can form, or the clay fails to meet the conditions and bloating pores will not form.

Previously published experimental work on the firing of clays has demonstrated the effect of firing conditions on the development of bloating. The heating rate, whether slow or fast, does not determine whether a clay can bloat, but a fast heating rate will lower the temperature at which bloating can begin to occur and increase the concentration of bloating pores produced due to the production of a localised internal reducing atmosphere [25–31]. The firing atmosphere of the kiln, although not related to the cause of bloating, will affect the development of bloating [31]. The use of a reducing atmosphere will cause both an increase in the amount of bloating and a reduction in the temperature at which bloating occurs [25, 30, 33]. Therefore the development of bloating pores can be affected by a combination of factors such as the firing temperature, the heating rate and the firing atmosphere.

The quality of the iron-reduction black slip on polychrome ware and the skill required to use the oxidising-reducing-oxidising (O-R-O) firing atmosphere effectively demonstrates the impressive control these potters had over the firing conditions. Although control of the firing atmosphere was crucial to the formation of the black slip, it was not responsible for the changes in the bloating patterns since all the black-slipped polychrome ware samples have been exposed to a similar reducing atmosphere. Instead, it is suggested that those samples exhibiting intense and common bloating at lower temperatures may have been exposed to a high rate of

Table 5 Cross-tabulation between the equivalent firing temperature estimate and the maximum size and quantity of bloating pores in the slip for the main Mesara production group. Incongruous bloating patterns are highlighted in bold. The Palaikastro production group exhibits a similar pattern.

	850–950°C	950–1050°C	1050–1080°C	>1080°C
None	0	0	**5**	0
Few FB	0	1	**12**	**2**
Frequent FB	0	2	2	**1**
Few CB	1	0	7	1
Frequent CB	0	0	7	1
Few vCB	0	0	2	0
Frequent vCB	0	0	7	2

Table 6 Cross-tabulation between the equivalent firing temperature estimate and the maximum size and quantity of bloating pores in the slip for the Central Cretan production group. Incongruous bloating patterns are highlighted in bold.

	750–850°C	850–950°C	950–1050°C	1050–1080°C	>1080°C
None	0	0	0	**1**	0
Few FB	0	2	6	**8**	0
Frequent FB	**3**	**1**	3	0	0
Few CB	0	0	3	6	1
Frequent CB	0	0	1	1	0
Frequent vCB	0	0	0	0	2

Table 7 Cross-tabulation between the equivalent firing temperature estimate and the maximum size and quantity of bloating pores in the slip for the Maliote production group. Incongruous bloating patterns are highlighted in bold.

	750–850°C	850–950°C	950–1050°C	1050–1080°C	>1080°C
None	0	3	0	0	**1**
Few FB	5	0	1	0	0
Frequent FB	**1**	1	0	0	0
Few CB	**1**	0	0	2	0

heating, while those samples exhibiting moderate and infrequent bloating at high temperatures may have been exposed to a slower rate of heating. Whilst the presence of individual samples exhibiting incongruous bloating in the repertoire of a production group is not evidence of a different firing practice, as they could be due to a one-off fluctuation in the heating rate, the identification of multiple examples from the same production group may allow us to suggest something other than just maximum firing temperature about the firing practices used in their production.

The samples were categorised into regional production groups on the basis of the chemical, mineralogical and microstructural data, with production identified in the Mesara, in Central Crete, around Malia and around Palaikastro [17–19], and when the patterns observed in the bloating pores are considered in these terms, there are contrasting patterns. The polychrome ware from the Mesara and Palaikastro both exhibit a similar pattern, with a significant proportion of examples with incongruously low bloating in the slip (Table 5). This suggests that the potters of these groups may have used a carefully controlled, relatively slower heating rate to achieve the maximum firing temperature.

In contrast, the polychrome ware originating from the production group in Central Crete (which, due to the very fine-grained nature of the fabric, does not contain any distinctive inclusions that would allow it to be tied down to a more precise source) shows a less accomplished (or less careful) control of the heating rate since there are many examples with both incongruously low and incongruously high bloating in the slip (Table 6). This might seem surprising since many of the best examples of polychrome ware, including most of the examples of eggshell ware analysed in this project, were made by this group. However, whilst the degree of control over the heating rate may be useful for defining the technological traditions of the potters, it may not have had a major impact on the final quality and finish of the pottery, which may instead reflect the quality of preparation of the raw materials, the firing temperature, etc., and may not have been an aspect of the firing process that the potters themselves considered important.

The slip microstructure of Maliote polychrome ware (Table 7), indicates a tendency towards incongruously high bloating, which would suggest the use of a relatively faster heating rate, but the small number of samples in this group makes it difficult to be more certain. However, when considered together with the example of Maliote red-slipped polychrome ware that exhibits a fast firing microstructure in the fabric, there may be an argument for a common use of controlled faster heating rates amongst the potters of this group.

Conclusions

An application of microstructural analysis based on the presence of bloating pores in iron-reduction slips will not be uniformly applicable to all archaeological ceramics. However, any technological behaviour is socially specific to an individual or group and thus investigations of crafts and technologies need to address the specific technological practices or the variety of choices faced by the individual potter. In this paper a comparison between the equivalent firing temperature estimate with the degree and amount of bloating present in the black slip has highlighted variants in the basic firing practice for Middle Minoan polychrome ware. Rather than just reflecting changes in the maximum firing temperature, it would appear that variances in the development of the complex bloating microstructure may reflect the use of different heating rates by individual production groups. This is useful for furthering our understanding of the technological traditions within a production group, where it has been shown that the ceramics were produced using similar raw materials, as well as specific decorative and firing practices [17–19]. Thus, at least in the case of Middle Minoan polychrome ware, the characterisation of the bloating pores in the slip may help to identify a further aspect of the technological practices used in the firing process and therefore is of significance for distinguishing the regional technological traditions.

Acknowledgements

I would like to take this opportunity to thank Peter Day and Vassilis Kilikoglou for their advice and guidance while undertaking this research. I am indebted to the following for permission to study and sample material referred to in this paper: G. Cadogan, F.M. Carinci, M.S.F. Hood, A. Kanta, V. La Rosa, C.F. Macdonald, J.A. MacGillivray, N. Momigliano, J.-C. Poursat, G. Rethemiotakis, L.H. Sackett, the Council of the British School at Athens, the 23rd Ephorate of Prehistoric and Classical Antiquities and the Conservation Directorate of the Greek Ministry of Culture, the 24th Ephorate of Prehistoric and Classical Antiquities and the Conservation Directorate of the Greek Ministry of Culture, and the 25th Ephorate of Prehistoric and Classical Antiquities and the Conservation Directorate of the Greek Ministry of Culture. The research presented here was supported by the EC TMR Research Network 'GEOPRO' (Integrating Geochemical and Mineralogical Techniques: A New Approach to Raw Materials and Archaeological Ceramic Provenance) (Grant number ERB FMRX CT 98-0165), a Wingate Scholarship from the Harold Hyam Wingate Foundation; and an educational grant from the Charles Henry Foyle Trust. The location map in Figure 1 was produced by David Taylor.

References

1. van der Leeuw, S.E., 'Giving the potter a choice: conceptual aspects of pottery technology', in *Technological Choices: Transformations in Material Culture since the Neolithic*, ed. P. Lemonnier, Routledge, London (1993) 238–288.
2. Pfaffenberger, B., 'Social anthropology of technology', *Annual Review of Anthropology* **21** (1992) 491–516.
3. Lemonnier, P., 'Introduction', in *Technological Choices: Transformations in Material Culture since the Neolithic*, ed. P. Lemonnier, Routledge, London (1993) 1–35.
4. Mahias, M.-C., 'Pottery techniques in India', in *Technological Choices: Transformations in Material Culture since the Neolithic*, ed. P. Lemonnier, Routledge, London (1993) 157–180.
5. Stark, M.T., 'Technical choices and social boundaries in material culture patterning: an introduction', in *The Archaeology of Social Boundaries*, ed. M.T. Stark, Smithsonian Institution Press, Washington, DC (1998) 1–11.
6. Bronitsky, G. and Hamer, R., 'Experiments in ceramic technology: the effects of various tempering materials on impact and thermal-shock resistance', *American Antiquity* **51** (1986) 89–101.
7. Gosselain, O.P., 'Bonfire of the enquiries. Pottery firing temperatures in archaeology: what for?', *Journal of Archaeological Science* **19** (1992) 243–259.
8. Tite, M.S., 'Firing temperature determinations – how and why?', in *The Aim of Laboratory Analyses of Ceramics in Archaeology*, ed. A. Lindahl and O. Stilborg, Kungliga Vitterhets Historie och Antikvitets Akademien, Stockholm (1995) 37–42.
9. Livingstone Smith, A., 'Bonfire II: the return of pottery firing temperatures', *Journal of Archaeological Science* **28** (2001) 991–1003.
10. Walberg, G., *Provincial Middle Minoan Pottery*, Verlag Philipp von Zabern, Mainz am Rhein (1983).
11. Walberg, G., *Kamares: A Study of the Character of Palatial Middle Minoan Pottery*, 2nd edn, Paul Åströms Förlag, Göteborg (1987).
12. MacGillivray, J.A., *Knossos: Pottery Groups of the Old Palace Period*, British School at Athens, London (1998).
13. Day, P.M. and Wilson, D.E., 'Consuming power: Kamares Ware in Protopalatial Crete', *Antiquity* **72** (1998) 350–358.
14. Knappett, C., *Thinking Through Material Culture: An Interdisciplinary Perspective*, University of Pennsylvania Press, Philadelphia (2005).
15. Manning, S.W., *The Absolute Chronology of the Aegean Bronze Age: Archaeology, Radiocarbon and History*, Sheffield Academic Press, Sheffield (1995) 217.
16. Ferrence, S., Swann, C.P. and Betancourt, P.P., 'PIXE analysis of white pigments on pottery from five Bronze Age Minoan archaeological sites', *Aegean Archaeology* **5** (2001) 47–54.
17. Day, P.M., Relaki, M. and Faber, E.W., 'Pottery making and social reproduction in Bronze Age Mesara, Crete', in *Pottery and Society: The Impact of Recent Studies in Minoan Pottery. Gold Medal Colloquium in Honor of Philip P. Betancourt. 104th Annual Meeting of the Archaeological Institute of America, New Orleans, Louisiana, 5 January 2003*, ed. M.H. Wiener, J.L. Warner, J. Polonsky and E.E. Hayes, Archaeological Institute of America, Boston (2006) 22–72.
18. Faber, E.W., *Middle Minoan Polychrome Pottery: An Integrated Microstructural, Geochemical and Mineralogical Investigation of its Production Technology and Provenance*, PhD thesis, University of Sheffield (2007).
19. Faber, E.W., Day, P.M. and Kilikoglou, V., 'Fine-grained Middle Bronze Age polychrome ware from Crete: combining petrographic and microstructural analysis', in *Interpreting Silent Artefacts: Petrographic Approaches to Archaeological Ceramics*, ed. P.S. Quinn, Archaeopress, Oxford (2009) 139–156.
20. Noll, W., Holm, R. and Born, L., 'Chemie und technik altkretischer vasenmalerei vom Kamares-Typ', *Naturwissenschaften* **58** (1971) 615–618.
21. Noll, W., Holm, R. and Born, L., 'Painting of ancient ceramics', *Angewandte Chemie International Edition* **14** (1975) 602–613.
22. Noll, W., 'Mineralogie und technik der keramiken Altkretas', *Neues Jahrbuch fur Mineralogie Abhandlungen* **142** (1982) 150–199.
23. Betancourt, P.P. and Swann, C.P., 'PIXE analysis of Middle Minoan pigments from Kommos', in *Archaeometry: Proceedings of the 25th International Symposium*, ed. Y. Maniatis, Elsevier, Amsterdam (1989) 177–181.
24. Poursat, J.-C. and Knappett, C., *La poterie du minoen moyen II: production et utilisation*, École Française d›Athènes, Paris (2005).
25. Maniatis, Y. and Tite, M.S., 'A scanning electron microscope examination of the bloating of fired clays', *Transactions and Journal of the British Ceramic Society* **74** (1975) 229–232.
26. Maniatis, Y. and Tite, M.S., 'Technological examination of Neolithic-Bronze Age pottery from central and southeast Europe and from the Near East', *Journal of Archaeological Science* **8** (1981) 59–76.
27. Tite, M.S. and Maniatis, Y., 'Examination of ancient pottery using the scanning electron microscope', *Nature* **257** (1975) 122–123.
28. Tite, M.S. and Maniatis, Y., 'Scanning electron microscopy of fired calcareous clays', *Transactions and Journal of the British Ceramic Society* **74** (1975) 19–22.
29. Kilikoglou, V., 'Scanning electron microscopy', in 'Ceramic regionalism in Prepalatial Crete: the Mesara imports in EM I to EM IIA Knossos', D.E. Wilson and P.M. Day, *Annual of the British School at Athens* **89** (1994) 70–75.
30. Austin, C.R., Nunes, J.L. and Sullivan, J.D., 'Basic factors involved in bloating of clays', *Mining Technology* **6** (1942) 1–11.
31. Riley, C.M., 'Relation of chemical properties to the bloating of clays', *Journal of the American Ceramic Society* **34** (1951) 121–128.
32. Sandrolini, F. and Palmonari, C., 'Role of iron oxides in the bloating of vitrified ceramic materials', *Transactions and Journal of the British Ceramic Society* **75** (1976) 25–32.
33. Hill, R.D. and Crook, D.N., 'Some causes of bloating in expanded clay and shale aggregates', *Australian Journal of Applied Science* **12** (1960) 374–384.

Author's address

Edward W. Faber, Department of Archaeology, University of Nottingham, Nottingham, UK (Edward.Faber@nottingham.ac.uk)

Examination of organic remains preserved by metal corrosion

Andrea Fischer

ABSTRACT Block-lifted metal artefacts from early medieval cemeteries were studied using variable pressure scanning electron microscopy (VP-SEM). Special emphasis was placed on highly degraded organic remains preserved by corrosion products. Considerable advantages were gained by being able to examine the material without laborious preparation and without a conductive coating. It was possible to visualise the state of preservation and preserved morphological characteristics of various organic materials *in situ*; identification was possible in many instances. To some extent, damp samples were able to be examined because water vapour could be introduced into the chamber and the temperature of a Peltier cooling stage controlled. In addition VP-SEM proved to be a useful tool for surveying conservation treatments for cleaning, consolidation and desalination, and for studying their potential effect on organic remains.

KEYWORDS burial, metal, organic material, mineralisation, identification

Introduction

The conservation and examination of grave goods lifted *in situ* is a big challenge for the conservator of archaeological artefacts. In most cases, the organic materials such as clothing, blankets, belts or sheaths are preserved solely as small traces because of the corrosion of the metal artefacts. The great importance of these organic materials to archaeology has been ably demonstrated by investigations in the research field of archaeological textiles (for example [1, 2]). Examining organic remains *in situ* may often enable one to understand the original function and use of textiles, leather and other organic materials. Since these remains are often very fragile, heavily degraded and barely visible to the naked eye, a careful way of working is needed in order to detect all preserved information. A systematic and methodical approach for examining and recording the block-lifted findings has been established in conservation practice. The position and stratification of all organic remains is depicted in order to reconstruct the details of the burial.

Corroding metals such as iron or copper protect organic materials. The processes of fibre decay and mineralisation are complex and involve many parameters, including the conditions of the burial environment, the microbiological activity, the pH of the soil and the composition of the organic materials as well as that of the metals.

Organic remains attached to metal artefacts were already mentioned in early conservation manuals [3]. Whilst Rathgen placed no value on the recording and preservation of this material within metal conservation treatments [4, 5], Rosenberg displayed more careful considerations [6], describing the organic remains as 'matières organiques enroulées'.

The study of archaeological textiles was established in the 1930s [2] with light microscopy and chemical analyses playing an important role in these studies [7, 8]. In 1964, Höhne published a comprehensive and very practical paper on the preparation and identification of archaeological textile finds [9]. Her work was pioneering for the conservation field because, for the first time, information about archaeological context, different mechanisms of preservation and technical analysis was combined with guidelines for excavation, cleaning and fibre identification.

Depending on their condition and state of preservation, it can be difficult to determine the nature of degraded and mineralised structures. Identification of the mineralised structures greatly improved with the introduction of scanning electron microscopy (SEM) in the 1970s. Since then, many studies on organic remains (including [10–12]) have improved the methodology so that this technique is now used routinely for identification [13], leading to a better understanding of the process of mineralisation. Considerable advantages arise from the higher magnification, the greater depth of focus and higher resolution of this method of investigation. Amongst others in the 1970s, Keepax investigated mineralised wood [10] and defined different types of mineralisation representing the progressive stages in coating and replacement of the organic matrix by iron corrosion products. In the following years, work was carried out to identify textile fibres [11, 14–16]), wood [10, 17], leather [18], bone, horn and ivory [19]. The associated mineralisation process was studied systematically by using complementary investigation methods. In particular, Fourier transform infrared spectroscopy (FTIR) provided valuable information about the degree of organic substance in mineralised fibres [20, 21]. However, the mechanism of fibre decay and mineralisation remains only partly understood.

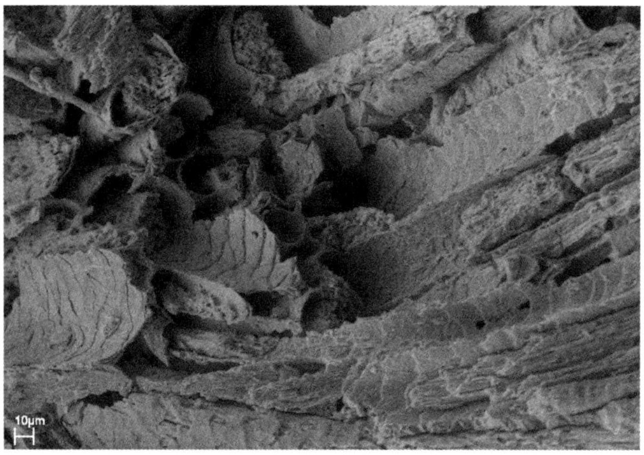

Figure 1 SEM SE image of a negative cast of keratin fibres, with remains of the fibrillated cortex (Hessigheim grave 166).

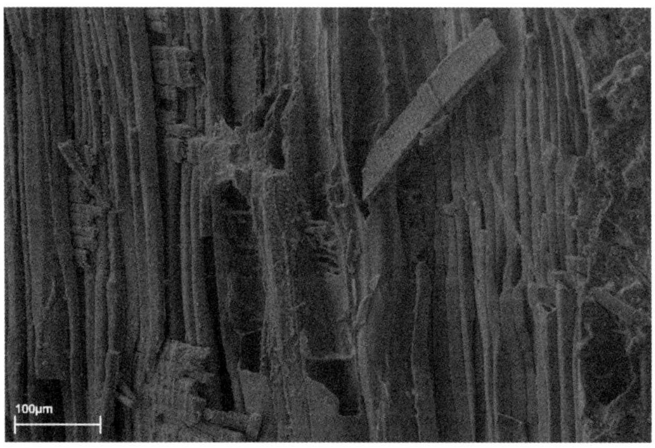

Figure 2 SEM SE image of a radial longitudinal section of birch wood (*Betula* sp.) with mineralised ray cells and scalariform perforation plates in vessels (Hessigheim grave 166).

Today, examinations of organic remains can be conducted under variable pressure (VP) conditions of SEM. Considerable advantages result from being able to examine the material without any preparation and conductive coating. Using a Zeiss EVO 60, investigations at the State Academy of Art and Design in Stuttgart have revealed the potential of VP-SEM in the study of organic remains encountered; alternative applications were tested on *in situ* lifted artefacts from early medieval cemeteries in Germany.

Materials and methodology

In many cases organic remains associated with metal artefacts from early medieval cemeteries are in poor condition, and it is difficult to differentiate between corrosion products and organic remains using optical microscopy. In close proximity to metals, the discoloration of the surrounding soil suggests the possibility of degraded textile, leather or wood, but it remains impossible to distinguish between these materials using this method and a non-destructive method of investigation was sought as the most appropriate alternative.

Since 2005, the Stuttgart Laboratory for Archaeometry and Conservation Science has been able to examine archaeological material using VP-SEM. In contrast to conventional SEM, it offers the possibility of working under different pressures. A small amount of remaining gas in the sample chamber prevents charging, which enables observation of non-conductive samples without the need for a conductive coating. The sample chamber can easily accommodate archaeological finds up to sizes of approximately 20 × 20 × 20 cm and it is possible to examine a small block of excavated soil.

The best means of evaluating VP-SEM for routine use was to select air-dried metal artefacts such as fibulae or knives with well-adhered and well-preserved organic remains for placing in the sample chamber. The mounting of the studied artefacts or archaeological finds in the sample chamber is relatively easy, as the chamber is equipped with two cameras, which allow for a safe and smooth alignment under the electron gun. To achieve optimal performance, the objects were first studied in detail using low magnification binocular microscopy before being placed in the chamber and areas of interest were indicated with small pressure-sensitive paper or copper foil arrows.

This preliminary methodological work revealed that the condition of the organic remains is essential in order to select the working parameters. If the organic substance is well preserved, the use of high acceleration voltage can generate charge. With an increasing degree of mineralisation, this problem occurs to a lesser extent. Working distances generally ranged from approximately 4 to 10 mm. While observing more degraded organic remains or soil surfaces, loose particles can be a risk for the detectors at a very short working distance and will contaminate the chamber.

The material in the case studies presented originates from three early medieval sites in Germany. The burial fields 'Lauchheim' and 'Hessigheim' are situated in Baden-Württemberg; the burial field 'Dunum' is located in East Frisia, Schleswig-Holstein. The environment of some of the graves at the site in Lauchheim is characterised by a clayey soil with a high water content, while the site in Dunum is characterised by a sandy soil and a high soil aeration. All SEM investigations were closely aligned to practical projects being carried out by students of the Objects Conservation Programme at the State Academy of Art and Design in Stuttgart.

Air-dried samples

The investigation of textile remains adhered to metal artefacts using VP-SEM shows that, even at low magnification, it is possible to image details of the weave structure and traces of use, which are often not very well defined using light microscopy. By moving the stage it is possible to characterise the material used for the warp and weft of the textile, any superposition of textile layers and, in some cases, details of cross-sections of fibres. The examination in the variable pressure mode of the SEM is quick and easy to perform. For highly degraded samples, the only evidence of textile remains can be the point at which two threads cross one another or from the negative cast of a fibre on a corroded metal surface. For the finds studied, the most common evidence of textiles was keratin fibres preserved by iron corrosion products (Figure 1).

The investigation of arrowhead shafts and several iron knife-handles was relatively straightforward. Although wood, horn or bone was commonly used for handles, the preserved mineralised remains were often too fragile and brittle to be sampled and analysed. Without sampling, already existing detachments of mineralised organic material were used, as they provided the requisite cross-sections for identification (Figure 2).

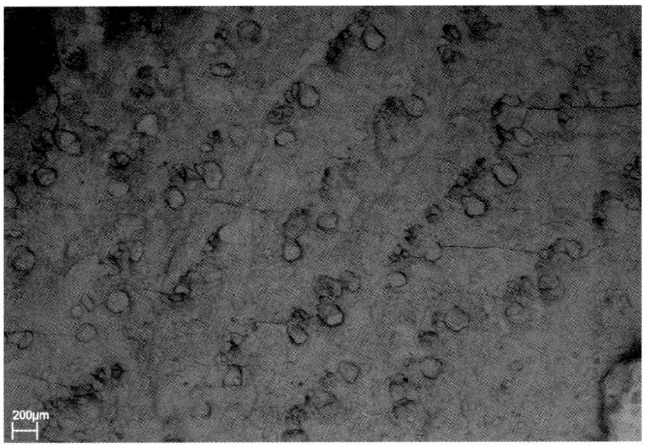

Figure 3 SEM BSE image of the surface of goatskin leather, preserved by ion corrosion products (Dunum grave 697).

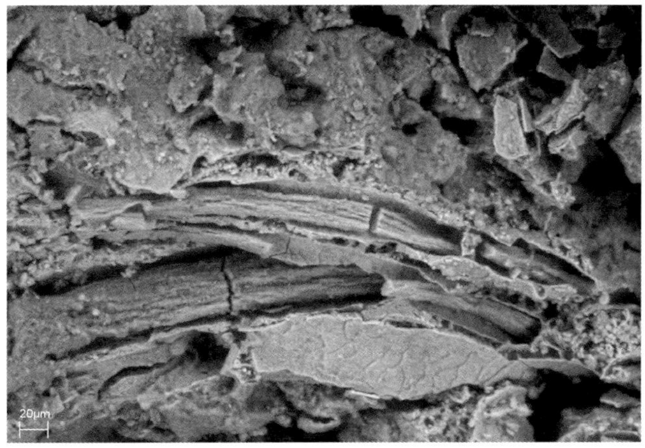

Figure 4 SEM BSE image of hydrated samples: detail of highly degraded keratin fibres (Lauchheim grave 271).

The identification of mineralised skin products and leather is still a problem. They can be identified if hairs or follicles survive [18], but often the surface has been removed by abrasion. Sometimes they can be observed as a powdery reddish-brown layer and can easily be mistaken for iron corrosion products. The use of VP-SEM enables a simple recording of the attached leather or skin product remains for incorporation into a reference database (Figure 3).

Hydrated material

As most block-lifted groups maintained their moist condition whilst being excavated in the conservation laboratory, the VP-SEM investigation of hydrated organic remains in their 'as found' state proved challenging. The study of hydrated samples requires the VP-SEM to operate in environmental mode, with a higher chamber gas pressure. If a Peltier cooling stage and a water control system is fitted to the VP-SEM, water vapour is used to discharge the sample surface. The temperature of the Peltier element can be controlled from −30°C to +35°C.

Preliminary investigations were carried out to explore the influence of humidity and temperature on the preservation and image quality of the samples. Small waterlogged textile fragments were selected initially because the aluminium sample holder of the cooling stage unit has a limited diameter of 1 cm. It was advantageous to freeze the samples before placing them on the Peltier couple. Appropriate working parameters in the sample chamber were: a chamber pressure of 500 Pa, a temperature of the cooling stage near the freezing point and a relative humidity just below 100%. Good quality images were achieved using an acceleration voltage of 20 kV and the backscatter detector (BSE).

A modified sample holder was needed to examine the fragile organic remains *in situ*, which were too large for the existing stage. A size of 5 cm in diameter was chosen for this sample holder in order to accommodate the small sections of the organic remains with some subsoil. It was made from an aluminium cylinder and sheet plate, fixed to the stub of the Peltier couple.

For the preparation, clay was used to embed the irregularly shaped and friable samples. However, using this method, drying of the sample surface was unavoidable, even by altering the operating parameters or by covering part of the specimen holder with foil. The capacity of the Peltier couple was a limiting factor. The processes of freezing and thawing were not always controllable, because the temperatures of the sample surface did not correspond to those of the Peltier couple. Observations of the drying process within the chamber showed that a quick working process was indispensible for the examination of these samples.

For the examination, several small sections of waterlogged organic remains were lifted and placed on the abovementioned sample holder. The investigation of degraded textile remains preserved under a copper alloy amulet disc from the site in Lauchheim (grave 271) illustrated the state of preservation: the keratin fibre was highly degraded and the cuticle was lost, but scale structure was preserved as a negative cast (Figure 4).

In many cases the examination of cross-sections of several layers of organic materials was of great value. The stratification of remains is vital to be able to draw conclusions about the nature and use of organic materials. A multilayered hydrated fragment of grave 737 in Lauchheim, probably belonging to a thong hanging from a belt, was investigated. The thong was made of two layers of skin material with a dense, non-oriented fibrous structure, as well as a third layer of another organic material.

The present work reaffirms that the imaging of waterlogged material *in situ* is possible, and provides an insight into structures that always benefit from further investigation. Even the identification of highly degraded material may be possible if features, such as scale structures, are preserved. In the present case studies, the preservation of the organic remains was influenced by a waterlogged environment as well as metal corrosion. It would be expedient to carry out a systematic study of degraded fibres in order to understand characteristic changes in morphology. For instance, the experimental procedure described here was used in studies of the freeze-drying process of leather finds [22] and in relation to the swelling process of canvas fibres [23]. However, it is necessary to remove a sample from the block of soil because the cooling stage has a limited capacity as far as both weight and size are concerned. Furthermore, as VP-SEM accelerates the drying process, the exposure time for *in situ* examinations of waterlogged organic remains is constrained.

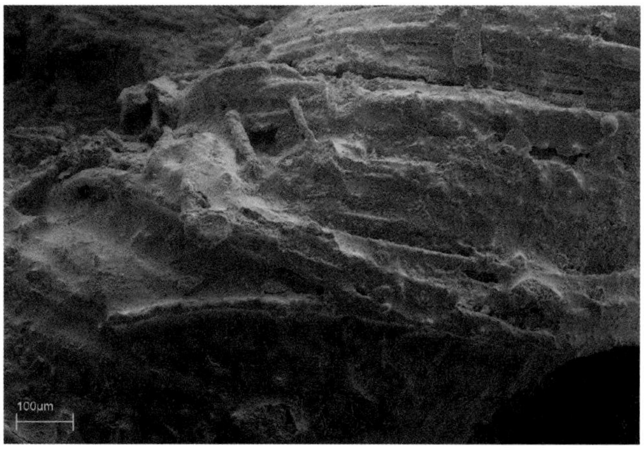

Figure 5 SEM SE image of mineralised textile before treatment (Lauchheim grave 271).

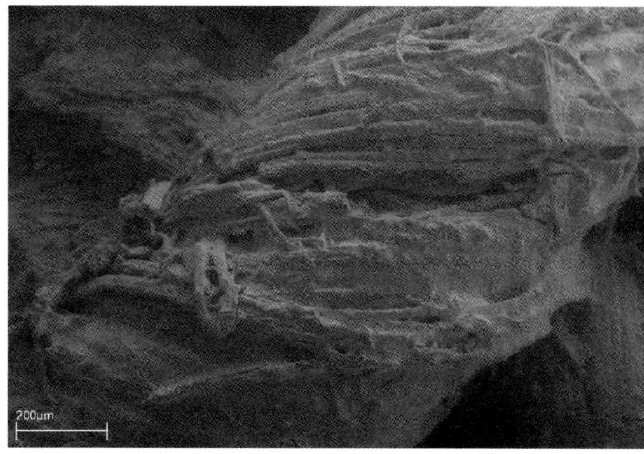

Figure 6 SEM SE image of mineralised textile after treatment in a disodium EDTA solution.

Conservation treatments

Several preliminary investigations were carried out to reveal the potential for recording and illustrating the processes involved in active conservation treatments of organic remains and any consequent changes.

Cleaning

Imaging the organic remains in the VP-SEM without prior dissection is a great advantage, particularly with regard to highly degraded samples, as each treatment can cause a loss of information. However, in the case of well-preserved organic remains, sample preparation or a cleaning treatment can increase the chances of observing characteristic features.

In one case it was possible to image the textile attached to a bronze ring after undergoing the following cleaning process. First, the artefact was treated with an ethanol/water solution to rinse away adhering soil and fibre fragments. The spin direction and the thread count were visible in some areas, but the single threads still appeared covered by soil after cleaning, as well as by a corrosion layer (Figure 5). In the second stage, cleaning continued using sequestrants. After a six-hour treatment in a 4% disodium ethylenediaminetetraacetic acid (EDTA) solution with a pH of 5.5, the corrosion layer was dissolved to some extent and single fibres started appearing (Figure 6).

In 1964, Höhne (later known as Farke) established the use of EDTA for the dissection of mineralised textiles [9, 24]. EDTA is an excellent sequestrant for most di- and trivalent metals. Today sequestrants are only applied in particular cases, because mordants of dyes or inorganic colouring agents may be chelated, precluding further analysis. Furthermore, the treatment itself poses a risk of damage to the mineralised organic remains depending on their state of preservation, as they are also composed of metal compounds. However, as the chemical process of chelation is quite slow and depends heavily on the pH and temperature of the solution, it is relatively easy to control.

We have already seen that VP-SEM is a useful tool for evaluating different cleaning methods at a microscopic level during the conservation treatment. Systematic series of tests are required to study the effect of the essential parameters (e.g. pH, concentration, immersion time, sequestering compound) involved in the formation process of chelate complexes and the mode of operation of different states of preservation. Other parameters than those used by Höhne [9, 24] could lead to better recognition of structural features.

Consolidation

In connection with any metal conservation treatment, all conservators have to deal at a given point with the problem of consolidating the fragile organic remains. Ideally the examination and identification of organic remains should be carried out before any consolidation is applied. It is seldom possible to identify mineralised organic material after consolidating with polymer resins, as the material is coated and sometimes impregnated and its structure is therefore no longer visible at a microscopic level. However, specialists for identification may not always be available during a conservation treatment. In such circumstances it would be relevant to investigate whether volatile binding media (e.g. cyclododecane) are a viable alternative to other materials commonly used for consolidation. The development of using volatile binding media is based mainly on the idea of temporarily consolidating fragile objects [25]. Depending on the conservation purpose and the condition of the material, they can be applied in different ways and sublimated within a reasonably short time without leaving any residue. Series of tests using VP-SEM would allow a preliminary evaluation as to whether the diagnostic microstructures of organic remains are changed by such volatile binding media.

Metal conservation treatments

The way in which conservation treatments developed for corroded metal artefacts affect organic remains is virtually unknown. The most common approach to the stabilisation of bronze artefacts is the inhibition of the active corrosion with benzotriazole (BTA). The use of this commercially well-known chemical was suggested by Brinch Madsen in 1967

[26]. BTA molecules form polymer complexes with copper and copper compounds, acting as a physical barrier for vapours and gases. Edwards [27] advises against the use of BTA if organic traces are adhering to the metal, because BTA is thought to crystallise and damage the mineralised organic material. Preliminary tests using VP-SEM were carried out to examine organic remains treated with a solution of BTA in ethanol. Edwards' doubts concerning BTA are yet to be confirmed; there was no noticeable evidence that revealed potential damage, so further tests are planned using a variety of samples.

For the long-term stabilisation of artefacts made of iron, the alkaline sulphite method, established by Rinuy and Schweizer in 1982 [28] is widely used. During the treatment, the objects remain in a strong alkaline solution maintained at 60°C for several weeks. From the subsequent macroscopic examination, however, it is evident that the fragile organic traces are damaged by the treatment. Experience has shown that, with large quantities of finds, it is scarcely possible to separately maintain and store all metals with attached organic remains. A few attempts were made to consolidate or to cover the organic traces. Several organic remains from the investigated iron finds appeared satisfactorily stable and were treated in an alkaline sulphite solution. Preliminary investigations using VP-SEM illustrated changes in the surface texture after the treatment. In some cases the surface of the iron and mineralised skin products were covered by fine needle-like crystals (which have not been analysed thus far).

The variety of observations made during and after treating metal artefacts with organic remains is a strong incentive for initiating several new research projects. The visualisation at a microscopic level is a useful aid in assessing and understanding detrimental alterations and risks. Based on the VP-SEM examinations, further research could include the use of complementary techniques such as FTIR, Raman spectroscopy, X-ray diffraction (XRD) and energy dispersive X-ray microanalyses (EDX).

Conclusions

This contribution has acknowledged further the benefit of direct observation of organic remains using VP-SEM. Metal artefacts with attached organic remains did not require coating for examinations at low vacuum or VP hence examinations proved to be very useful while excavating soil blocks in the conservation laboratory. These organic remains can provide extremely valuable evidence in the archaeological context, reinforcing the importance of leaving them unaltered for study. Non-destructive observations provided a more thorough insight into this material, showing the state of preservation of disintegrated substances and preserved morphological characteristics, thereby assisting in their identification. Imaging hydrated samples in their wet state was also possible. In the present study, the size of the samples and the operating time were limited by the capacity of the available Peltier cooling stage. Finally, the effects of various conservation techniques such as cleaning, consolidation or active metal conservation treatments on organic remains can easily be recorded and illustrated.

Acknowledgements

I wish to express my thanks to Christoph Krekel, head of the Laboratory for Archaeometry and Conservation Science; Gerhard Eggert, head of the Objects Conservation Programme; and the students of the Objects Conservation Programme at the State Academy of Art and Design in Stuttgart. I would also like to thank the Landesamt für Denkmalpflege Baden- Württemberg for providing the block-lifted findings from the sites in Hessigheim and Lauchheim and the Niedersächsisches Institut für Historische Küstenforschung for providing the artefacts from Dunum.

References

1. Hägg, I., 'Historische Textilforschung auf neuen Wegen', *Archäologisches Korrospondenzblatt* **19**(4) (1989) 431–439.
2. Walton Rogers, P., *Cloth and Clothing in Early Anglo-Saxon England, AD 450–700*, CBA Research Report 145, York (2007).
3. Voss, A., *Merkbuch, Alterthümer aufzugraben und aufzubewahren: eine Anleitung für das Verfahren bei Ausgrabungen, sowie zum Konserviren vor- und frühgeschichtlicher Alterthümer*, ed. Ministerium der Geistlichen, Unterrichts- und Medizinalangelegenheiten, Berlin (1888).
4. Rathgen, F., *Die Konservirung von Alterthumsfunden*, Berlin (1898).
5. Rathgen, F., *Die Konservierung von Altertumsfunden*, 3rd edn, Berlin and Leipzig (1926).
6. Rosenberg, G. *Antiquités en fer et en bronze: leur transformation dans la terre contenant de l'acide carbonique et des chlorures et leur conservation*, Copenhagen (1917)
7. Stockar, W. von, 'Zur mikroskopischen Untersuchung vorgeschichtlicher Gewebearbeiten', *Manus* **26** (1934) 300–320.
8. Geijer, A., *Textilfunde aus den Gräbern (Birka III)*, Uppsala (1938).
9. Höhne, H., 'Präparation und Bestimmung vorgeschichtlicher Textilreste', *Neue Museumskunde* **7** (1964) 211–228, 306–319.
10. Keepax, C., 'Scanning electron microscopy of wood replaced by iron corrosion products', *Journal of Archaeological Science* **2** (1975) 145–150.
11. Janaway, R.C., 'Textile fibre characteristics preserved by metal corrosion: the potential of SEM studies', *The Conservator* **7** (1983) 29–34.
12. Watson, J., 'The identification of organic materials preserved by metal corrosion products', in *Scanning Electron Microscopy in Archaeology*, ed. S. Olson, BAR International Series 452, Oxford (1988) 65–76.
13. Fischer, A., *Reste von organischen Materialien an Bodenfunden aus Metall – Identifizierung und Erhaltung für die archäologische Forschung*, ed. K. W. Bachmann, Schriftenreihe des Instituts für Museumskunde an der Staatlichen Akademie der Bildenden Künste Stuttgart 13 (1997).
14. Jakes, K.A. and Sibley, L.R., 'Survival of cellulosic fibres in the archaeological context', *Science and Archaeology* **25** (1983) 31–38.
15. Jakes, K.A. and Sibley, L.R., 'Survival of protein fibres in archaeological contexts', *Science and Archaeology* **26** (1984) 17–27.
16. Jakes, K.A. and Sibley, L.R., 'An examination of the phenomenon of textile fabric pseudomorphism', in *Archaeological Chemistry III.*, ed. J.B. Lambert, Advances in Chemistry Series 205, Washington, DC (1984) 403–424.
17. Keepax, C., 'Corrosion preserved wood: advances since 1975', in *Evidence Preserved in Corrosion Products: A New Field in Artefact Studies*, ed. R. Janaway and B. Scott, United Kingdom Institute for Conservation of Historic and Artistic Works Occasional Papers 8, London (1989), 15–20.
18. Cameron, E., 'Identification of skin and leather preserved by iron corrosion products', *Journal of Archaeological Science* **18** (1991) 25–33.
19. Moraitou, G., *Structure, Deterioration and Identification of Bone, Antler, Horn and Ivory*, unpublished BSc dissertation, University of London (1983).

20. Gillard, R.D., Hartman, S.M., Thomas, R.G. and Watkinson D.E., 'The mineralisation of fibres in burial environments', *Studies in Conservation* **39**(2) (1994) 132–140.
21. Chen, H.L., Jakes, K.A. and Foreman D.W., 'SEM, EDS, and FTIR examination of archaeological mineralized plant fibres', *Textile Research Journal* **66** (1996) 219–224.
22. Wiesner, I. and Krekel, C., 'Low vacuum scanning electron microscopy of waterlogged archaeological leather', in *Proceedings of the 10th ICOM Group on Wet Organic Archaeological Materials Conference, Amsterdam*, ed. K. Strœtkvern and D.J. Huisman, Nederlandse archeologische rapporten 37, Amersfort (2009), 741–760.
23. Haller, U., 'Gemälde im VP-REM: Beobachtungen zum Feuchteverhalten von Werkstoffen der Malerei', in *Beiträge zur 20. Tagung des Österreichischen Restauratorenverbandes* (2008) 20–27.
24. Farke, H., 'Der Einsatz von Komplexon in der Textilkonservierung', *Arbeitsblätter für Restauratoren* **25**(2), Gruppe 10 (1992) 176–178.
25. Hangleiter, H., Jägers, E. and Jägers, E., 'Flüchtige Bindemittel', *Zeitschrift für Kunsttechnologie und Konservierung* **9** (1995) 385–392.
26. Brinch Madsen, H., 'A preliminary note on the use of benzotriazole for stabilizing bronze objects', *Studies in Conservation* **12**(4) (1967) 163–167.
27. Edwards, G., 'Guidelines for dealing with material from sites where organic remains have been preserved by metal corrosion products', in *Evidence Preserved in Corrosion Products: A New Field in Artefact Studies*, ed. R. Janaway and B. Scott, United Kingdom Institute for Conservation of Historic and Artistic Works Occasional Papers 8, London (1989) 3–7.
28. Rinuy, A. and Schweizer, F., 'Entsalzung von Bodenfunden in alkalischer Sulfitlösung', *Arbeitsblätter für Restauratoren* **15**(1), Gruppe 2 (1982) 160–174.

Author's address

Andrea Fischer, Staatliche Akademie der Bildenden Künste, Stuttgart, Germany (a.fischer@abk-stuttgart.de)

A study of pre-Columbian gold beads from Panama

Ainslie Harrison, Kim Cullen Cobb, Harriet F. Beaubien, Paul Jett and Julia Mayo

ABSTRACT Research into the fabrication of pre-Columbian gold beads from Panama was carried out as part of an ongoing study on pre-Columbian gold at the Smithsonian Institution. Roughly 2100 beads were included in the study, originating from the collections of the National Museum of the American Indian (NMAI), the National Museum of Natural History (NMNH), and from recent excavations in Panama. Technical examination and X-ray fluorescence (XRF) analysis were carried out in the first phase, which permitted the beads to be divided into compositional, formal and technological categories. The most common type, the rolled and joined sheet bead, was chosen for more in-depth study. Variable-pressure scanning electron microscopy with energy dispersive X-ray analysis (VP-SEM-EDX) was carried out on selected beads in order to identify the joining method used. No elemental variation was detected along the exterior surfaces or over cross-sections of the join regions, suggesting that solder was not used as a bonding method for this type of bead.

KEYWORDS SEM-EDX, metallography, gold, beads, Panama

Introduction

While the techniques of gold jewellery fabrication in Europe have been extensively studied, to date there has been relatively little research on the production of pre-Columbian gold ornaments, and more specifically, the manufacturing techniques of gold beads. Ongoing research at the Smithsonian Institution on the history of pre-Columbian gold working in Panama provided an excellent opportunity to undertake an in-depth study of bead fabrication. This study is the latest phase of a larger research project that began in 2007, when researchers from the Smithsonian's Museum Conservation Institute (MCI), in collaboration with archaeologists at the Smithsonian Tropical Research Institute (STRI), conducted a technological study of two large collections of pre-Columbian gold in Panama. The second phase, which provided the context for the current study, focused on examination and analysis of the Panamanian gold in two Smithsonian collections in Washington, DC, those of the National Museum of the American Indian (NMAI) and the National Museum of Natural History (NMNH), in addition to well-contextualised finds from recent excavations in Panama curated at STRI. Access to the recently excavated material for analysis was approved by the Patrimonio Histórico of Panama through partnership with archaeologists at STRI.

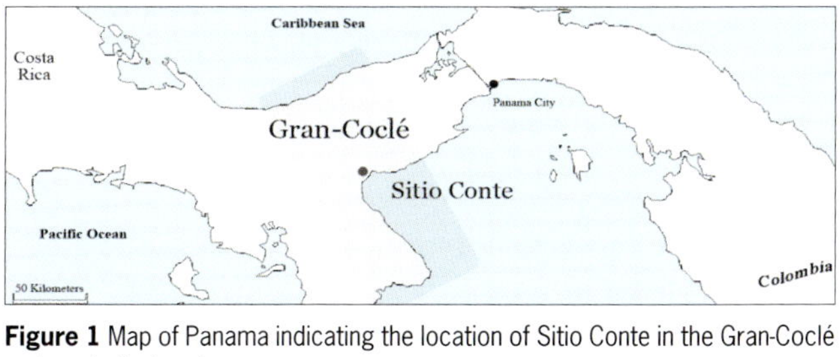

Figure 1 Map of Panama indicating the location of Sitio Conte in the Gran-Coclé archaeological region.

Background

Metallurgy arrived in Panama in the first several centuries AD through contact with the cultures of Colombia and Ecuador,[1] [1, 2]. While methods for smelting and casting had originated in South America, spreading from Peru to neighbouring regions and then into Central America, there remained a great deal of diversity in technology and style among cultures occupying this vast area. Panamanian goldwork shares many traits with the goldwork from Colombia, and the close relationship with Costa Rican material has led some researchers to consider these Isthmian countries as part of one larger metalworking region [3, 4].

Gold and copper were the metals of primary importance in Panama and were often combined in an alloy known as *tumbaga*. Many pre-Columbian gold objects, however, are made from a natural alloy of gold, silver and copper, which results from the use of gold grains and nuggets from secondary alluvial deposits that commonly contain small quantities of copper and silver. Gold fabrication techniques known to the pre-Columbian inhabitants of Panama include sheet forming, annealing, repoussé, chasing, lost-wax casting (closed core and open-back casting), soldering and surface enrichment, also known as *mise en couleur* or depletion gilding [4, 5].

Sitio Conte, a funerary site excavated by Harvard and University of Pennsylvania archaeologists in the 1930s, is the most well-known site for gold finds in Panama. It is located in Gran-Coclé, the region considered by archaeologists to encompass the culturally related pre-Columbian societies of central Panama (Figure 1). The gold objects excavated at Sitio Conte, which reportedly date to between AD 750 and AD 950, have provided the clearest evidence of complex, ranked societies from that time period [1]. The recently excavated beads discussed in this paper are roughly contemporary to

Table 1 Categorisation of beads included in study by fabrication method and type.

Fabrication	Type	Number	%	Total%
Cast beads	Small circular beads	4	0.2	1.32
	Large tubular beads	21	1.03	
	Animal/anthropomorphic effigy beads	2	0.1	
Sheet beads	Coiled with no join	46	2.26	98.68
	Biconical	2	0.1	
	Sheathing (complex shapes)	151	7.41	
	Rolled and joined	1812	88.91	

those from Sitio Conte and are believed to have been produced by a culturally related group in Gran-Coclé.

Gold beads provide an excellent starting point for looking at specific fabrication technologies used by Panamanian goldsmiths as they are a common and widespread class of artefact in Panama. Beads also represent a range of forming technologies from hammered sheet to lost-wax casting, allowing the study of a variety of production methods within one category of artefact. The large quantity of beads that have been found in Panama to date also provides a considerable sample for analysis.

Despite their prevalence, the references to Panamanian gold beads in the literature are rare and often contradictory. According to Emmerich, the necklaces excavated at Sitio Conte were composed of hundreds of identical cast gold beads, many of which were 'cast over inner cores' [6]. Conversely, the archaeologist at Sitio Conte, Samuel Lothrop, described the majority of beads as 'hammered or pressed' to shape over a perishable substrate [5]. While Stone and Balser also mention hammered beads among those that they examined, they suggest that the majority was made up of different types of cast gold beads, including cast tubes cut at regular intervals to produce identical smaller beads [4]. None of these assertions, however, is supported by analytical investigation.

Methodology

Examinations and analyses were carried out on the pre-Columbian gold beads from Panama in the two Smithsonian museum collections and from recent excavations in order to identify the fabrication techniques used in their manufacture. The study was divided into two phases: (1) preliminary investigation to determine the final sample set, and (2) analysis of the selected objects. Phase 1 included technical examination and X-ray fluorescence (XRF) analysis, which allowed the identification of different categories of beads by fabrication type, shape and features. Using this information, representative beads were selected from the most common categories for further analysis in Phase 2. In this second phase the selected beads underwent imaging and microanalysis using variable-pressure scanning electron microscopy with energy dispersive X-ray analysis (VP-SEM-EDX).

Preliminary investigation to select sample set

Roughly 2100 beads were included in the first phase of the study, comprising over 1700 beads at the National Museum of the American Indian, 60 beads at the National Museum of Natural History, and 231 beads from recent excavations in Panama. Of the recently excavated beads, 223 originated from the site of El Caño in the Gran-Coclé region. In cases where multiple beads were catalogued as a single item, such as the strung necklaces and bracelets, several representative beads were selected for examination and XRF analysis.

A total of 113 beads was analysed by calibrated XRF, using a portable bench-top ElvaX X-ray fluorescence analyser. Analyses were conducted at 45 kV for 100 seconds of live time in heavy element mode, which allows detection of elements from chlorine (Cl, atomic number 17) to uranium (U, atomic number 92). A set of approximately 35 matrix-matched gold reference standards (Royal Canadian Mint BCR 8079) were analysed by Jeff Speakman at MCI to create a material-specific calibration in the ElvaX software, reporting results in wt%. A selection of these standards was also analysed periodically during testing to adjust the calibration and monitor the instrumental precision and accuracy. The limits of detection for each element vary somewhat according to the matrix in which they are present. For copper and silver in a gold matrix, they are approximately 0.1–0.2% and c.0.25% respectively and for gold in a copper matrix, c.0.25–0.5%.

Using information gathered during technical examination, the beads were divided by fabrication technique into two main categories: cast and sheet fabrication (Table 1). Based on the presence of several visible casting features, such as dendrites, casting flaws and porous surface texture, only 27 beads were identified as cast, which represented 1.3% of the total beads examined. The cast beads ranged from large tubular beads to animal effigy beads and small plain circular beads. The beads fabricated from sheet were found to be far more numerous at just over 2000. Common features identified on these beads included flanges and burrs at edges, chisel lines around bead openings, rough texture or unidirectional striations on the interior surface, and crimped and creased edges. These beads were either coiled with no joined edge or alternatively joined with a bonded overlap (Figure 2). The beads with a bonded join were by far the most common type at 88.9% of all beads examined. While many of the beads examined were identified as rolled and joined sheet, a join line was visible on only a fraction of these beads. Finally, 151 of the sheet beads were considered to fall under the sub-category of sheathing, i.e. thin hammered sheet that had been formed by burnishing and crimping over organic cores in complex shapes such as shells and teeth.

Based on the information provided by the preliminary investigation, the sample set for further analysis was narrowed down to the most common category of bead – those formed from rolled and joined sheet. By focusing on the most frequently found bead type, it might be possible to characterise

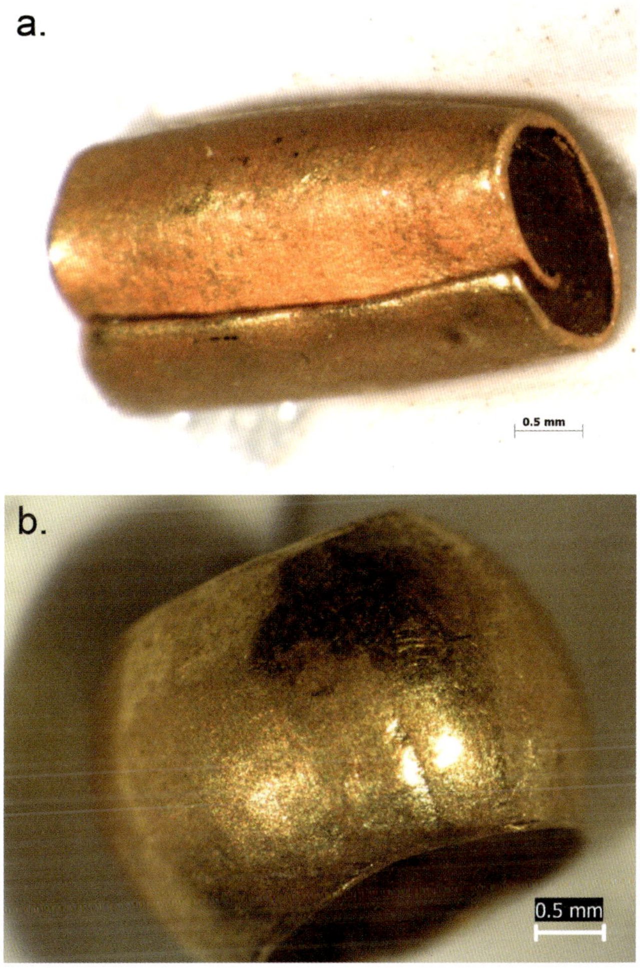

Figure 2 (a) Bead that has been coiled with no join (NA20 R5387). (b) Bead that has been rolled and joined (PAPG H7).

a large category of artefact from Panama through analysis of a small representative sample. Similarly, as the most numerous type of bead from recent excavations, the rolled and joined sheet beads had greater potential for the approval of sampling requests. This particular category of bead also raised several interesting research questions including the types of joining techniques used to form the bonded overlap.

Joining methods

The identification of methods used to bond the overlap of rolled sheet gold alloy beads is fundamental to our understanding of fabrication. The procedures for joining metals considered in this study include hot-working processes such as fusing, soldering, the formation of eutectic systems and sintering, and mechanical, cold-working methods such as crimping and burnishing.

Fusing is a metal bonding process using heat to unite two homogeneous metals without the addition of other materials. Adjacent surfaces are heated until they become fluid and diffuse into each other. This process takes place at the liquidus or melting temperature of the metal. Heat is rapidly transmitted through gold, silver and copper, owing to their superior conducting properties, so the possibility of melting the entire metal piece during the fusing process is greatly increased [7]. Although lack of variation in the metallic bond may make detection difficult, telltale signs of a roughened surface or wavy edges can be reasonable visual indicators of a fused seam.

Soldering is a method of joining metals by using heat to flow an alloy of a lower melting temperature metal between the two surfaces of a metal with a higher melting temperature. For our purposes, the solder would likely be formulated of gold, copper and silver in a ratio that reduces the melting temperature below that of the original alloy sufficient to flow the solder without melting the object. The solder required to achieve this disparity in melting temperatures may not be visible under magnification. However the compositional variation between the solder and original alloy may be measured by a variety of analytical techniques including SEM-EDX [8–10].

Another method of joining gold elements is the use of a eutectic system to form a bond. Eutectics are also referred to as proto-brazing [11], reaction-soldering [12] or – when attaching small granules of metal – granulation [13]. Gold alloys with a low percentage of copper will form a eutectic when copper salts in an organic adhesive are applied between the surfaces to be joined, followed by heating. Carbonisation of the adhesive reduces the salt to metallic copper, which forms an alloy with the original metal. The localised reduction in melting temperature results in bonding at the points of contact. The slight change in composition of the eutectic system has been identified in several previous studies using both surface analysis with SEM-EDX and analysis of cross-sections [14, 15].

A bond can be formed using powdered metal of the same composition as the bulk metal, which is packed into the seam with an organic adhesive. This process is called sintering. When the metal powder is heated to 65–80% of its solidus temperature the grains coalesce, closing the seam. Sintering may be distinguished by the granular appearance at the surface of the join; however, metallography highlighting the dendritic structure of the metal in the seam is a more reliable indicator [12].

Finally, cold-working approaches for joining gold beads include crimping and burnishing. For our purposes, crimping is the process of pushing one metal surface into another, thereby deforming one or both surfaces to form a join. Depending on the force applied, the metal surfaces may come into close contact along most of the seam. Crimped seams are more likely to be visible to the naked eye and will be obvious under magnification. Regions of work hardening may be visible in cross-sectional analysis.

Burnishing extends this process, bringing the two metal surfaces into intimate contact by applying sufficient force to the front of an object while the back is supported against a rigid surface. The metal surface is burnished or rubbed with a hard, rigid material, such as a polished stone, and the combination of pressure and friction compacts the two surfaces together. Burnished seams may not be visible to the naked eye although they may be visible under magnification, and regions of work hardening may be visible in cross-sectional analysis.

Beads selected for further analysis

To better understand the joining technology, more nuanced compositional information was required. SEM-EDX analysis of whole and sectioned beads was determined to be the best method available at MCI for this second phase of the study. As none of the Smithsonian museum material was permitted to undergo destructive sampling, 29 beads from the NMAI and NMNH collections (15 and 14, respectively) were selected to undergo SEM-EDX of the whole beads only. These 29 beads represent eight catalogued objects such as necklaces and bracelets, each with between 1 and 550 beads. Each of the selected beads exhibited obvious overlaps or visible seams and were clearly fabricated from sheet.

A representative sample of eight rolled and joined beads was selected from the 223 beads excavated at El Caño, for which export and analysis permission was granted. Four of these underwent further analysis at MCI, the results of which are featured in this paper. The four beads, R4853, R5101.2, R5341 and R5352, were all manufactured from hammered sheet with an overlap or seam visible on either the interior, exterior, or on both sides. The bead diameters measured between 2.15 mm and 4.05 mm and the lengths measured between 2.15 mm and 2.8 mm. While three of the beads are round in cross-section and in profile, one of the selected beads, R5352, is round in cross-section but tubular in profile. Bevels, burrs, chisel marks or a combination of these features was noted on the bead openings, or apertures. Three of the four beads also had evidence of abrasion in the form of a rough texture on the interior surfaces. XRF measurements indicated that all were high in gold, ranging from c.81 to c.93 wt%. Copper content was c.1–2 wt% while silver content ranged from c.5 to c.19 wt% (Table 2).

Table 2 XRF data of El Caño beads selected for further analysis.

Object	Au (%)	Ag (%)	Cu (%)
R5352.2	92.9	5.4	1.7
R4853	90.9	8.2	0.8
R5101.2	87.5	6.7	1.2
R5341	80.5	18.7	0.8

Analysis of selected beads: SEM-EDX and metallography

Prior to analysis, the selected El Caño beads were cleaned in an ultrasonic bath; cleaning was not permitted for the NMAI and NMNH beads. SEM-EDX was carried out first on the whole beads, and then subsequently on the mounted and sectioned beads from El Caño, using a Hitachi S-3700N VP-SEM with Bruker EDX. All beads were analysed at 15 kV at approximately 10–15 mm working distance. Analysis of the whole beads was carried out under full vacuum while the sectioned beads were analysed in VP mode. Elemental mapping was conducted over the surface of the joins in order to determine whether any compositional variation from joining techniques could be detected. Elemental maps were acquired in the Hypermap mode of the Quantax software for approximately 15 minutes.

Surface analysis of whole beads

Elemental mapping of gold, copper and silver concentrations showed homogeneous composition over the exterior surface of visible joins on all beads that underwent SEM-EDX. The elemental maps of bead A396672 (Figure 3) in the NMNH collection are typical in that they reveal no compositional variation across the join region. Comparison of the elemental count rates in areas across and adjacent to the join quantitatively confirmed the uniformity of composition. While no solder or eutectic bonding was detected in this initial step, no conclusions could be reached regarding the presence of these materials inside the seam as only the surface composition of these beads was measured. Therefore, variation in composition at the interior of the join required further investigation. Although the Smithsonian beads could not undergo destructive sampling, confirmation of the initial results was possible through cross-sectioning of the representative beads from El Caño.

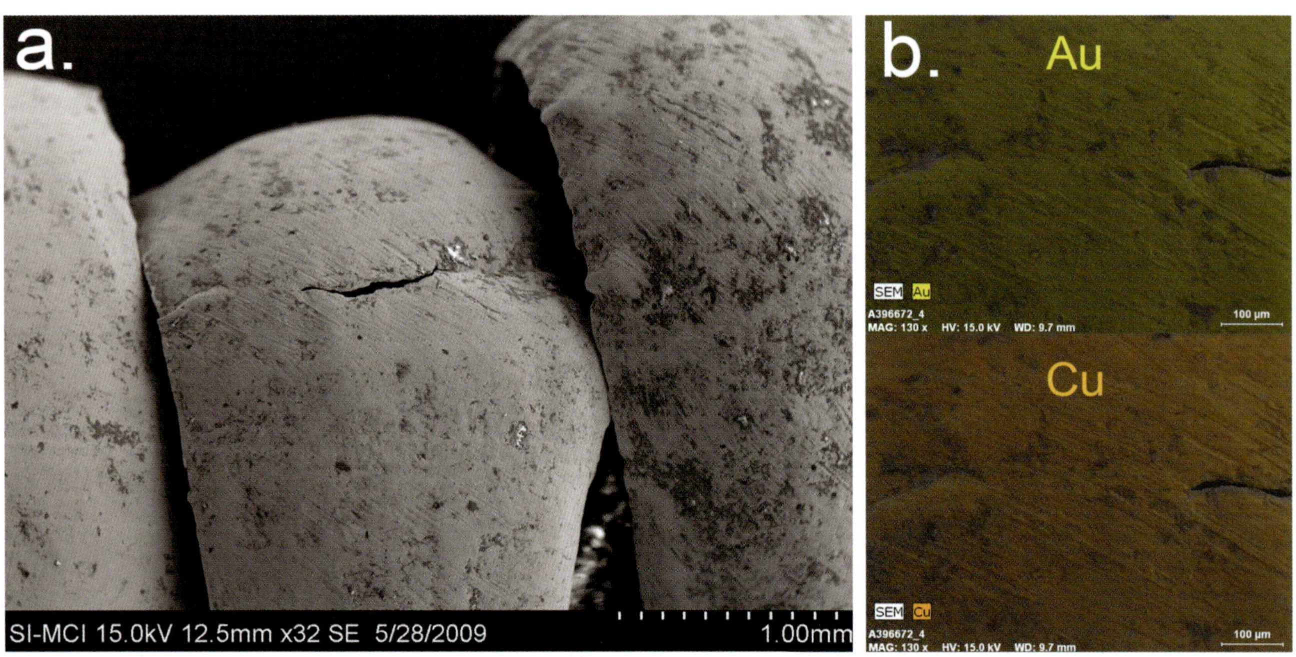

Figure 3 (a) SEM image of bead A396672 (NMNH). (b) Elemental maps of Au and Cu in the joint.

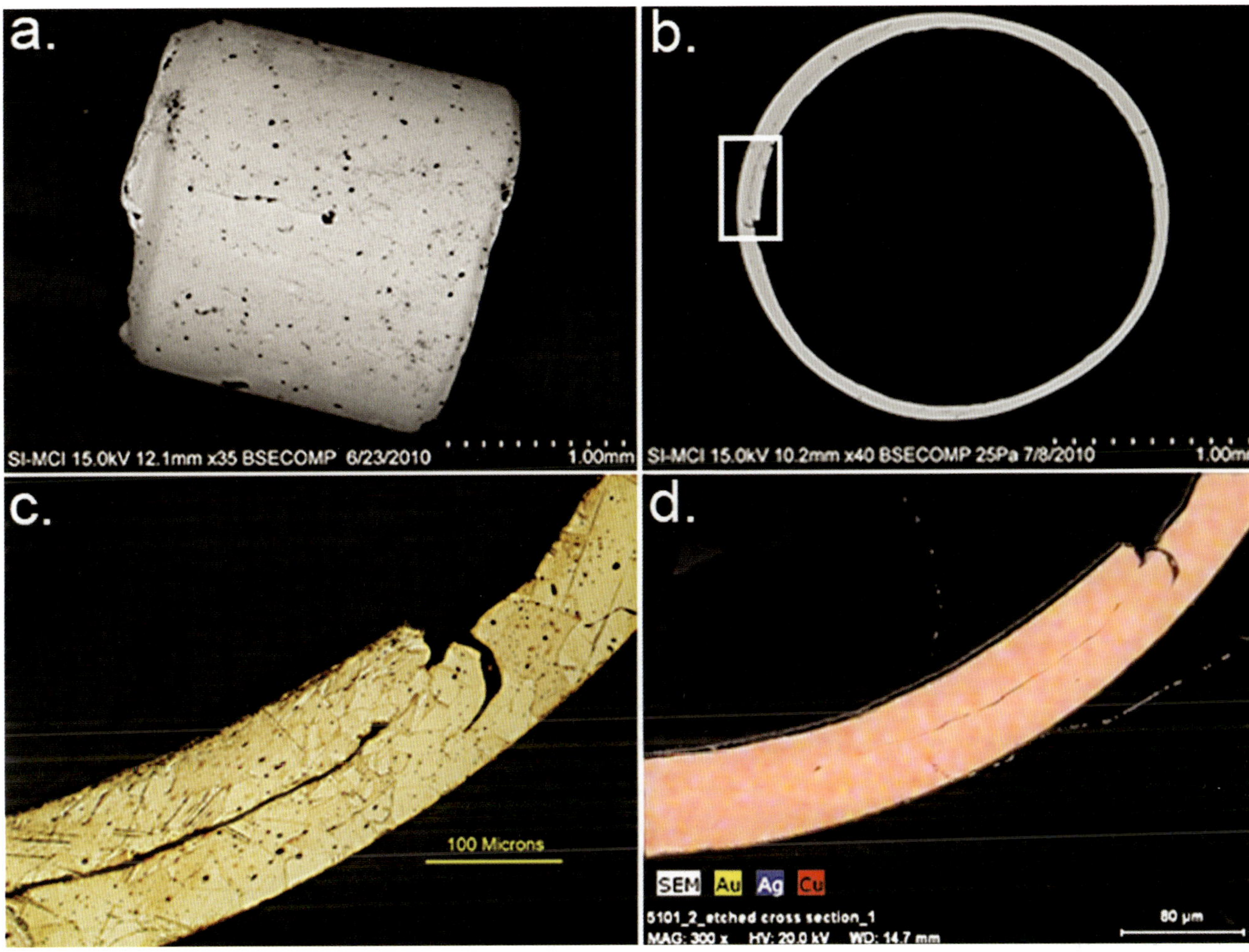

Figure 4 (a) SEM-BSE image of bead R5101.2. (b) SEM-BSE image of bead R5101.2 in cross-section. (c) Cross-section of R5101.2 etched in aqua regia. (d) Elemental map of Cu, Ag, Au concentrations in the join region of R5101.2.

Cross-sectional analysis

The four recently excavated beads (R4853, R5101.2, R5341 and R5352) were mounted in epoxy and polished down to reveal the cross-section along one edge. Bead number R5341 was also sectioned longitudinally with a low speed diamond saw to reveal the profile of the bead in addition to the remaining half of the cross-section. All of the samples were polished with a series of increasingly finer grit grinding papers up to 600 grit. The final polish was achieved on a lap wheel with 0.05 Alpha Alumina and water on a polishing cloth. The polished cross-sections were examined under the metallographic microscope, photographed and analysed by VP-SEM-EDX. After documentation and analysis of the un-etched samples, the cross-sections were etched in aqua regia and re-examined under the metallographic microscope, followed by carbon coating and analysis once again by VP-SEM-EDX.

Results and discussion

R5101.2

SEM-EDX analysis of bead number R5101.2 was first undertaken on the whole bead to measure the composition across the surface of the join. This small tubular bead, which measures just 2 mm long and 2 mm in diameter, has a join visible on both the exterior and interior surfaces, the overlap of which measures approximately 1.5 mm (Figures 4a and 4b). The inner surface of the bead is rough in texture while the exterior is polished with chisel marks present around the circumference of the bead aperture. An irregular flange is also visible on the interior of the bead openings. Elemental mapping at the surface of the join at ×160 revealed no variation in gold, copper and silver composition.

Examination of the cross-section with the metallographic microscope after etching in aqua regia revealed an irregularly shaped grain pattern from cold working (Figure 4c). The straight twinned grains throughout indicate that the sheet was annealed after working. Some of the twin lines at the join are slightly curved, however, suggesting additional localised deformation after the final annealing. The zone directly between the two layers at the very end of the overlap is tightly aligned across the grain boundary and a crack is present in the metal sheet adjacent to the end of the overlap, possibly having widened due to working at the join.

No variations in microstructure or colour were noted across the join region under the metallographic microscope, suggesting no addition of solder or other material. These initial findings were confirmed using EDX analysis. Elemental mapping of the overlapping region showed no compositional variation across the join other than that related to the variation in the microstructure of the metal itself (Figure 4d). The combined evidence from metallography and SEM-EDX suggests that the bead was joined without any solder or use of any other additive material. The chisel marks around the bead apertures indicate that the bead was cut from a longer tube.

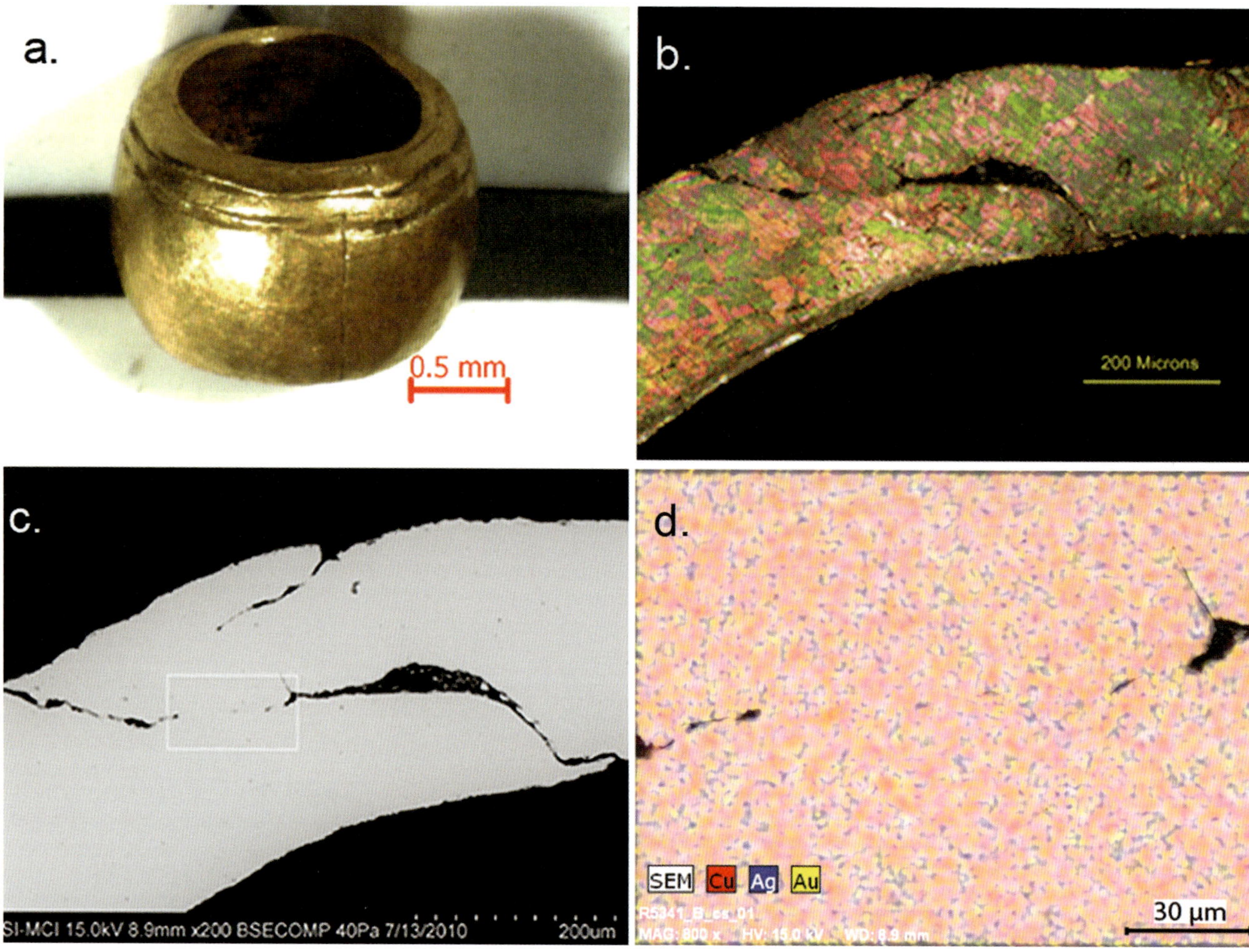

Figure 5 (a) Bead R5341. (b) Cross-section of R5341 etched in aqua regia. (c) SEM-BSE image of cross-section of the join region. (d) Elemental map of Cu, Ag, Au concentrations in the join region of R5341

R5341

Bead R5341 is fairly similar in shape to many of the other beads from El Caño; however, analysis by XRF revealed an unusually high percentage of silver, at 19 wt% with 80 wt% gold and approximately 1 wt% copper. The bead is round in cross-section and profile with an overlap visible on the interior and exterior of the bead (Figure 5a). The edges of the apertures are bevelled and a double chisel line is present around the circumference of one side.

As in the case of the NMAI and NMNH beads, SEM-EDX analysis across the exterior join revealed no anomalous variation in composition. Metallography exposed a distinct seam with no solder present but a small area where the grain boundaries appear tightly aligned across the join (Figure 5b). Elemental mapping using SEM-EDX provided further evidence of the homogeneity across the join region (Figure 5c). Count rates in areas at and adjacent to the join were also checked in the Quantax software to quantitatively confirm the visual interpretation of the elemental maps (Figure 5d).

R5352.1 and R4853

Analysis of the remaining two beads from El Caño produced very similar results to those already described. The sectioned beads revealed an overlapping join with no solder or compositional variation along the seam. The join regions were not found to be fused; however, the surfaces on either side were very closely aligned.

Discussion

Based on the results of our investigation, all of the beads from El Caño appear to have been fabricated using the same manufacturing technique: they were made from sheet that was formed into a tube with overlapping edges, joined, and then cut at regular intervals. The lack of compositional variation along the seams indicates that solder was not used as the joining method. Sintering with powdered gold alloy would leave no appreciable compositional variation but was ruled out as the characteristic microstructural features of this method were not present in our samples.

Eutectic bonding is also considered unlikely as we would anticipate a distinguishable difference in elemental distribution at the join. The compositional variation resulting from eutectic bonding has been identified through analysis in other studies; however, it should be noted that there are additional steps in fabrication that could prevent detection of this joining method. The diffusion mechanism that initially allowed the formation of a eutectic may continue with prolonged or repeated heating, causing continued diffusion of copper into the object [16]. The increased homogeneity in composition resulting from this mechanism could thus complicate detection of the eutectic zone by SEM-EDX analysis.

While burnishing of the seam over a substrate is considered a plausible joining method, there is some question as to whether this method alone can achieve a secure bond. Replication experiments in this area have had varied results suggesting that the specifics of the replication procedure probably play a crucial role. During experimental reproduction of burnished joins on sheet gold, Armbruster [16] was not able to produce a lasting join while Harling's experiments proved successful [17]. Further research involving experimental reproduction with all of these joining techniques for comparison with the archaeological beads will be necessary in order to determine the exact method of manufacture.

Conclusions

Initial results from examination and analysis of the Panamanian gold beads in the Smithsonian collections and from recent excavations in Panama suggest that the majority of these beads were manufactured from hammered sheet, formed into a tube with overlapping ends, joined along the seam and cut at regular intervals. While XRF and microscopic examination were effective methods for the categorisation of beads by type and composition, cross-sectional analysis coupled with SEM-EDX was essential in the investigation of joining methods. Although the beads in the NMAI and NMNH collections did not undergo destructive sampling, cross-sectional analysis of four representative beads recently excavated in Panama provided sufficient evidence to support our initial hypotheses regarding the fabrication of this category of bead. While results to date suggest that neither solder nor sintering methods were used to effect joins, experimental reproduction of beads using variations on the hypothesised method will be the next step in this research in order to fully work out the sequence, techniques and tools used in the manufacture of these beads.

Acknowledgements

We would like to thank Richard Cooke (STRI) and Arq. Jaime Zárate (Patrimonio Histórico, INAC) for their collaboration and for providing access in Panama to the recently excavated material. We are incredibly indebted to Jeff Speakman for providing the calibration and training for use of the XRF, to Judy Watson for sharing her SEM expertise, and to Javier Iñañez for his assistance in SEM sample preparation. We are also grateful to Pat Nietfeld (NMAI), Linda Greatorex (NMAI), Janet Pasciuk (NMAI), Deb Hull-Walski (NMNH), and David Rosenthal (NMNH) for providing access to the beads in the Smithsonian collections.

Note

1. The earliest known metal artefacts excavated in Panama, including both copper and gold alloys, were discovered at Cerro Juan Díaz in a grave with a calibrated radiocarbon date of AD 130–370. R. Cooke, Smithsonian Tropical Research Institute, personal communication (May 2009).

References

1. Cooke, R.G., Sánchez, L.A. and Udagawa, K., 'Contextualized goldwork from "Gran Cocle", Panama: an update based on recent excavations and new radiocarbon dates for associated pottery styles', in *Precolumbian Gold: Technology, Style and Iconography*, ed. C. McEwan, British Museum, London (2000) 154–176.
2. Cooke, R.G., Isaza, I., Griggs, J., Desjardins, B. and Sánchez, L.A., 'Who crafted, exchanged, and displayed gold in pre-Columbian Panama?', in *Gold and Power in Ancient Costa Rica, Panama, and Colombia*, ed. J. Quilter and J.W. Hoopes, Dumbarton Oaks, Washington, DC (2003) 91–158.
3. Scott, D.A., 'Goldwork of pre-Columbian Costa Rica and Panama: a technical study', in *Materials Issues in Art and Archaeology IV: Symposium held 16–21 May 1994, Cancun, Mexico*, Materials Research Society 352 (1995) 499–526.
4. Stone, D. and Balser, C., *Aboriginal Metalwork in Lower Central America*, Editorial Antonio Lehmann, San José (1958).
5. Lothrop, S.K., *Coclé: An Archaeological Study of Central Panama*, vol. 1, Cambridge, MA (1937).
6. Emmerich, A., 'Master goldsmiths of Sitio Conte', *Natural History* **74**(8) (1965) 18–25.
7. Brepohl, E., *The Theory and Practice of Goldsmithing*, Brynmorgen Press, Portland, ME (2003).
8. Scott, D.A. and Doehne, E., 'Soldering with gold alloys in ancient South America: examination of two small gold studs from Ecuador', *Archaeometry* **32**(2) (1990) 183–190.
9. Duval, A.R. and Eluere, C., 'Solder characterization on ancient gold artifacts with the electron microprobe', *Scanning Electron Microscopy* **4** (1986) 1331–1335.
10. Demortier, G., 'The soldering of gold in antiquity', in *Materials Research Society Symposium Proceedings No. 123, Symposium held 6–8 April 1988, Reno, Nevada*, ed. E. Sayre, P. Vandiver, J. Druzik and C. Stevenson, Materials Research Society, Pittsburgh (1988) 193–198.
11. Griffin, J.G., 'An investigation in joining metals', in *Metalurgia de América Precolombina*, ed. C. Plazas, Banco de la República, Bogotá (1986) 369–376.
12. Echt, R. and Thiele, W.R., 'Sintering, welding, brazing and soldering as bonding techniques in Etruscan and Celtic goldsmithing', in *Prehistoric Gold in Europe: Mines, Metallurgy and Manufacture*, ed. G. Morteani and J.P. Northover, Kluwer Academic Publishers, Boston (1995) 435–450.
13. Wolters, J., 'The ancient craft of granulation: a re-assessment of established concepts', *Gold Bulletin* **14**(3) (1981) 119–129.
14. Eluere, C., 'A "gold connection" between the Etruscans and early Celts?', *Gold Bulletin* **22**(2) (1989) 48–55.
15. Duval, A.R., Eluere, C., Hurtel, L.P. and Menu, M., 'The use of scanning electron microscopy in the study of gold granulation', in *Archaeometry: Proceedings of the 25th International Symposium* (1989) 325–333.
16. B. Armbruster, University of Toulouse, personal communication (September 2010).
17. D. Harling, Kentucky School of Craft, Hazard Community and Technical College, personal communication (July 2010).

Authors' addresses

- Ainslie Harrison, Metropolitan Museum of Art, Objects Conservation Department, New York, USA (ainslie.c.harrison@gmail.com)
- Kim Cullen Cobb, Museum Conservation Institute, Smithsonian Institution, Suitland, MD, USA (cobbk@si.edu)
- Harriet F. Beaubien, Museum Conservation Institute, Smithsonian Institution, Suitland, MD, USA (beaubienh@si.edu)
- Paul Jett, Freer Gallery of Art and Arthur M. Sackler Gallery, Smithsonian Institution, Washington, DC, USA
- Julia Mayo, Smithsonian Tropical Research Institute, Center for Archaeology and Paleoecology, Panama (mayoj@si.edu)

Subsurface analysis by application of FIB-SEM to samples of geological and historical importance

Diane Johnson, Stuart Kearns and Monica M. Grady

ABSTRACT The successful application of focused ion beam scanning electron microscopy (FIB-SEM) to a range of geological and historical objects has shown that important three-dimensional information can be revealed, enabling the characterisation of the fine structure and chemical composition contained within the samples. Fossil feathers from theropod dinosaurs showed submicron-scale details suggestive of cellular to sub-cellular level preservation. Ancient coins revealed surface finishing processes such as gilding. Further applications of the FIB-SEM technique were explored, most specifically in the study of meteorite weathering processes occurring during curation in the restricted environment of a museum collection. Examination of the internal chemical distribution of rusting patches on and within iron-nickel alloy regions of meteorites help to define best storage conditions for samples containing significant metal areas. One of the advantages of the technique is that samples may be analysed intact and without any preparation. Charge neutralisation can also be employed for analysis of non-conductive samples, allowing stable milling of a sample without charge build.

KEYWORDS FIB, microstructure, *in situ*, archaeometry, conservation

Introduction

Analysis of precious materials has always led to concern over sample damage and loss of material through cutting or polishing in preparation for microscopy examination. When electron microscopy or ion beam technology is employed, there is often the additional complication of coating a sample with a conductive layer, potentially introducing contamination; coating is also almost always unacceptable for historical artefacts.

A FIB-SEM is an instrument with two separate beams: a scanning electron (SE) microbeam and focused ion beam (FIB); the two beams share a common sample chamber. As a combined technique, the SE microbeam is used to view and analyse a surface, whilst the FIB can mill away small areas, allowing direct analysis of the freshly revealed subsurface by the SE microbeam. If this can be performed without macroscopically cutting or coating the sample, any damage caused by FIB milling is restricted to a scale of microns, and thus is almost non-existent. The ethical issues associated with the conservation and analysis of cultural heritage materials are frequently complex and require careful consideration [1, 2]. This FIB-SEM study attempts to minimise damage to the artefact making the risk associated with analysis acceptable.

When the FIB-SEM was originally developed its use became common in the semiconductor industry and was applied primarily to locate faults and modify circuits on silicon devices [3]. However, since then it has also become a routine research analysis tool in the field of materials science, being widely employed to characterise samples principally through the manufacture of transmission electron microscopy (TEM) sections from precise locations. FIB-SEM has also found applications within specialist areas of scientific research, such as the study of space materials. For example, when the NASA Stardust mission recovered particles from the coma of Comet Wild-2, the submicron size of the samples and the small impact structures they produced on parts of the collector array meant that FIB-SEM was the ideal tool to study these samples, allowing their extraction from very precise locations for subsequent analysis [4, 5]. Another recent research application within the space sciences field is the analysis of interstellar dust grains that have been identified within primitive meteorites. Because of their nanometre-scale dimensions, the FIB-SEM is an important tool for extracting the grains for further detailed structural, compositional and isotopic analysis [6].

The FIB-SEM results discussed here were obtained using an FEI Quanta 200 3D FIB-SEM at the Open University, UK. The instrument incorporates an *in-situ* gas injection system (GIS), allowing deposition of platinum on the sample surface. It also has a lift-out device attached to the inside of the sample chamber, allowing extraction of TEM wafers under vacuum. Secondary (SE) and backscatter (BSE) electron detectors and an energy dispersive X-ray analyser (EDX) enable imaging and qualitative elemental analysis of extracted wafers.

Samples

A range of samples was chosen for FIB-SEM characterisation on the basis of specific analytical requirements and to illustrate the variety of sample structures and compositions. For the historical materials, if sample preparation could not be avoided, it was kept to an absolute minimum in order to reduce sample damage and contamination. The damage caused by FIB milling is localised to the site of interest; it is typically restricted to a micron scale and is one of the unique attributes of this technique.

Fossils

Samples of fossilised therapod dinosaurs and early bird species from the 125 million-year-old Jehol Group, north-east China [7], which consisted of small, typically sub-centimetre sized, fragments of fossil feathers, were mounted on Al pin stubs using silver paint or double-sided carbon tape, then coated in a thin layer of gold prior to SEM and FIB-SEM analysis (Table 1).

Ushabti

Ushabtis are ancient Egyptian funerary figures that were made in a variety of materials including wood, stone, and commonly, glazed faience. Faience is a mixture of crushed quartz, calcite, lime and alkali mixtures such as natron [8, 9]. Copper oxide was also a common component of the faience, producing a strong blue coloration when fired in an oven. Ushabtis were usually placed close to the sarcophagus inside a tomb, typically small in size (less than 10 cm length), and often produced as multiples. The Ushabti was believed by ancient Egyptians to animate after the judgment of the dead and then work as a servant or protector of the deceased [10].The example reported here is faience with a blue lustre glaze and is dated at 400–600 BC. This sample was analysed intact (Table 1).

Shipwreck coin

The story of the *El Cazador* (*The Hunter*) shipwreck is an important part of American history. In the late 1700s paper currency was devalued by economic failure and Spain controlled almost a million square miles of North America. King Carlos III decreed that newly minted Mexican silver coinage would replace what had become almost worthless paper currency. The Spanish warship *El Cazador* was transporting hundreds of thousands of these silver coins when it was lost at sea, last sighted in June 1784. The loss of this cargo destabilised the economy further and Spain gave this territory back to France, who sold it to America at a very low price in 1803. In 1993, while trawler fishing, a boat called the *Mistake* accidently caught silver coins in its net; this led to the discovery of the *El Cazador* shipwreck site [11] and the recovery of hundreds of thousands of silver Reale coins. The ship's bell and a number of cannons were also recovered. The coin discussed here is an 8 Reale coin, known commonly as 'pieces of eight'. It was franked in 1783, with assayer initials of FF, and is approximately 4 cm in diameter. Most of the coinage when recovered was naturally fused into large clumps because of long-term seawater exposure. This coin was already separated and cleaned of decay when examined (Table 1) and required no further preparation.

Roman coin

The coin analysed is a silver denarius of Severus Alexander that dates from approximately AD 222–228. The obverse of the coin features the laureate head looking right; on the reverse Mars is standing looking right holding a shield and spear (Table 1). This coin was already cleaned free of contamination so no further sample preparation was required; it was analysed intact.

Meteorite

The meteorite analysed is classified as a pallasite and named Brenham, which is the find location in Kiowa County, Kansas, USA. The meteorite was first discovered in 1882, and samples are still being recovered – over 1 tonne of material so far including a 635 kg oriented individual sample found in 2005 that formed the leading edge of the main meteoroid body as it passed through the atmosphere approaching the earth. Brenham is classified as a pallasite; these are stony-iron meteorites that are generally believed to have formed at the core–mantle boundary region of a large asteroid. It is composed

Table 1 Summary of the variety of samples examined by FIB-SEM: compositions, ages and structures.

Sample	8 Reale *El Cazador* shipwreck	Roman coins	Ushabti	Dinosaur fossils	Meteorite
Composition	Metal	Metal	Silicates	Silicates	Metal and Silicates
	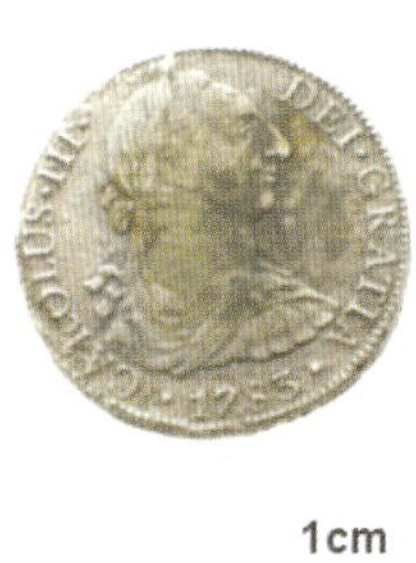	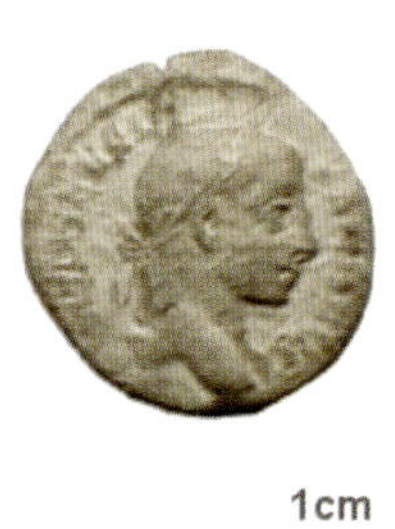	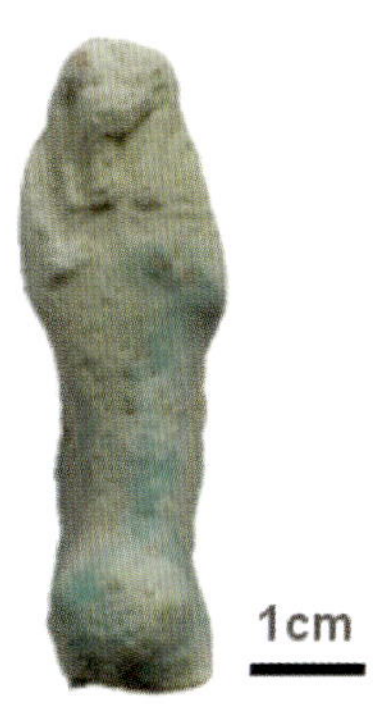	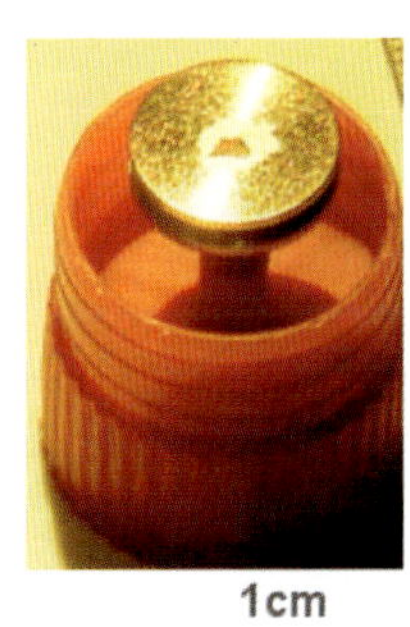	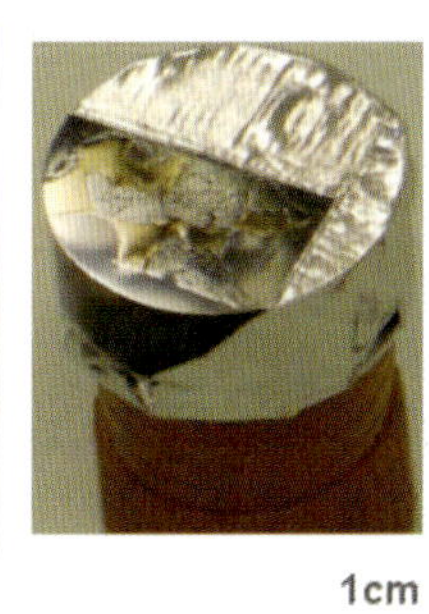
Age	227yrs	1790yrs	2410yrs	120Myrs	4500Myrs

predominantly of large centimetre-sized olivine crystals and an iron-nickel alloy. It also contains small quantities of iron sulphide (FeS), chromite ($FeCr_2O_4$) and schreibersite ($(Fe,Ni)_3P$). Meteorites are witnesses to the birth of the solar system and contain important information about its subsequent evolution. This sample was cut with a circular saw and embedded in epoxy, polished and carbon coated prior to analysis (Table 1).

Methods

FIB-SEM is a technique that allows both an electron beam and a focused ion beam to share access to a sample through a common sample chamber. Both beams are aligned and focused at the eucentric height of the sample centred on the site of interest. The SEM can be used to characterise the sample producing BSE and SE and EDX analysis via point spectra and element mapping. The electron imaging in this study was always performed at 20 kV, beam current 0.6 nA, FIB at 30 kV (beam current ranging from 1 nA to 20 nA).

The FIB has a gallium liquid metal ion source, having a spot size relating to beam current magnitude; this is rastered over the sample surface to produce an image similar to the electron beam. When the high energy Ga beam hits the sample surface a range of processes take place. Inelastic scattering causes the production of secondary and auger electrons and photons in addition to other processes taking place such as gallium ion impregnation within the surface of the sample. Hence an image recorded using the FIB and the SE detector is composed of a broad range of signals. It can therefore appear different to that of an electron beam secondary detector image of the same area. When elastic collisions take place between the FIB and target sample atoms, momentum is transferred and results in sputtering of the target atoms – this is the physical principle of the FIB milling technique. If the beam current is sufficiently high and scan speed slow then the sample can start to mill slightly during imaging so appropriate settings must be chosen carefully and adjusted to the individual sample and its electrical properties.

When milling significant volumes of material from the sample, such as a trench structure to reveal cross-sectional information, the volume milled away relates directly to beam current. Often a trench is milled in stages with the majority of the material removed with a high beam current. Then a final cleaning trench is milled across the face of the trench with a low beam current. However, it can sometimes be difficult to select an ideal beam current as the beam can cause complicated unwanted structures. With high beam currents this frequently takes the form of large columns of material or vertically oriented lines dispersed across the trench face, often called the 'curtain effect', as can be seen in Figure 1. This is due to variations in the elemental composition of specimens: elements with low atomic numbers are preferentially milled relative to elements with high atomic numbers, producing grooves and ridges. Deposition of a layer of platinum above the trench helps to eliminate this effect [12]. A second problem is that if the beam current is too low, sample atoms can redeposit across the face such that the trench has well-defined edges and structure, but the face takes on a hazy appearance that blocks out details of the underlying sample structure. The problems of the curtain effect and sample redeposition can be overcome by choosing a more appropriate beam current mainly determined by experimentation, although sometimes a high beam current is appropriate with electron beam charge neutralisation.

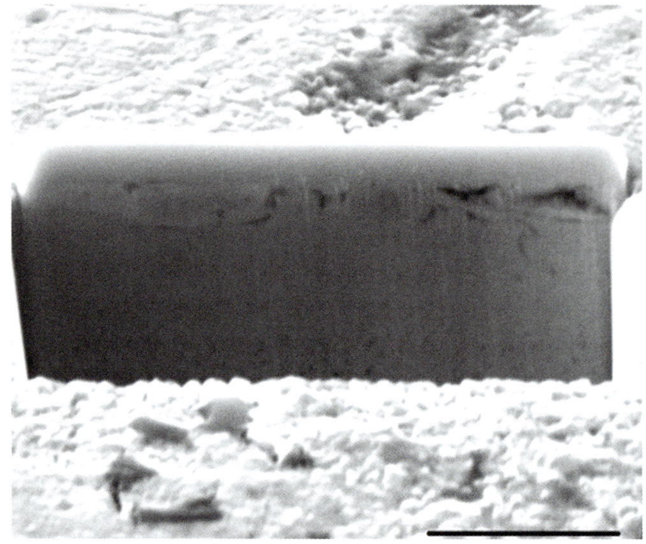

Figure 1 SE image of trench FIB milled into 8 Reale coin revealing mercury patches at the surface. Hole structures are also visible in the lower half of the trench, as are the thin vertically oriented lines down the trench face formed by the interaction of the FIB with the sample of varying localised composition, often known as the curtaining effect. Scale bar = 5 µm.

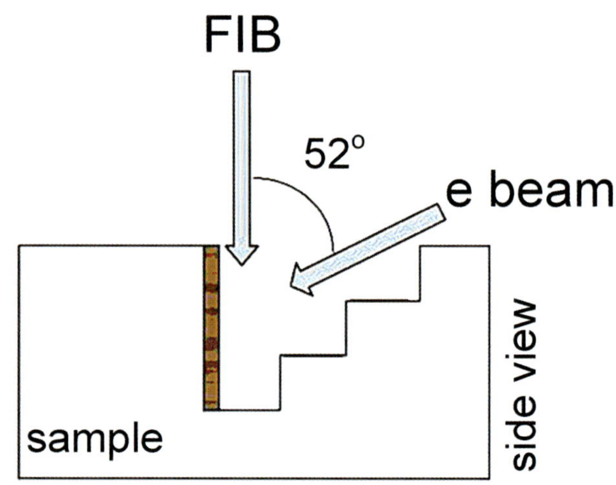

Figure 2 Diagram of the geometry of a typical FIB milled trench structure allowing direct electron imaging of the sample subsurface.

For each of the samples in the current study, a surface was initially imaged with SEM to identify a feature of interest. The FIB was then used to deposit a 1 µm wide and 1 µm deep strip of platinum across the specific feature that was to be milled; the strip runs along the length of the trench to minimise charge build up. A trench was then milled into the site of interest; this has a step structure into a surface that terminates in a flat vertical clean face. It is this face that is illuminated by the electron beam to view a cross-section through the subsurface (Figure 2). In samples of poor conductivity, additional measures are required to stabilise the sample during the milling process. In some samples, it is acceptable to cut, polish and coat material, which creates an ideal substrate for milling. But if we wish to avoid destroying and contaminating samples, more conservative *in-situ* methods can be applied, such as charge neutralisation. This is a technique that neutralises the positive charge of the FIB (which can, in some cases, build up on the sample) with a

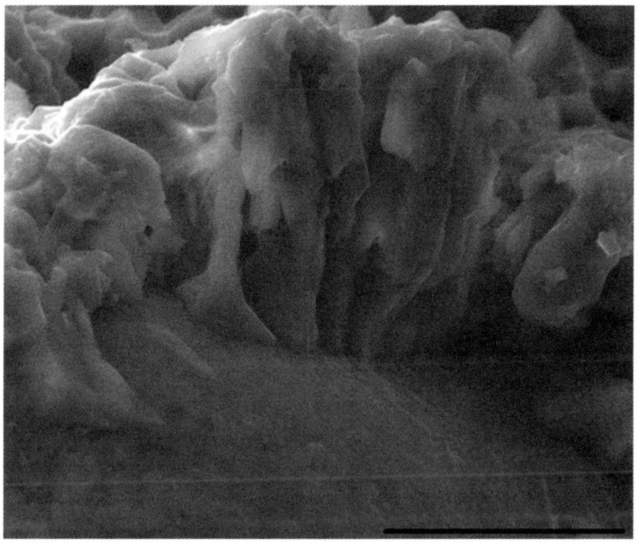

Figure 3 SE image showing detail of Ushabti surface. Scale bar = 30 μm.

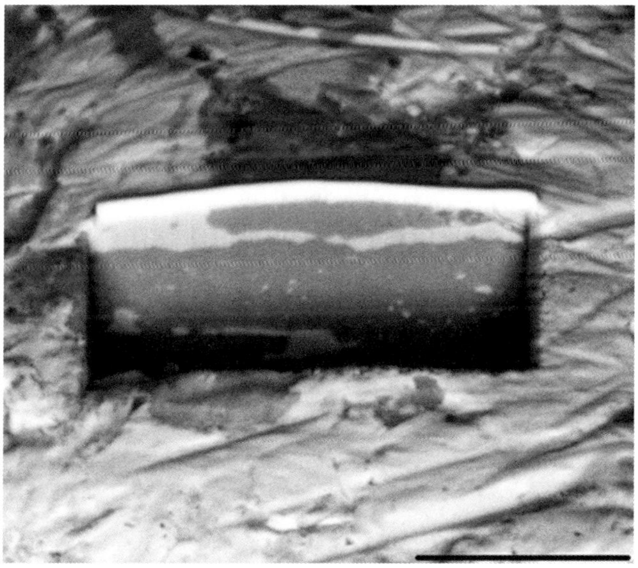

Figure 4 SE image of trench FIB milled into Roman denarius coin revealing layers of silver and copper intermixed that crystallised out due to immiscibility. Scale bar = 10 μm.

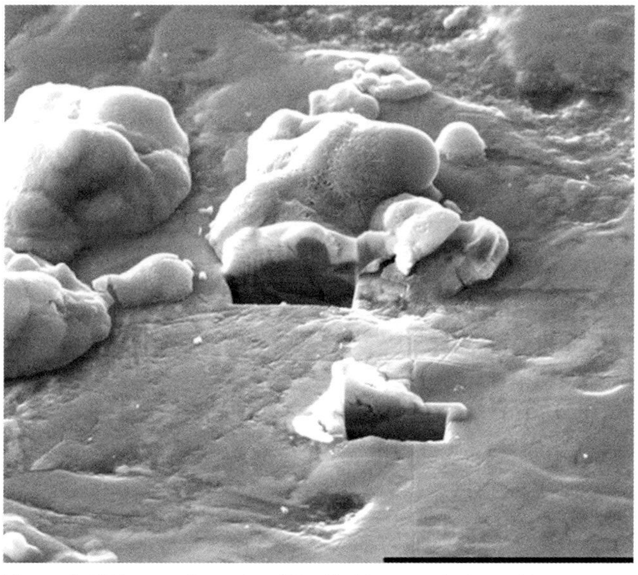

Figure 5 SE image of trenches FIB milled into weathering oxide structures grown on a metal area of a meteorite in the laboratory over one month's air exposure. Scale bar = 20 μm.

localised cloud of electrons produced by the SEM. The exact settings for the electron beam must be determined for each individual sample; accelerating voltage and beam current, along with an option to defocus the electron beam by a fixed amount, allows appropriate values to be finely set.

Results and discussion

Fossils

The feathered dinosaur fossils were characterised in a previous study and found to contain micron- to submicron-scale cellular structures interpreted as melanosomes [13], which are the colour-bearing melanin-containing cells. These structures were classed into two types: eumelanosomes, rod-shaped cells producing dark grey to black colours and phaeomelanosomes, sub-spherical cells producing browns and reds. The morphology and distribution of the melanosomes is identical to those of modern-day bird feather melanosomes. Interpretation of the structural form and distribution of the melanosomes was used to decipher the fossil colours and produce true colour reconstruction images of feathered dinosaurs. These fossils have such exceptional preservation that we are now making efforts to examine them for possible sub-cellular structures such as cell walls by milling out thin wafers of fossil from strategic points for further analysis by TEM studies.

Ushabti

SEM images revealed that the surface has large grains embedded within it that are coated with a smooth compact layer of material. In places this layer has rough regions consisting of angular fragments largely composed of calcium carbonate and small amounts of quartz. They appeared to be oriented linearly away from the surface to a thickness of 50 μm. The area selected for FIB milling was a compact smooth region; an example of such an area is shown in the lower half of Figure 3. This sample required the use of charge neutralisation with the electron beam during FIB milling. A well-defined trench was produced that clearly shows a subsurface interfacial layer (c.2 μm thick) buried approximately 1 μm under the surface. X-ray analysis revealed a Cu content of this layer of 13 wt%, much higher than in the main body of the Usahabti. The strong contrast revealed on the image produced with the FIB may also indicate some localised crystallography, possibly related to the heating and cooling process involved in the original production of the Ushabti. The contrast can appear on FIB images as a result of the relative alignment or disalignment of the FIB beam with the sample crystallographic planes: when aligned, the beam penetrates more deeply because of fewer collisions with sample atoms [14].

Shipwreck coin

A series of trenches was milled under identical beam conditions, revealing that for the greater part of the coin, the outermost 10 μm of the surface was composed of silver. Small areas of mercury were observed always at the surface level within 2 μm depth of the surface (Figure 1). The coin surface is generally smooth and continuous, with some fracture lines present, although there are areas that have degraded into patches of roughened oxidised damage. The trenches milled all indicated uniform silver composition except for occasional mercury patches or quartz grains. An interesting structural feature is observed 5 μm below the surface, where seemingly randomly positioned holes of less than 100 nm diameter occur across the trench face (Figure 1). It is impossible to determine from these trenches if the porous structure formed as a consequence of the time that the coin spent under the sea, or if it is an original feature of the coin production. As a more compact layer overlies the porous layer, this is suggestive of a manufacturing process that produced the bulk of the structure followed by a surface finishing process, perhaps mercury gilding, to leave a more robust silver layer finish.

Roman coin

The overall surface structure was observed to be quite roughened with significant topographic variation on a scale of tens of microns. Some areas have substrate materials splitting away from the surface in localised patches. When viewed with BSE, significant contrast is seen between small spots of less than 10 μm in diameter, which X-ray analysis revealed to be largely of copper, and the majority of the surface that has a composition of 80 wt% Ag with minor amounts of copper, aluminium and palladium. A series of trenches was milled into the surface of the coin, and layers of silver and copper revealed, with some layers mainly of copper with minor inclusions of silver (Figure 4). The structure is interpreted as resulting from the immiscibility of silver and copper: as the metal started to cool, the elements began to separate out in layers with areas of dendritic crystal growth.

Meteorite

The Brenham pallasite has a well-documented history of oxide and rust formation, but study of this alteration process in meteorite samples held in the restricted environment of a meteorite collection has been limited. A randomly chosen sample of Brenham was subjected to air exposure over a period of one month and monitored for oxide formation on the surface. Significant changes were observed to take place between 14 and 21 days in the form of thin layers of oxide associated with chlorine, becoming even more pronounced by 28 days, where large domes, spheres and tube structures formed (Figure 5). FIB-SEM was applied to mill away a small volume of one of these altered dome structures to allow a direct view of the internal chemistry of the structure. The milled surface was imaged and element maps produced across the face of the resulting trench. Iron was present throughout the area of the sample, as expected, as was a band of oxygen in the sample subsurface, feeding the larger protruding surface structure. This prominent surface feature was also found to contain approximately 10 wt% chlorine. This result is not unexpected, as chlorine is often encountered as an industrial atmospheric pollutant to which meteorites are exposed upon atmospheric entry. However, it might be advantageous to explore more thoroughly the storage conditions of metal-rich meteorites in order to identify the environmental conditions that allow maximum preservation for a range of different meteorite compositions.

Conclusions

FIB-SEM as a general technique for materials analysis is a very powerful tool with great potential for many new applications. It has the ability to make precise cuts into a surface on a small but very exact scale. When a simple trench is milled into a sample surface, a cross-sectional view of the subsurface environment can be directly imaged, with either an electron beam or focused ion beam, which can be complemented by X-ray analysis. It is then relatively easy, if required, to remove a thin slice of the sample centred on the feature of interest for additional analysis by other techniques. The FIB-SEM allows high precision, three-dimensional characterisation of micron-scale structures. The standard approach is to cut the sample with a macroscopic saw, polish then coat with a conductive layer of carbon or metal, but with use of the FIB-SEM these steps may be avoided. Specifically, if a sample is uncoated, then a positive charge can build up over its surface as a result of the FIB milling process. Deposition of a platinum strip over the area where a trench is to be milled supports charge dispersion, allowing stable FIB positioning. A charge neutralisation technique employing the broad, defocused negatively charged SEM electron beam can also be applied when appropriate. The operational parameters of the two beams require careful adjustment, but with patience, suitable conditions can be found for most samples, allowing stable milling in either a coated or uncoated state, with significant topography or flat polished surfaces. Here we have illustrated that the applications of FIB-SEM extend beyond the semiconductor industry and materials science research. Precious samples can be analysed in detail with minimal preparation, allowing preservation of important and valuable materials, and reducing damage and contamination.

Acknowledgements

We would like to acknowledge STFC (grant: ST/F003102/1) and NERC (grant: Ne/E011055/1). We would also like to thank the Institute of Vertebrate Palaeontology and Palaeoanthropology, Chinese Academy of Sciences, Beijing, China for allowing access to fossil samples and the sharing of expertise.

Suppliers

- Quanta 200 3D FIB-SEM and platinum gas injection system: FEI, Oregon (www.fei.com/)
- Liftout device Xtreme access: Ascend Instruments, Omniprobe, Dallas, Texas (www.omniprobe.com/)
- Silicon Drift Detector, 80 mm X-Max: Oxford Instruments, Abingdon, Oxfordshire (www.oxford-instruments.com/)

References

1. Ashley-Smith, J., *Risk Assessment for Object Conservation*, Butterworth-Heinemann, London (1999).
2. Richmond, A., 'The ethics checklist – ten years on', *V&A Conservation Journal* **50** (2005) 11–14.
3. Zschech, E., Langer, E., Engelmann, H.J. and Dittmar, K., 'Physical failure analysis in the semiconductor industry', *Materials Science in Semiconductor Processing* **5**(4–5) (2002) 457–464.
4. Zega, T.J., Nittler, L.R., Busemann, H., Hoppe, P. and Stroud, R.M., 'Co-ordinated isotopic and mineralogic analyses of planetary materials enabled by in-situ lift out with a focused ion beam scanning electron microscope', *Meteoritics and Planetary Science* **42** (7/8) (2007) 1373–1386.
5. Graham, G.A., Teslich, N.E., Kearsley, A.T., Stadermann, F.J., Stroud, R.M., Dai, Z., Ishii, H.A., Hutcheon, I.D., Bajt, S., Snead, C.J., Weber, P.K. and Bradley, J.P., 'Applied FIB techniques for sample preparation of astromaterials for integrated nanoanalysis', *Meteoritics and Planetary Science* **43**(3) (2008) 561–569.
6. Vollmer, C., Brenker, F.E., Hoppe, P. and Stroud, R.M., 'Direct laboratory analysis of silicate stardust from red giant stars', *The Astrophysical Journal* **700**(1) (2009) 774–782.
7. Chen, P.J. (ed.), *The Jehol Fossils: The Emergence of Feathered Dinosaurs, Beaked Birds and Flowering Plants*, Academic Press, Shanghai (2008).
8. Nicholson, P.T. and Peltenberg, E., 'Egyptian faience', in *Ancient Egyptian Materials and Technology*, ed. P.T. Nicholson and I. Shaw, Cambridge University Press, Cambridge (2000) 177–194.
9. Vandiver, P., 'Appendix A', in *Ancient Egyptian Faience: Analytical Survey of Egyptian Faience from Pre-dynastic to Roman Times*, ed. A. Kaczmarczyk and R.E. Hedges, Aris and Phillips Ltd., London (1983) A1–A4.
10. Brier, B., *The Encyclopedia of Mummies*, Checkmark Books, New York (1998) 186.
11. Broad, W.J., '1784 Spanish ship is found in gulf,' *New York Times* (19 December 1993).
12. Giannuzzi, L.A. and Stevie, F.A. (eds), *Introduction to Focused Ion Beam Instrumentation, Theory, Techniques and Practice*, Springer, New York (2005) 178.
13. Zhang, F., Kearns, S.L., Orr, P.J., Benton, M.J., Zhou, Z., Johnson, D., Xu, X. and Wang, X., 'Melanosomes and the colour of Cretaceous dinosaurs and birds', *Nature* **463** (2010) 1075–1078.
14. Giannuzzi, L.A. and Stevie, F.A. (eds), *Introduction to Focused Ion Beam, Instrumentation, Theory, Techniques and Practice*, Springer, New York (2005) 43–46.

Authors' addresses

- *Corresponding author:* Diane Johnson, Planetary and Space Sciences Research Institute, Open University, Milton Keynes, UK (D.Johnson@open.ac.uk)
- Stuart Kearns, Department of Earth Sciences, University of Bristol, Bristol, UK (Stuart.Kearns@bristol.ac.uk)
- Monica M Grady, Planetary and Space Sciences Research Institute, Open University, Milton Keynes, UK, and Natural History Museum, London, UK (M.M.Grady@open.ac.uk)

Scanning electron microscopy imaging of tool marks on Qin bronze weapons using silicone rubber impressions

Xiuzhen Janice Li, Marcos Martinón-Torres, Nigel Meeks and Yin Xia

ABSTRACT Scanning electron microscopy (SEM) allows the detailed examination of tool and wear marks on ancient objects under high magnification. However, the SEM study of archaeological artefacts is limited by the chamber size, and by the potential difficulties of having to move objects to laboratories for analysis. This paper presents a pilot study in which dental silicone rubber (vinyl polysiloxane) was employed to obtain accurate impressions of tool marks on the surfaces of metal artefacts for SEM examination, overcoming the problems of transportation and size. Methodological protocols, potentials and limitations are discussed. The features examined were the inscriptions, filing, grinding and polishing marks, but no obvious wear traces, on the Qin bronze weapons unearthed from the pits of the Terracotta Army. The method allowed the identification of the chiselling techniques used by craftspeople to create inscriptions, as well as the use of hand filing to remove excess metal from casting seams. The evidence suggests that rotary mechanical devices were used on a large scale to grind and polish the bronze weapons during their mass production.

KEYWORDS SEM, silicone rubber, tool marks, handheld files, rotary tools

Introduction

Scanning electron microscopy (SEM) has been used extensively for the technical examination of ancient artefacts. However, any such SEM study is limited by chamber size. For example, the chamber size of the Hitachi S-3400N SEM used in this study is 200 mm in length, and the samples are normally put on aluminium stubs varying in diameter from 3.1 to 51 mm. The Hitachi S-3700N offers a significant improvement, having a larger chamber and stage traverse, with a maximum specimen size of 300 mm in diameter and an observable area of about 200 mm in diameter, with rotation and a height of up to 60 mm. However, many archaeological artefacts are even larger than this; furthermore, it is often difficult to move unique or delicate objects to laboratories for SEM examination.

Both of these problems were encountered in our ongoing project, focused on the bronze weapons unearthed from the pits of the Terracotta Warriors within the Emperor Qin Shihuang's tomb complex, built in Lintong, Xian, China between 246 BC and 206 BC. The thousands of terracotta warriors discovered in the 1970s were equipped with a large quantity of functional bronze weapons. The excavated assemblage includes over 40,000 arrowheads as well as hundreds of crossbow triggers, swords, lances, spears, halberds, hooks, honour weapons (*Su*) and the ferrules that were affixed to the ends of wooden shafts. Scratches and marks visible on some of the weapons suggested that grinding or polishing took place during their production processes after casting. Furthermore, some of the weapons displayed incised inscriptions, which vary from small symbols to relatively long sentences referring to the workshop or craftspeople involved in their production. One specific objective of this study was to investigate the techniques used to inscribe and polish bronze during the Qin period, including the shapes and materials of the tools, and the way these tools were used. Particular attention was paid to the potential identification of the use of any mechanical devices, including rotary tools. However, due to the limitations in chamber size and conservation concerns, it was difficult to move the actual weapons to the laboratory to be observed under SEM. For example, the swords are 810–948 mm in length [1] and are therefore unsuitable for the SEM chamber. Even for smaller weapons such as bronze arrows and triggers, moving them from their secure storage facilities was deemed too risky.

In order to overcome these problems, silicone rubber impressions were taken from the surface of the objects, which were examined subsequently using SEM. The initial results of our study are discussed at length elsewhere [2] and will only be outlined here. In this paper, we detail our methodological procedures with a view to assessing the potential and the limitations of this approach for this and similar studies.

Methods

The examination of tool marks is part of a larger project investigating the craft organisation and specialisation during the production of the bronze weapons [3, 4]. Except for a few weapons displayed in exhibitions, most of these are now

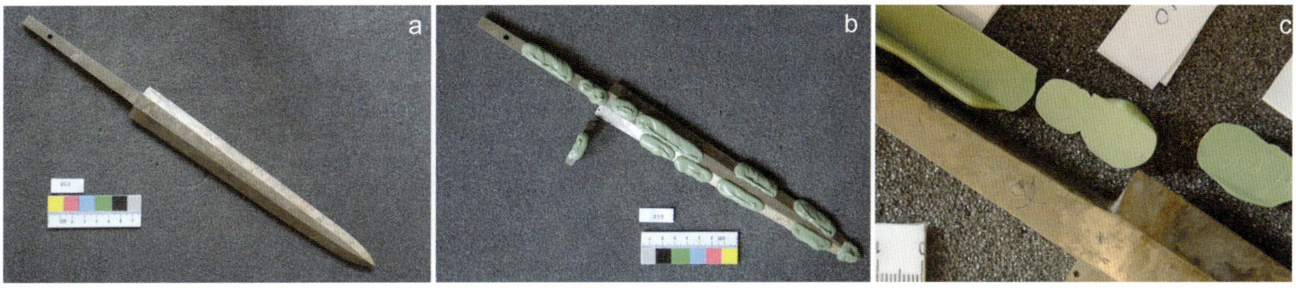

Figure 1 Obtaining silicone impressions: (a) well-preserved lance in its current condition; (b) the vinyl polysiloxane impression material is applied on the area of interest and allowed to set for 3–5 minutes; (c) silicone rubber impressions are easily detached from the lance.

housed at the Conservation Department of the Museum of Emperor Qin Shihuang's Terracotta Army, China. In addition to an initial typological analysis, including detailed measurements, statistical and spatial analyses of the distribution of the weapons in the pit, the swords, lances, arrows and triggers were taken out of their storage and the macroscopic features of their inscriptions and their grinding and polishing marks were noted. Many of these weapons were examined and photographed under a stereomicroscope in order to identify areas of interest. Subsequently, a subset of these, including fragments of 20 triggers, two swords, two lances and six arrowheads, was selected to obtain silicone rubber impressions for more detailed SEM examination.

Mould material

The moulding material employed for this project is a silicone rubber paste known as vinyl polysiloxane (VPS) impression material used in dentistry. It is a low viscosity/light body hydrophilic syringeable material designed for making precise dental impressions. An auto-mix system simultaneously mixes the silicone liquid and catalyst (normally Pt-catalyst), and then delivers the paste directly onto the targets via a cartridge. It normally takes 3–5 minutes to set. The excellent elastic properties of this silicone rubber make the removal of the mould from the artefacts very easy.

Previous use of moulding material

The use of silicone rubber impressions to study tool marks is not new in archaeology. Sax and Meeks used this method to investigate engraving techniques on British Museum (BM) cylinder seals [5], which were commonly made from natural minerals or stones and used in Mesopotamia from 3500 BC until 400 BC. Silicone rubber moulds were cast from the seals and observed using SEM; techniques such as micro-chipping, filing, wheel cutting and drilling could be identified [5, 6]. More recently, this method was also employed to examine Chinese jade carving [7, 8], and it is currently being used at University College London (UCL) to study the manufacture of beads made of stone (Roseleen Bains) and glass (Yixian Lin). However, as far as we are aware, the method has not been used on metal artefacts despite the interest shared by many scholars in the carving and finishing marks left on the surfaces of early Chinese bronze objects [9, 10].

Application and problems encountered

Considerable effort was exerted to perfect the moulding methods on these bronze weapons, particularly as corrosion or dust on their surfaces could mislead the results. Given that the weapons are generally very well preserved, initially we selected the most promising areas and took silicone rubber impressions directly on uncleaned surfaces. However, examination of these moulds in the SEM showed that a few dust particles had become attached to the silicone rubber and affected the quality of the impressions. Subsequently, cleaning with a soft brush or a gas duster was sufficient to remove loose particles from the samples prior to taking the impressions. Given our interest in very fine polishing marks, it was important not to use cloths or similar materials that might add new scratches to the metal surfaces.

The moulding material was applied to the tool marks on the surface of the object (Figure 1). Care was taken during the application of the moulding material to avoid air bubbles and other moulding faults. A thin layer (2–3 mm) of the silicone mix was first applied on the areas of interest and then more was applied directly to build up sufficient depth to give the mould adequate rigidity without wrinkling. Therefore, for a more efficient process, sometimes several artefacts were lined up on a table for the impression material to be applied in a quick sequence before it solidified and blocked the mixing tip.

SEM observations

Owing to the elasticity of the silicone rubber, removing the moulds from the weapons was always easy, and we never experienced any problems with residues being left on the artefacts or, conversely, any corrosion products other than minute loose particles being detached from the metal and trapped in the silicone rubber. The impressions were mounted on 51 mm diameter aluminium stubs prior to being coated with gold for SEM observation. Most analyses were carried out using an Hitachi S-3400N at the Institute of Archaeology, UCL, with a few pilot tests carried out on an Hitachi S-3700N at the BM. Both instruments have virtually identical detectors and software and the results are therefore comparable. A large area solid-state backscattered electron (BSE) detector was employed for imaging. This imaging mode gives the best contrast and visual display of depth for inscribed features compared with conventional secondary electron (SE) imaging (Figure 2). The moulds could be rotated

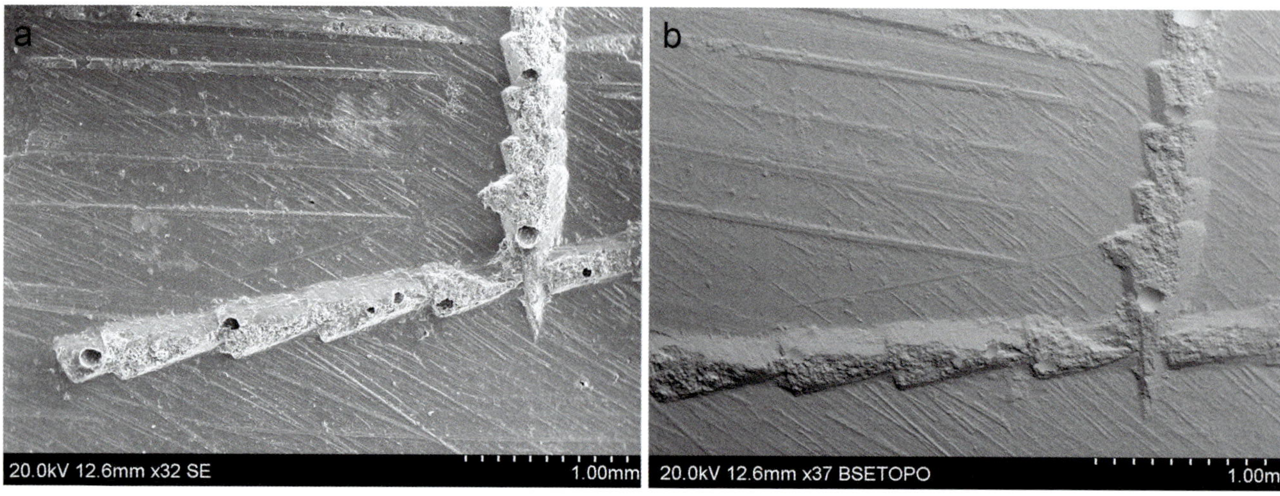

Figure 2 Comparative images of the same tool marks using SE (a) and BSE detectors (b) on gold-coated silicone rubber impressions. Note the improved contrast and resolution of the latter approach.

and tilted in the SEM to give the best angles of observation, and five radial detector segments could be turned on and off as appropriate to create topographic shadows and allow the photomicrographs to represent accurate geometric shapes and topographic features [5]. SE imaging was also used in our initial studies for some features such as the very fine polishing marks.

It is important to note that the tool marks are present on the silicone impressions as positive features, that is, the SEM images show the marks as negative compared to those on the weapons. This offers several advantages as it facilitates the observation of the profile and depth of inscribed features and of details within these features.

Recasting with resin

Most of the silicone impressions could be observed directly with SEM, but some very deep casts of silicone rubber, such as the bolt heads of the triggers, had to be recast in resin in order to make the tool marks accessible for observation. In this case, drawing on Roseleen Bain's experience with stone beads [11], the silicone rubber impression was used as a mould, and a mixture of epoxy resin and hardener (in a weight ratio of 4:1) was poured into it and allowed to set for 24 hours. The resin cast was then carefully separated from the elastic mould avoiding the formation of cracks that could compromise the results. As the original bolt heads had grinding and polishing marks on the top and side surfaces, the silicone moulding material concealed the side marks after casting; the recasting process was used to expose traces on all surfaces, and the resin cast examined as above.

Results

Examination of the silicone rubber impressions under the SEM allowed us to investigate two types of tool marks: (1) the chiselling marks composing the inscriptions and (2) the surface finishing traces after casting, including filing, grinding and polishing.

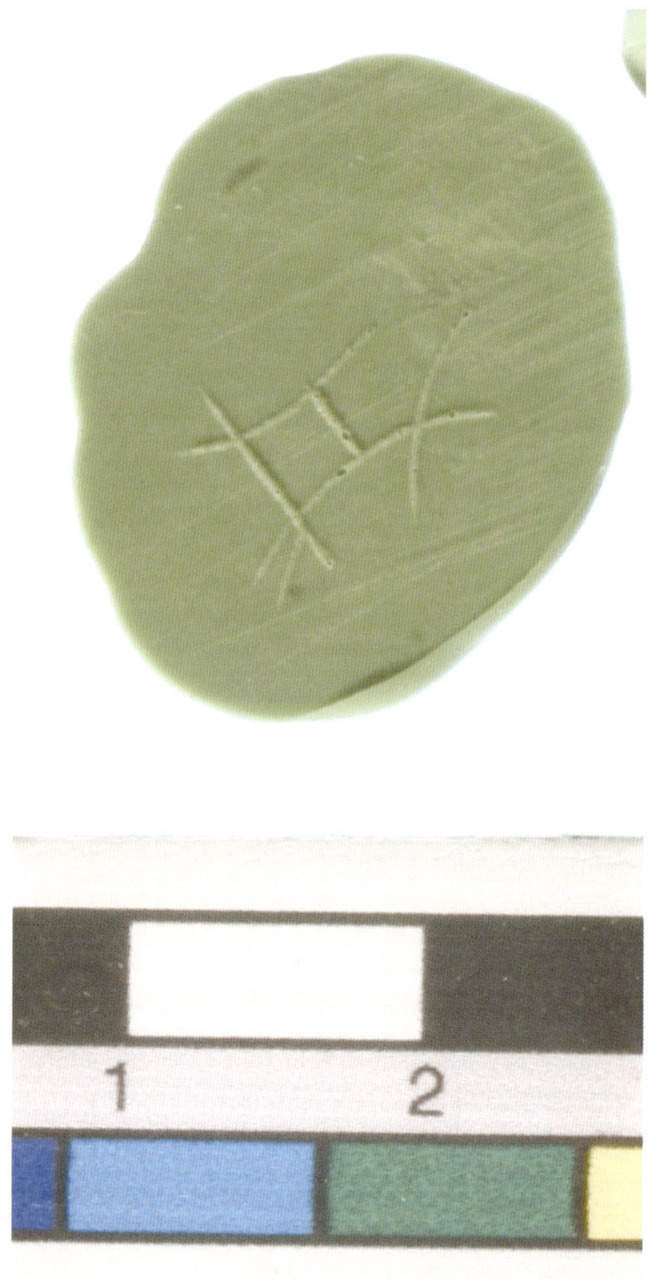

Figure 3 Silicone rubber impression of the inscription on a bronze trigger (the Chinese character *Wu*).

Chiselling techniques for creating inscriptions

The chiselling techniques used to work inscriptions are illustrated by the examination of the mould details of a bronze trigger (Figure 3). The SEM images (Figure 2b) show the components of the characters to be formed from the juxtaposition of overlapping cut marks, resulting in ragged edges. Each of these marks has an elongated triangular shape, indicating that the line was chiselled blow by blow, with a finely pointed sharp tool ending with a wedge-shaped tip in a V-shape. The overlapping of the marks indicates that the chisel was hammered into and along the bronze at an angle, starting from where the base of the triangle is now, and finishing at the sharp end of the mark. A comparison of such features in weapons is beginning to allow the identification of different tool types, and even of slight variations that may have been caused by different chiselling techniques or craftspeople [2].

The chisels would have to be made from a material both harder and tougher than the bronze weapons in order to produce the sharp marks without becoming blunt or shattering after a few strokes. Whilst stones can be harder than bronze, they are relatively brittle and thus not suitable for chiselling. It seems more likely that either high-tin bronze, cast iron or quenched steel were used. During the Warring States period (475–221 BC), iron tools and implements were already used, and indeed some iron weapons, such as an iron spear and three iron arrows, have been unearthed from the pits of the Qin terracotta warriors [12], as well as some iron implements discovered from the Emperor Qin Shihuang's mausoleum complex itself [13]. Although the chisels and other iron tools found in the mausoleum complex are too large to have been used on these weapons, their presence indicates that such technology and tools were familiar to Qin metalworkers. Unfortunately, lacking analytical data, it is not possible to determine whether they are made from bloomery or wrought iron, or to assess their carbon content to determine their hardness and toughness. Evidence of quenched steel and cast iron has been discovered in many contemporary archaeological sites in central China [14, 15], and some even show earlier iron production activities, which can be traced back to the eighth century BC [16].

The presence of symbols or texts on the surface of Chinese bronzes is widely documented in Shang (1600–1100 BC) and Zhou (1100–221 BC) bronzes. However, the predominant methods of application in the earlier periods focused on the design of the moulds, either by adding thin coils or strips of clay that would appear as depressions in the cast artefacts, or by carving the inscriptions directly on the model or mould that would appear as ridges on the cast artefacts [17–19]. Technically different from the cast inscriptions, the use of chisels directly onto the surface of the bronzes, as documented in the First Emperor's weapons, becomes common only during the Warring States period [9]. Such a technical change may be related to both sociological and technological developments; one might even conjecture that the chiselling technique is associated with the availability of cast iron and steel tools.

The final potential of this method of investigating tool marks can be highlighted here, that is the observation of stratigraphical relationships between different marks (in the impressions). The fact that the inscriptions break through the filing and grinding traces (discussed in the next section) suggests that the inscriptions were added after the filing and grinding, perhaps when the finish of the weapons was deemed suitable for them to receive the official mark of quality control or to aid their assembly.

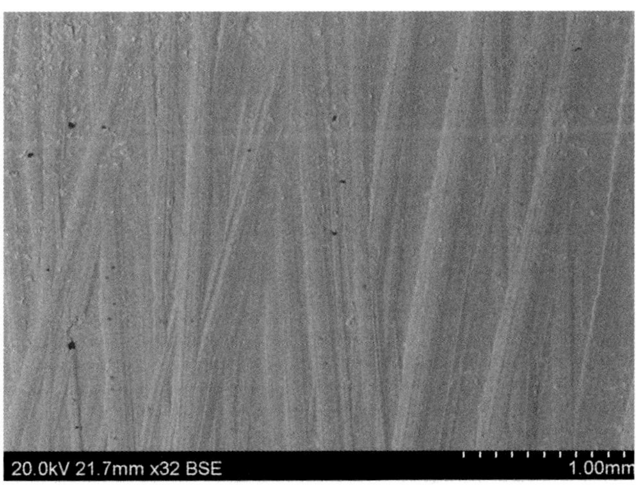

Figure 4 BSE micrograph of filing marks on the trigger on a silicone rubber impression showing coarse overlapping marks from manual filing.

Filing, grinding and polishing marks

Crossbow triggers, lances, swords and arrows were examined for signs of tool marks. Filing, grinding and polishing marks were discovered on a large number of the weapons unearthed from the tomb complex. Some appear rather coarse and easily discernable under the naked eye, whilst other marks are much finer and only perceptible, under magnification, thanks to the generally exceptional state of preservation of most of these artefacts.

Filing marks

Filing marks are particularly obvious on the composite bronze crossbow triggers. Each trigger comprises three mechanical parts joined by two bolts, and each part has its own shape and function [2]. The five parts were all cast separately and assembled before attaching them to the crossbow. The introduction of the trigger mechanisms during the Warring States period revolutionised military warfare, as it ensured that the crossbow required less skill and strength to use than a composite bow and was said to be able to fire heavier arrows over a longer distance [1, 20, 21]. Most of these marks surround the holes of the three mechanical parts, where they would have slid over each other as they moved. These traces are probably the results of filing off excess metal from the surfaces to make them fit together during assembly, to ensure an effective functioning of the crossbow. However, similar traces have also been found on some of the bolt heads and on other parts of the triggers, where they appear to serve no obvious technical purpose.

Visually, the orientation of the marks on the triggers appears parallel. However, under the SEM, it is possible to see that these marks are typically multidirectional and that they cross and overlap each other (Figure 4). These features are diagnostic of filing by hand, which involves back and forth movements of the file on the surface of the object. Based on measurements of the individual grooves, the diameter of the ridges or points on the file would have been approximately 100–200 μm.

Although we lack direct evidence for filing tools employed during the Qin period, the relative depth and the sub-parallel arrangement of the filing marks suggest that handheld files made of a hard material, such as stone, steel or cast iron, could have been used. Experimenting with a handheld polishing stone in a back-and-forth motion on a brass surface, we produced sub-parallel filing marks that, albeit thinner, resembled the arrangement found on the archaeological artefacts [2].

Similar marks from filing or a relatively coarse grinding are also noticeable on the tangs of arrowheads. They typically appear broadly parallel to each other and are oriented diagonally. Although they appear predominantly on the casting seams, seemingly as a result of trying to remove imperfections, in some cases the marks encompass the entire circumference of the tangs. Under the SEM, the marks appear quite similar to those identified on triggers, suggesting that a similar kind of filing tool was used [2].

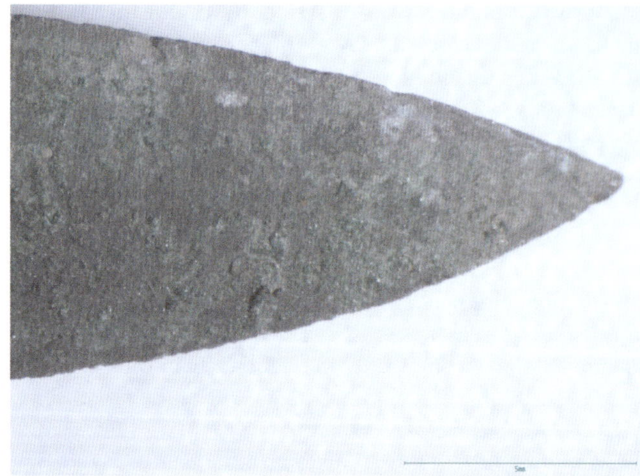

Figure 5 Detail of the grinding marks on an arrowhead, as seen under the stereomicroscope.

Grinding and polishing marks

Much finer grinding or polishing marks were found all over the surfaces of swords and lances, as well as on the tens of thousands of arrowheads. In these examples, it is assumed that the objective was to obtain a smooth and shiny surface, as well as a sharp blade. On the swords and lances, the marks are generally longitudinal on the spine section, and transverse on the blade section. On the arrow surfaces, these marks are typically transverse on all sides. In some instances, it is possible to discern two stages of fine grinding/polishing on the arrowhead surfaces, one finer than the other, carried out in orientations perpendicular to each other. Sometimes these finer marks extend onto the surface of the arrow tangs, removing the evidence of the coarser filing described in the previous section (Figure 5).

When examined under the SEM, the grinding marks on all three types of artefacts (swords, lances and arrowheads) appeared as extremely fine, densely packed and parallel lines (Figure 6). The shallowness of the marks indicates that only a very small amount of material was removed from the weapon surface during this finishing stage. Such characteristics appear to originate from a mechanical grinding device, probably a rotary tool. This interpretation is supported by our experiment using a small, hard stone rotary wheel on a brass surface, which created similar marks [2]; conversely, we failed in our attempts at using handheld tools to create marks that were equally parallel under the microscope.

A relatively hard stone was most probably the material used for this finishing process, as naturally abrasive materials were easy to acquire and sufficiently hard to grind the bronze weapons. It is also possible that relatively soft organic materials, such as linen, could have been employed for the last and finest stage, which gave the weapons their characteristic sheen [22]; perhaps a lubricant and very fine loose abrasive material could have been used. Tools for this purpose might be comparable to bristle brushes or buffing wheels used in modern crafts. The softness and flexibility of these materials would explain the continuity of polishing marks overlapping different planes of the same weapons, and their very presence in small, barely accessible, intricate parts of the arrowheads, where creating such parallel marks with a handheld tool would seem difficult. Further research will concentrate on the potential differences between grinding and polishing marks, and the possible materials and tools used during these processes.

Another interesting finding of this work is the noteworthy similarity among the grinding marks detected on the different artefacts with regard to the thickness, density and relative shallowness of the marks. Although the exact dates of manufacture

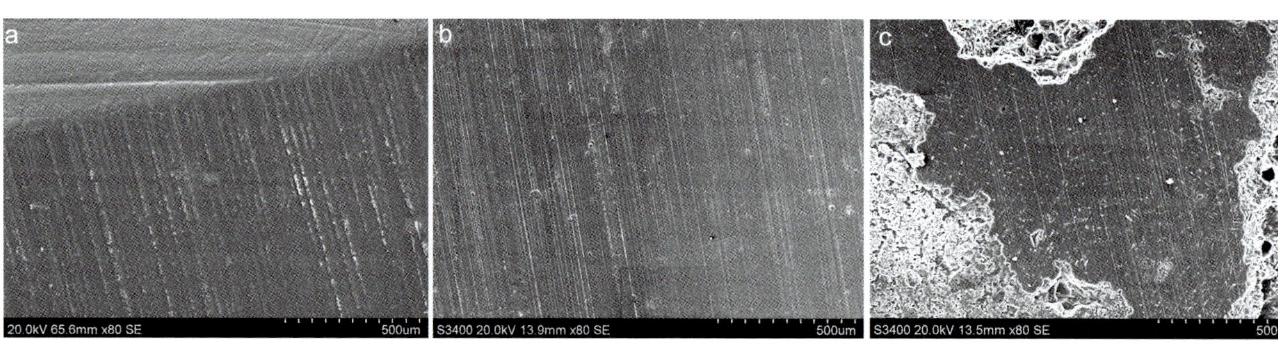

Figure 6 SE images from silicone rubber impressions of the fine parallel grinding marks on lances, swords and arrowheads, taken at the same magnification to facilitate comparison: (a) lance 860, (b) sword 854, and (c) arrow 2118 (the bright crusts on the latter are impressions of corrosion products). Compare to coarser filing marks on the trigger in Figure 4.

of the swords and arrowheads are unknown, the inscriptions on the lances indicate a narrow chronological range of 232–228 BC. The technical similarity of the grinding marks on all three artefact categories may be taken as an indication that all the weapons could have been ground employing very similar techniques, although further investigations are needed to confirm the chronological or organisational relationships between the producers of different weapons.

Anecdotal visual evidence on Shang and Zhou bronze vessels indicates that they were subject to filing and grinding finishing techniques. However, whilst their casting technology has been discussed abundantly in the literature [14, 23, 24], these finishing marks have rarely been studied, which makes a detailed history of this technology difficult to reconstruct. Until further analytical and experimental data are available it may not be possible to ascertain whether iron or steel files were employed on these bronze weapons, but these marks were almost certainly not made by loose abrasive materials.

Of particular importance is the positive identification of the use of mechanical tools on a systematic and industrial scale, for grinding the tens of thousands of bronze weapons. Some simple rotary techniques appear to have been used in jade carving during the Neolithic period and developed during the Bronze Age, however, systemic use of lathe-mounted rotary wheels for incising and grinding jade was not made until the Tang Dynasty (AD 618–906) [8, 9].

As regards bronzes, it appears that rotary wheels were used on metals as early as the Shang dynasties [9]. A bronze mask recovered from Guanghan Sanxingdui with a rectangular hole on the forehead is thought to have been cut using a rotary wheel, because the four corners of the hole display intrusive crossing lines typical of wheel cutting [9, 25]. This would constitute the earliest example of mechanical methods discovered so far on Chinese bronzes.

Rotary grinding wheels appear also to have been used on bronze weapons from the Warring States era based on the observations of surface traces. Dong [26] has argued that the Yuewang Goujian sword, reputed to have been owned by the king of the Yue state, probably had its blades trimmed and ground using rotary wheels because the lines are all parallel. However, the weapons from the Terracotta Army may constitute the earliest evidence of the use of such rotary methods in mass production.

Conclusions

Vinyl polysiloxane dental impression materials have been used in the past to obtain high-resolution impressions of stone archaeological artefacts that could be examined for tool marks and manufacturing traces in the SEM. This study has shown that the same approach can be used to examine areas of the surfaces of metal artefacts, overcoming the potential limitations of SEM chamber size and avoiding the need to move artefacts to laboratories.

Silicone rubber impressions are easy to produce, relatively inexpensive and inflict no damage to the artefacts' surfaces. They bear a high fidelity to the original surface textures, topography and the very fine striations associated with manufacture or use. They are very durable and can be re-examined if required. Tool and wear marks on metal objects can be examined directly as positive features on the silicone rubber, or nearly exact replicas can be examined by casting epoxy resin in the moulds and observing the resulting pieces under the SEM. Our experience indicates that the best images are obtained using a large area solid-state BSE detector with radial segments that can be switched on and off to adjust the type and directionality of the contrast, creating shadows of the features of interest but minimising the visual effects of sample charging or surface dirt.

In the present study, examination of impressions of the surfaces of bronze weapons from the Terracotta Army has given an insight into the techniques used in chiselling, filing, and even the fine grinding and polishing of the artefacts. Besides the technical reconstruction of the tool types and shapes and the operational sequences, this research provides indirect evidence of the use of iron/steel tools and rotary wheels on an industrial scale during the finishing processes of the production of bronze weapons during the Qin period.

Using the same approach for investigating manufacturing traits on unused weapons, impressions could also be used for detailed examination of use or wear marks on other metal artefacts, which could also be compared to traces produced experimentally. In addition, the growing availability of variable-pressure SEMs (VP-SEMs) will allow the examination of impressions without the need for carbon or gold coating, thus overcoming the potential problems caused by the cracking or soiling of these coatings in the long term. With ongoing studies expanding the applicability of this analytical method to a variety of stone, ceramic and glass artefacts, we encourage researchers to continue considering this approach as an alternative when moving archaeological objects out of secure locations may entail risks to their integrity or unnecessary expenses.

Acknowledgements

We are very grateful to the Museum of the Emperor Qin Shihuang's Terracotta Army and the Institute of Archaeology, UCL, as we all benefit from the cooperative aspects of this project, especially Peter Ucko, who made this collaboration possible. Many thanks are due to the Museum Director, Wu Yongqi, for valuable suggestions regarding the making of these tool marks, and to Thilo Rehren and Andrew Bevan for their practical advice and constructive comments. By courtesy of the British Museum, working experience was gained by Xiuzhen Janice Li using the SEM in the Department of Conservation and Scientific Research. Thanks are due to Roseleen Bains for generously sharing her gun, silicone and experience on mould casting. We would also like to thank Zhao Kun, Chang Qiuyue, Yan Hongxia and other colleagues in the Conservation Department of the Terracotta Army Museum for their support during the examination of the weapons and the casting of the silicone rubber moulds, as well as many other friends who support this project. A special appreciation is given by Xiuzhen Janice Li to Shiona Airlie for her time and comments on the use of English.

Material and supplier

Vinyl Polysiloxane Impression Material produced by 3M ESPE AG, Coltène/Whaledent, Switzerland and supplied by Dental A2Z (www.dentala2z.co.uk)

References

1. Yuan, Z., *Research on the Terracotta Warriors and Horses from the Mausoleum of Emperor Qin Shihuang*, Culture Relics Press, Beijing (1990).
2. Li, X.J., Martinón-Torres, M., Meeks, N.D., Xia, Y. and Zhao, K., 'Inscriptions, filling and polishing marks on the bronze weapons from the Qin Terracotta Army in China', *Journal of Archaeological Science* **38** (2011) 492–501.
3. Martinón-Torres, M., Li, X.J., Xia, Y., Zhao, K. and Rehren, Th., 'Making weapons for the Terracotta Army', *Archaeology International* **13/14** (2011) 65–75.
4. Li, X.J. *Standardisation, Labour Organisation and the Bronze Weapons of the Qin Terracotta Warriors*, PhD thesis, UCL Institute of Archaeology, London (2012).
5. Sax, M. and Meeks, N.D., 'Methods of engraving Mesopotamian quartz cylinder seals', *Archaeometry* **37**(1) (1995) 25–36.
6. Sax, M., McNabb, J. and Meeks, N.D., 'Methods of engraving Mesopotamian cylinder seals: experimental confirmation', *Archaeometry* **40**(1) (1998) 1–21.
7. Sax, M. and Meeks, N.D., 'The identification of carving techniques on Chinese jade', *Journal of Archaeological Science* **31** (2004) 1413–1428.
8. Sax, M., Meeks, N.D., Ambers, J. and Michaelson, C., 'The introduction of rotary incising wheels for working jade in China', in *Scientifc Research on the Sculptural Arts of Asia*, ed. J.G. Douglas, P. Jett and J. Winter, Archetype Publications, London (2007) 20–25.
9. Dong, Y., *Mould Casting Bronzes*, Beijing Art and Science Press, Beijing (2006).
10. Strahan, D and Fenn, M., 'A transfer of technology: jade abrasive methods used to create inscriptions in ancient Chinese bronzes', *Scientific Research on the Sculptural Arts of Asia*, ed. J.G. Douglas, P. Jett and J. Winter, Archetype Publications, London (2007) 26–36.
11. Roseleen Bains, personal communication (June 2009).
12. Institute of Shaanxi Archaeology and Team of Qinling Archaeology, *Excavation Report of Pit 1 of the Terracotta Warriors and Horses (1974–1984)*, Culture Relics Press, Beijing (1988).
13. Yuan, Z., *Research into the Archaeological Discoveries from the Emperor Qin Shihuang's Tomb Complex*, Shaanxi People Press, Xian (2002).
14. Wagner, D., *Science and Civilisation in China. Vol. 5: Chemistry and Chemical Technology. Part 11: Ferrous Metallurgy*, Cambridge University Press, Cambridge (2008) 98–128.
15. Han, R. and Ke, J., *Mining and Metallurgy: A History of Chinese Science and Technology*, Science Press, Beijing (2007).
16. Han, R. and Duan, H., 'Metallurgy and an early iron-using centre in the ancient Jin state region (8th–3rd century BC)', in *Metallurgy and Civilisations: Eurasia and Beyond*, ed. J. Mei and Th. Rehren, Archetype Publications, London (2009) 99–106.
17. Li, X., *The Wonder of Chinese Bronzes,* Foreign Language Press, Beijing (1980).
18. Ma, C., *Ancient Chinese Bronzes*, Oxford University Press, Oxford (1986).
19. Bagley, R., 'Anyang mold-making and the decorated model', *Artibus Asiae* **69**(1) (2009) 39–90.
20. Yang, H., *Research on the Ancient Chinese Weapons*, Cultural Relics Press, Beijing (1980).
21. Yates, R.D, 'The rise of Qin and the military conquest of the Warring States', in *The First Emperor: China's Terracotta Army*, ed. J. Portal, The British Museum Press, London (2007) 30–57.
22. Yuan, W., Yuan, Z., Zhao, P., Zuo, Z. and Wang, X., 'Interpreting standardisation of production of bronze weapons unearthed from the pit of the Terracotta Warriors', *Shaanxi Standardisation* **2** (1981) 21–26.
23. Liu, Y., 'Origins and evolution of the casting technology of Anyang bronze ritual vessels: an exploratory survey', in *Metallurgy and Civilisations: Eurasia and Beyond*, ed. J. Mei and Th. Rehren, Archetype Publications, London (2009) 55–61.
24. Barnard, N., 'Bronze casting and bronze alloys in ancient China', in *Monumenta Serica Monograph 14*, Australia National University, Canberra (1961).
25. Chen, D., *San Xing Dui*, Sichuan People Press, Chengdu (2000).
26. Dong, Y., 'An investigation into the origin of mechanical means based on the ancient Chinese bronze objects', *Journal of Ezhou University* **34–38** (1999).

Authors' addresses

- Xiuzhen Janice Li, Museum of Emperor Qin Shihuang's Mausoleum, Xian, China and Institute of Archaeology, University College London, London, UK (xiuzhen.li@ucl.ac.uk)
- Marcos Martinón-Torres, Institute of Archaeology, University College London, London, UK
- Nigel D. Meeks, Department of Conservation and Scientific Research, The British Museum, London, UK
- Yin Xia, Museum of Emperor Qin Shihuang's Terracotta Army, Xian, China

A forgotten tradition; the rediscovery of Mexican feathered textiles

Hector Manuel Meneses Lozano

ABSTRACT This paper highlights some rare Mexican feathered textiles of late seventeenth or early eighteenth century date. They display an unusual manufacturing technique, being made of feathered cotton yarns woven on a back strap loom. This technique, since disappeared, differs greatly to the one used to make the only feathered textiles in Mexico today, and is unlike the way in which feathers have been incorporated into some Amazonian and Andean textiles.

For a better understanding of the textile processes that are no longer practised by indigenous weavers, an important colonial feathered textile was studied in the Museo Textil de Oaxaca and treated in conjunction with the Escuela Nacional de Conservación, Restauración y Museografía 'Manuel del Castillo Negrete' in Mexico City. Fundamental to the research was an evaluation of the implications of using scanning electron microscopy (SEM) for the identification of down feathers. SEM permitted detailed observation of the different structures of down feathers according to species. Eight samples were taken from the textile and two structures of down feathers were identified by analysing barbules length, the forms and numbers of nodes as well as their distribution along the barbules. Two ducks were identified: *Cairina moschata* and *Anas platyrhyncos*. SEM also enabled a better understanding of the structure of the feathered yarns: two cotton plies with down feathers encircling them, instead of feathers and cotton spun at the same time. Researchers have commonly referred to this technique as 'spun feather', however, after SEM imaging, a new term was coined: 'twisted feather'. The reproduction of this yarn by weavers and textile artists has also confirmed the SEM findings

KEYWORDS textile, feathers, Mexico, colonial, identification

Introduction

The Mexican art of featherwork has long delighted people from around the world. Since the conquistadores arrived in the New World, much has been said about the beauty that these objects possess; a beauty that is gracefully achieved by the soft texture and unique brightness of feathers. Feathered mosaics are the most common examples that have survived throughout the centuries. In contrast, Mexican feathered textiles are far less common: there are only six known textiles of this kind, all of which belong to the late seventeenth or early eighteenth century.

Feather creations in Mesoamerica were highly esteemed, not only because they were precious materials, but also because they had a symbolic meaning. Feathers were reserved for the elite because they were associated with power and divinity [1]. Fray Bernardino de Sahagún, a Spanish Franciscan missionary who arrived in Mexico in 1529, recorded four different ways of using feathers [2, 3]. The most famous are the feather mosaics, which caught Sahagún's attention to the extent that he described their manufacturing process in detail. The other feather manifestations consisted of objects made with a series of knotted feathers, such as headdresses and fans, and small feathered animals, which have not yet been documented in collections, and textiles, which are the subject of this study.

Due to the environmental conditions that exist in Mexico, only a few pre-Columbian textiles have been found in archaeological excavations, none of which has shown the use of feathers. To date, the study of feathered textiles in pre-Columbian societies has only been documented in surviving codices and manuscripts [3–5]. Unfortunately, as rich as these documents are as information sources, they provide little, if any, precise descriptions when it comes to the use of feathers in textiles. It seems that the Spaniards were so completely astounded by the splendour of feathered garments that they focused mainly on the beauty of these textiles. There are only a few mentions of technical aspects related to these textiles and these will be referred to below.

Fortunately a few feathered textiles have survived from the colonial period, six of which have been documented: a tapestry band at the Cooper-Hewitt, National Museum of Design in New York [6]; a cape known as Tlamachayatl at the Museo Nazionale Preistorico Etnografico 'Luigi Pigorini' in Rome; two blankets, one at the Museo Nacional del Virreinato and the other at the Museo de Bellas Artes de Toluca, both in Estado de México; a woman's robe called a *huipil* at the Museo Nacional de Antropología in Mexico City and a fragment of a *huipil*, referred to as Tlamachte:ntli de Madeline, at the Museo Textil de Oaxaca [1]. This latter fragment was studied and treated in collaboration with the ENCRyM in 2006. This paper focuses on the identification of feathers and the understanding of the feathered yarn's manufacturing technique. However, a detailed analysis of the *huipil* fragment was undertaken at the time.

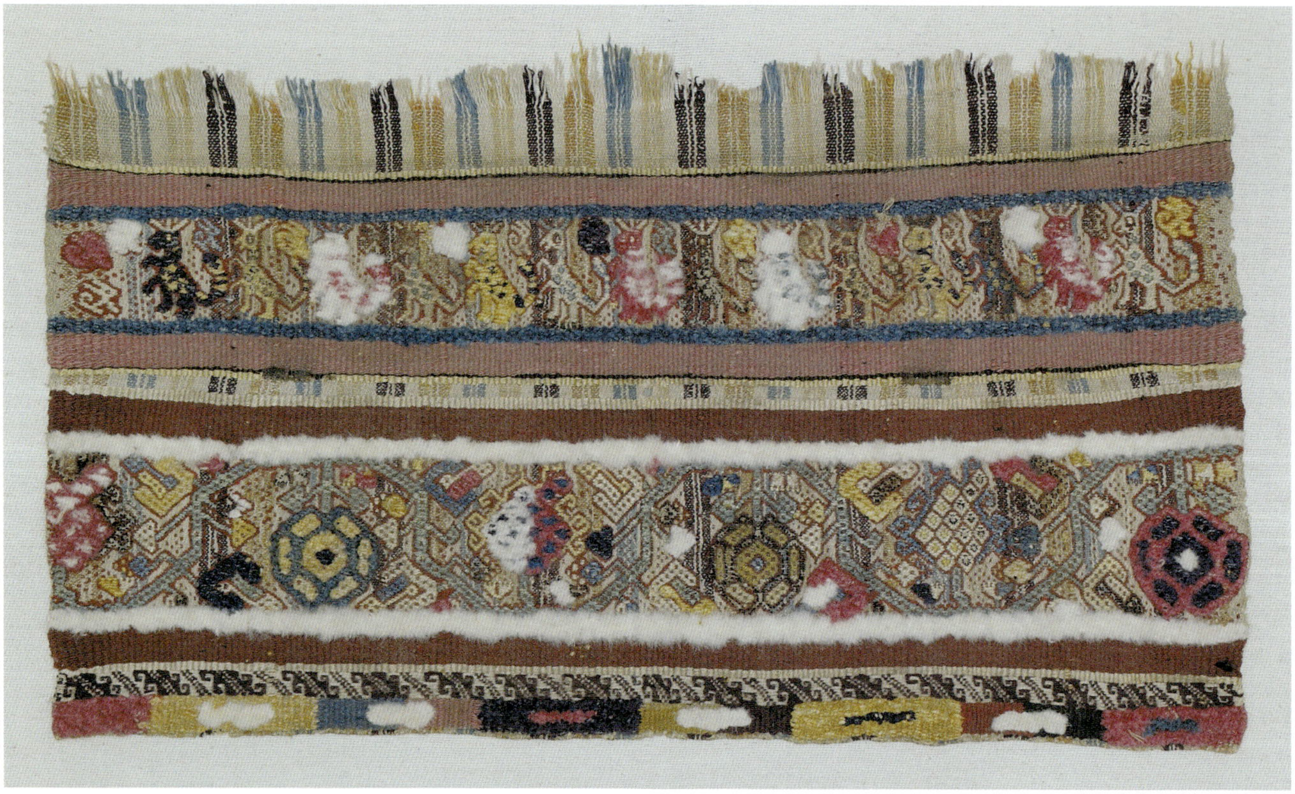

Figure 1 Tlamachte:ntli de Madeline, late seventeenth to early eighteenth century. Museo Textil de Oaxaca. Inv. no. HUI-0513.

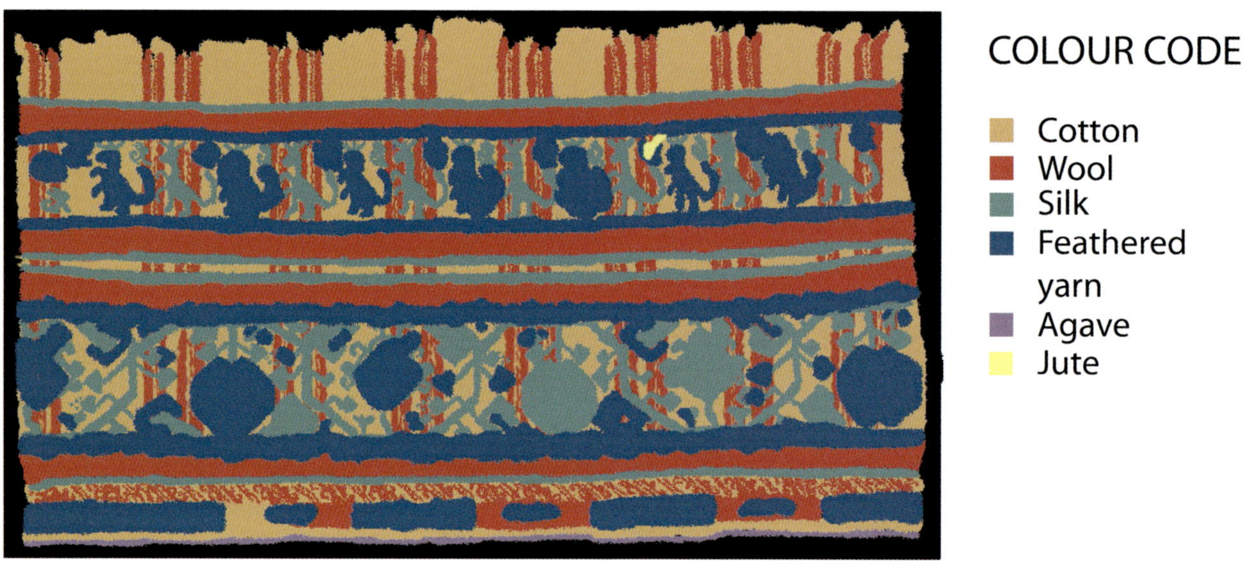

Figure 2 Diagram of fibre distribution.

Although the exact origin of these textiles is still unknown, they are all contemporary – late seventeenth to early eighteenth century – and all of them share similar technical characteristics. They appear to form a one-of-a-kind ensemble as the technique used in their manufacture has not been documented elsewhere. This technique appears to have fallen out of use in the eighteenth or nineteenth century. The only feathered textiles that exist nowadays in Mexico come from Zinacantan, Chiapas, located in southern Mexico [7]. These textiles, used for wedding gowns, show a completely different manufacturing technique to that observed in the colonial examples. This technique incorporates feathers into the weaving, securing them in between each line of weft instead of using a feathered yarn. It is still unknown if both techniques coexisted in the past or if the surviving practice is the result of an evolutionary process. The technique is also different to the way that some other cultures have incorporated feathers into their textiles, such as in the Amazonian and Andean regions [8]. At some point during the eighteenth or nineteenth century, the making of feathered yarns disappeared in Mexico.

Study object

The Tlamachte:ntli de Madeline entered the Textiles Conservation Seminar at the ENCRyM in Mexico City in 2006 (Figure 1). The object measures 55 cm in length and 31 cm in width. Being so remarkable, it was subjected to several studies:

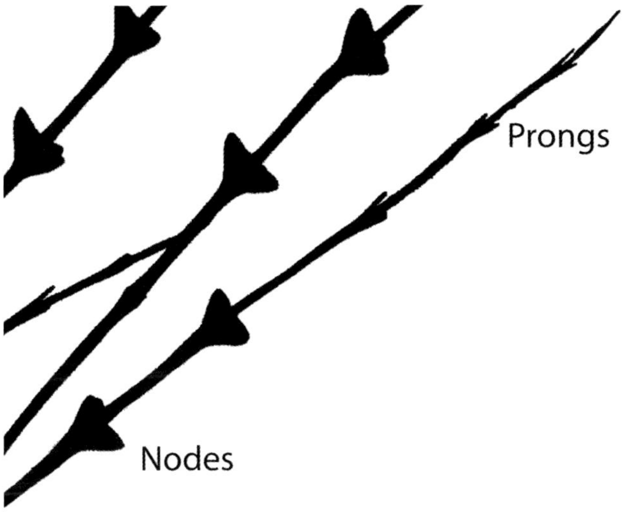

Figure 3 Nodes and prongs on down barbule from muscovy duck.

polarised light microscopy (PLM) was used for fibre identification, and microchemical analysis, thin layer chromatography (TLC) and ultraviolet/visible (UV/VIS) light spectrometry were used for dye identification. The results of these complementary analyses helped to provide detailed documentation of this textile.

The ground fabric was woven with hand-spun cotton (warp and weft) and wool (warp). The cotton warps include the use of the natural brown cotton known as *coyuchi* (colour of the coyote) in Mexico, as well as cotton dyed blue. Woollen warps were dyed orange, ochre and purple. Most of the ground fabric was woven using a plain weave. There are bands of half-basket weave using degummed silk and wool as weft yarns. Silk was dyed in a yellow colour – now extremely faded – and dark green. Pink- and dark red-dyed wool, as well as undyed and blue-dyed feathered yarns were also used in a half-basket weave. Close to the bottom border, a fret was woven with undyed cotton and purple-dyed wool. The bottom border shows dovetail tapestry technique using a combination of cotton, wool and feathered yarns. Supplementary wefts were also used to create different motifs in silk, wool and feathered yarns. Figure 2 shows a diagram of the distribution of the fibres present on the textile, which emphasises the complexity of the weaving.

All reds and pinks were obtained with cochineal. Indigo was the source for obtaining blue and purples were achieved by using both cochineal and indigo. Yellow dyes, however, could not be identified. Although it was not possible to fully identify the yellow dyes with UV/VIS light spectrometry, it was observed that common yellow dyes were not used to dye the down feathers of this textile. Some of these common yellow dyes are *zacatlaxcalli* (*Cuscuta americana* L.), *xochipalli* (*Cosmos sulphurous*) and fustic (*Clorophora tinctoria*) [9]. The exact composition of green dyes could not be identified either, but microchemical analysis indicated the presence of indigo as one of the components. The dark green dye used on some of the silk yarns has been partially characterised: it was obtained through a common combination of tannin-based dye and iron. X-ray fluorescence (XRF) showed that the green silk yarns contain eight times more iron than any other section in the textile. The combination of the tannin compounds and the iron particles would explain the severe depolymerisation of the silk fibres, which results in a very brittle yarn [10].

Down feather identification

Preliminary approach to down identification

The feathers found on colonial feathered textiles belong to the 'down feather' category [11]. Down feather may grow in either the apteria or pterylae, or both. It is composed of a very fine rachis with filamentous barbs [12]. The thinness of the rachis provides down with great flexibility, which is highly advantageous in the yarn-making process.

The first attempt at down identification was made by ornithologist Lourdes Navarijo [13], from the Universidad Nacional Autónoma de México (UNAM), who stated that birds are identified according to their vaned feathers, i.e. those that cover the exterior of their bodies. These feathers have particular characteristics that are useful for identification purposes, such as size, form and colour. This methodology, however, cannot be applied to down feathers for several reasons. Navarijo [13] stated that the size of the down feathers makes it difficult to appreciate significant differences in form. Also, the colours observed on down from the colonial feathered textiles are misleading, as down feathers were commonly dyed before making the yarn. Therefore, this common procedure for feather identification could not be performed.

DNA-sequence analysis was thus considered as a possible alternative. However, the oldest samples that the Instituto de Biología (Institute of Biology (IB), UNAM) has identified correctly using DNA-sequence analysis came from 30-year-old specimens. This posed a significant restriction, since the samples studied here were close to 200 years old. It was then decided to explore another possibility that might give more reliable results.

After discussion with UNAM ornithologist Patricia Escalante [14], it was decided to investigate the potential of scanning electron microscopy (SEM). As this was new to the IB, a reference library with which to compare results was not available. There was therefore a need not only to create a small reference library, but also to determine whether, indeed, there was variability amongst and/or within down feathers from different species.

Down identification using SEM

Escalante [14] and Alejandro de Ávila [15], curator at the Textile Museum of Oaxaca, suggested five birds as reference models. The first choice was made after reviewing historical references. The geographical chronicles of the sixteenth century in Mexico mention that a black-coloured duck with a red beak was raised in Oaxaca for its feathers, which were used in the manufacture of textiles [16]. This description belongs to the muscovy duck (*Cairina moschata*), which was domesticated and widely distributed in pre-Columbian Mexico. The four other reference species consisted of birds also common in Mexico and whose feathers were used for different applications, such as fans and headdresses according to Sahagún's descriptions; these were: mallard (*Anas platyrhynchos*), snow goose (*Anser caerulescens*), snowy egret (*Egretta thula*) and sea hawk (*Pandion haliaëtus*) [17]. Samples consisted of true down

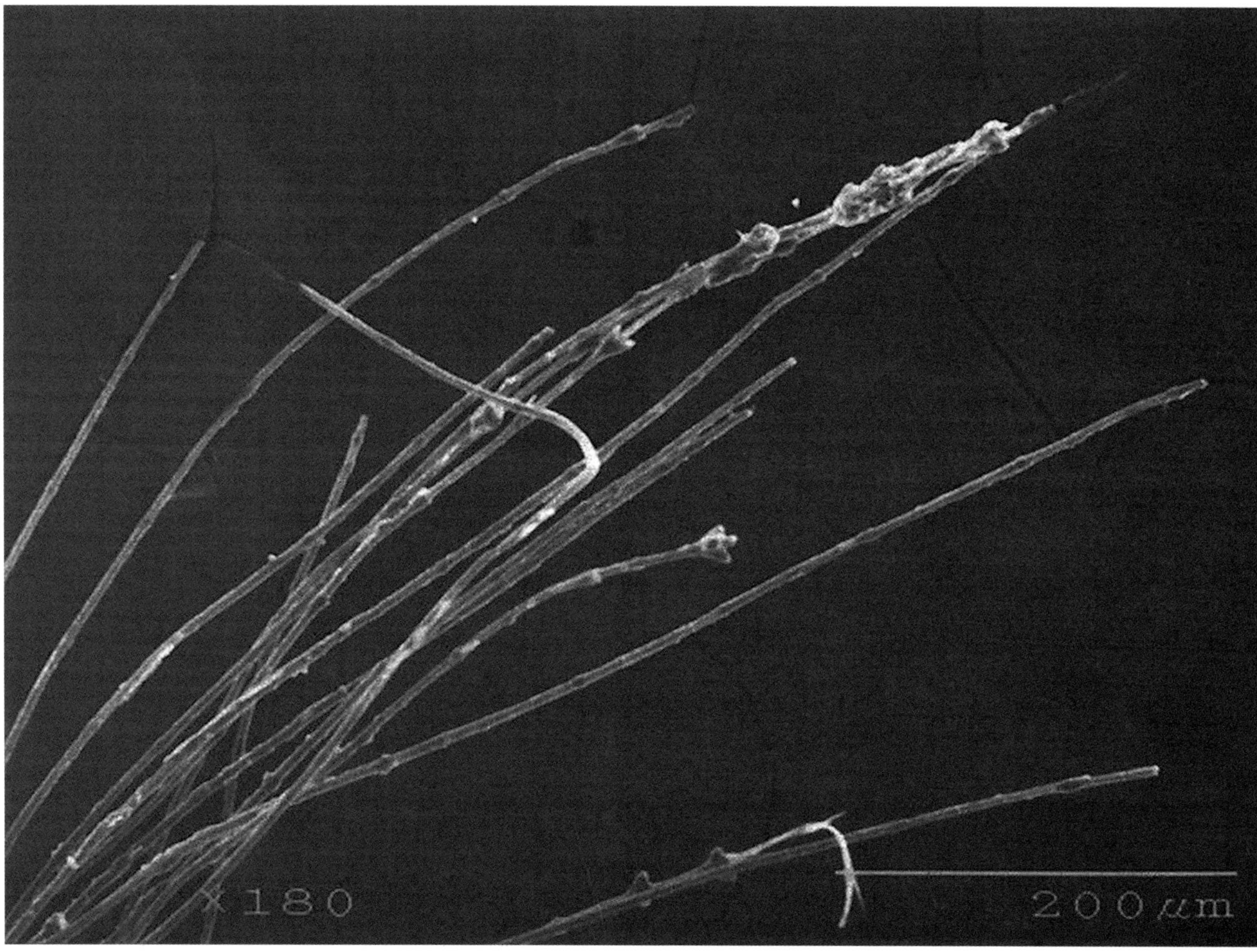

Figure 4 Broken barbules from a down feather dyed with a yellow natural dye.

obtained from the breasts of bird specimens belonging to the IB collections.

The SEM used for this study was a Hitachi S-2460N, with a voltage of 15 kV. Samples were prepared on carbon-coated adhesive discs and coated with gold for use in high vacuum (HV) mode.

The approach to examining down feathers using SEM consisted of three initial steps with the first two steps only carried out on samples from *Cairina moschata*. The first step was to become familiar with the structure of down feathers [12, 18]. Barbs, barbules, nodes and spines/prongs (spines are shorter than prongs) were found as indicated in the literature (Figure 3). The second step was to verify if there were any significant differences in the down of the same bird, from the breast and under the wing. The structure of both samples was almost identical, with the length of the barbules being the only difference. The final step was to confirm actual differences among species, using only down feathers from the breast of each bird. SEM enabled the acquisition of very detailed three-dimensional images of barbule structures and revealed variations in the distribution, form and number of nodes as well as the distribution of spines and/or prongs along the barbules. These variations permitted differentiation amongst all five samples, which was a crucial factor for proposing this analytical methodology as a tool for down identification. Length was not considered a determinant factor at the time, although further studies are required since it is still unknown if the length of down barbs and barbules is a vital factor for characterising feathers.

Reference library

Details of the average length of barbules from each sample, as well as a description of the nodes, spines and/or prongs observed, are explained in Table 1. Once the reference library had been created, eight down samples from the Tlamachte:ntli de Madeline were analysed. Samples were taken from the reverse of the textile, where loose ends of yarn can be found. Individual down feathers, measuring from 0.3 to 0.8 cm in length, were sampled. Six different down structures were documented, out of which only two were identified. One of the results was expected, and corresponded to the muscovy duck (*Cairina moschata*). The other structure that could be identified belongs to the mallard (*Anas platyrhynchos*). A larger reference library is needed to compare the other four structures found on the textile. Table 2 describes the physical characteristics of the analysed samples from the object.

It is important to note that the images provided by SEM also gave additional information on the process of degradation of down. On the macroscopic scale, all down dyed in yellow shows a significant decrease in volume and texture. My experience with five of the six colonial feathered textiles has made it possible to determine that in all cases the yellow-dyed down exhibits this effect. SEM images clearly show all the ruptures on the barbules. Figures 3 and 4 illustrate the contrast of the end tips of barbules: normally tapered on Figure 3 and broken on Figure 4. Further material characterisation is needed to determine which step of the dyeing process has provoked such damage.

Table 1 Characteristics of down barbule structures.

Sample	Barbules: average length (in μm)	Nodes	Spines/prongs	Additional remarks
Cairina moschata	800–900	Groups of two or three triangular nodes located close to the tips of all barbules.	Groups of three or four prongs located at the tip of each barbule.	Nodes and barbules are not intercalated among each other.
Anas platyrhynchos	700	Triangular nodes are not present in all barbules. When nodes are found on barbules, it is possible to appreciate one or two. These are located near the tips of barbules.	Varying number of prongs located close to the tip of each barbule.	Nodes are intercalated among prongs.
Anser caerulescens	300–350	No nodes present.	Prongs are located from the mid-section up to the tip of each barbule.	–
Pandion haliaëtus	400–500	No nodes present.	Prongs are located from close to the base up to the tip of each barbule.	–
Egretta thula	1400–1500	Variable number of oblong nodes located from the mid-section up to the tip of each barbule.	Spines are located close to the tip of each barbule.	–

Study of the feathered yarn

The six colonial feathered textiles show the same structure regarding the feathered yarns, with only minor variations. The descriptions in codices do not provide precise information on how these yarns were made. Moreover, the technique for creating these yarns was unknown to contemporary weaving indigenous communities in Mexico. Hence, detailed SEM observation has been the only available resource (thus far) to decipher the technique used for these feathered yarns.

The first examination of the yarn's structure was made using magnifying lenses. This revealed that all feathered yarns used for the Tlamachte:ntli de Madeline comprised a two-ply cotton yarn and down feathers. Each basic component of the yarn is clearly separate, i.e. down feathers are not mixed with the cotton fibres. The yarn was used as a structural weft yarn and also as a supplementary weft yarn to create the brocaded motifs.

Methodology of the study of the feathered yarn

The study of the feathered yarn was carried out at the Instituto Nacional de Investigaciones Nucleares (ININ). To suit the available equipment at the ININ, it was decided to mount the samples of the feathered yarn on a carbon-coated adhesive disc and place them into the chamber of a Jeol 5900, under a pressure of 19 Pa. Samples were taken from the reverse of the textile, where loose ends of the supplementary wefts can be found. The sizes of the samples varied between 0.5 and 1 cm to ensure that they remained twisted.

Contribution of SEM to the study of the feathered yarn

SEM enabled the acquisition of images of the structure of the feathered yarns at higher magnification and resolution. The images enabled clear differentiation of each cotton ply from the down feathers. Cotton fibres are twisted in a 'Z' direction to form each ply and then twisted together with down feathers, to form a final yarn twisted in an 'S' direction. The images clearly show the way in which down revolves around each ply: the more down, the more it covers both cotton plies (Figure 5).

The results of this analysis challenge information commonly found in the literature. When studying codices and/or actual feathered textiles, some authors have referred to these being made with the 'spun feather' technique, which implies that the cotton fibres were combined with down while it was being spun [1, 19, 20]. This, however, would create a one-ply yarn in which cotton fibres would be completely mixed – and entangled – with down feathers.

The SEM images show that this is not the case: the down feathers are not mixed with cotton fibres. Therefore, spinners must have twisted already-spun cotton yarns with down, creating enough torsion in the yarns to prevent the down feathers from coming away. The selection of down is important not only for the texture it creates, but also because of its flexibility which enables it to revolve around an already-spun yarn. Furthermore, the number of barbules present in these feathers, together with the length of barbs and barbules, serves to create better anchorage.

Since down is not spun along the cotton fibres to create a yarn, the new term of 'twisted feather' is proposed as a more appropriate description of this particular way of

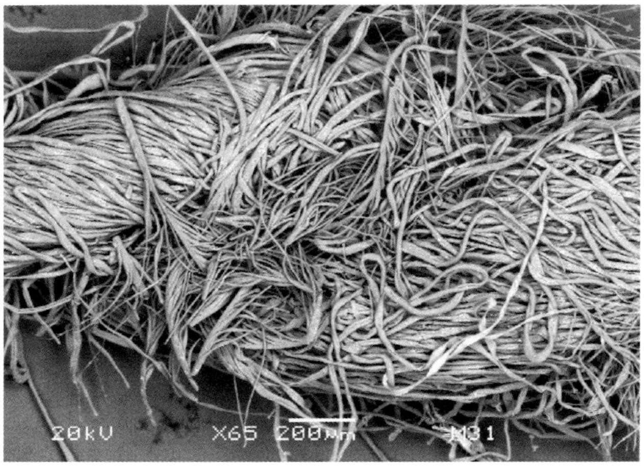

Figure 5 SEM image of a feathered yarn (detail).

Table 2 Feather samples from the Tlamachte:ntli de Madeline.

Sample no.	Identified bird	Nodes	Spines/prongs	Additional remarks
1	*Cairina moschata*	Groups of two or three triangular nodes located close to the tips of all barbules.	Groups of three or four prongs located at the tip of each barbule.	Nodes and barbules are not intercalated among each other.
2	*Anas platyrhynchos*	Triangular nodes are not present in all barbules. When nodes are found on barbules, it is possible to appreciate one or two. These are located near the tips of barbules.	Varying number of prongs located close to the tip of each barbule.	Nodes are intercalated among each other.
3	*Cairina moschata*	Groups of two or three triangular nodes located close to the tips of all barbules.	Groups of three or four prongs located at the tip of each barbule.	Nodes and barbules are not intercalated among each other.
4	Unidentified	Groups of two, three or four triangular nodes located at the mid-sections of all barbules.	Groups of three, four or five prongs located close to the tip of each barbule.	
5	Unidentified	Groups of two, three or four triangular nodes located at the mid-section of the distal and proximal barbule sections of the rachilla.	Groups of five prongs (at least) located from the mid-section and up to the tip of the distal and proximal barbule sections of the rachilla. The rest of the barbules present these prongs starting from the base.	
6	*Cairina moschata*	Groups of two or three triangular nodes located close to the tips of all barbules.	Groups of three or four prongs located at the tip of each barbule.	Nodes and barbules are not intercalated among each other.
7	Unidentified	Groups of four or five triangular nodes located close to the tips of all barbules.	Groups of four or five prongs located at the tip of each barbule.	Nodes and barbules are not intercalated among each other.
8	Unidentified	Groups of two or three triangular nodes located at the mid-sections of all barbules.	Groups of up to eight prongs located from the mid-section to the tip of all barbules.	

incorporating feathers into textiles. As all six colonial textiles seem to show this same technique and feathered garments were found in many Mesoamerican societies, it is possible that the twisted-feather technique is linked closely to pre-Columbian times.

Conclusions

The technical analyses on the Tlamachte:ntli de Madeline provide a better understanding of Mexican textile development. In addition to clarifying the manufacturing technique of feathered yarns, these analyses were fundamental to establish the basic knowledge to be able to reproduce such technique in the conservation laboratory. Collaborations with artists and designers in Mexico City offered a glimpse of one possible way in which this yarn may have been made in the seventeenth century or earlier. This in turn has led to a series of workshops with weavers from different communities in Mexico to try to revive the twisted-feather technique, which seems to have only been performed in this country. The Museo Textil de Oaxaca now houses the first two textiles with these types of feathered yarns woven in 2008 by Román Gutiérrez, a weaver from Teotitlán del Valle, Oaxaca.

The observation of down feathers using SEM assisted in creating a reference library of diagnostic images. Although in need of further development, it has been demonstrated here that SEM examination may be applied for the purposes of down identification. To date, down samples from the feathered blanket at the Museo de Bellas Artes de Toluca have been identified following the same methodology described. Down samples from the Tlamachayatl, in Rome will be analysed next.

Understanding and documenting the technique of the feathered yarn was useful and helped to establish links among the six colonial feathered textiles. Modern reproductions made it possible to visualise what feathered textiles could have looked like in earlier Mesoamerican societies. Furthermore, it laid the foundations for the technique's revival at the beginning of the twenty-first century.

The combination of science, conservation and the humanities allows a better interpretation and understanding of the technique with which these feathered textiles were made. Even more enriching has been the opportunity to communicate these findings to contemporary Mexican weaving communities. This in turn has provided the opportunity to delve more deeply into traditional textile techniques that might have been used by pre-Columbian societies.

Acknowledgements

I would like to thank Lorena Román (ENCRyM) and Alejandro de Ávila (MTO) for giving me the opportunity to study this wonderful textile. I would also like to thank everyone who helped me in my quest of down feather identification, as well as the reproduction of the twisted-feather technique: Patricia Escalante (Instituto de Biología, UNAM), Manuel Espinosa (Instituto de Investigaciones Nucleares), Oliva Fuentes (Museo de Bellas Artes de Toluca), Berenit Mendoza (Instituto de Biología, UNAM), Lourdes Navarijo (Instituto de Biología, UNAM), Arturo Ramírez (ENCRyM), Lilia Ramírez (ENCRyM), Leonel Sánchez (Museo de Bellas Artes de Toluca) and Donatella Saviola (Museo Nazionale Preistorico Etnografico 'Luigi Pigorini').

References

1. Johnson, I., 'Telas emplumadas en la época virreinal', in *El Arte Plumaria en México*, T. Castelló, Fomento Cultural Banamex, Mexico (1993) 79–99.
2. Anders, F., 'Las artes menores', *Artes de México* **17**(137) (1970) 4–45.
3. Sahagún, B., *Historia General de las Cosas de Nueva España*, 4th edn, Porrúa, Mexico (1981).
4. Cervantes, F., *Crónica de la Nueva España*, Porrúa, Mexico (1985).
5. Tableros, L., *Los Elementos de Pluma Contenidos en los Códicos Matrícula de Tributos y Códice Mendocino*, unpublished BA dissertation, Escuela Nacional de Antropología e Historia (1985).
6. Phipps, E. and Commoner, L., 'Investigation of a colonial Latin American textile', in *Proceedings of the Textile Society of America 10th Biennial Symposium, Toronto*, Textile Society of America (2006) 485–493.
7. Martínez, R., *K'uk'umal Chilil – El Huipil Emplumado de Zinacantán*, Casa de las Artesanías de Chiapas, Mexico (1990).
8. Varela, C., *Catálogo Arte Plumario Amazónico Museo de América*, Ministerio de Cultura, España (1993).
9. Roquero, A., *Tintes y tintoreros de América*, Ministerio de Cultura, Madrid (2006).
10. Tímár-Balázsy, Á. and Eastop, D., *Chemical Principles of Textile Conservation*, Butterworth-Heinemann, London (1998).
11. Moreno, M.O., *Conservación de Arte Plumario en México*, Unpublished thesis, Escuela Nacional de Conservación, Restauración y Museografía, 'Manuel del Castillo Negrete', Mexico (1983).
12. Dove, C., *A Descriptive and Phylogenetic Analysis of Plumulaceous Feather Characters in Charadriiformes*, Ornithological Monographs vol. 51, American Ornithologists' Union (2000).
13. Lourdes Navarijo, personal communication (September 2006).
14. Patricia Escalante, personal communication (November 2006).
15. Alejandro de Ávila, personal communication (November 2006).
16. Acuña, R. (ed.), *Relaciones Geográficas del Siglo XVI: México*, Universidad Nacional Autónoma de México – Instituto de Investigaciones Antropológicas, Mexico (1985).
17. Howell, S.N.G. and Webb, S., *A Guide to the Birds of Mexico and Northern Central America*, Oxford University Press, Oxford (1995).
18. Pettingill, O.S., *Ornithology in Laboratory and Field*, 5th edn, Academic Press, New York (1985).
19. Román, L. and Gutiérrez A., 'Conservation treatment of an eighteenth-century Mexican textile woven with feathers', in *Conservation Combinations: Proceedings of the North American Textile Conservation Conference*, North American Textile Conservation Conference (2000) 184–185.
20. Sayer, C., *Costumes of Mexico*, University of Texas Press, Austin (1985).

Author's address

Hector Manuel Meneses Lozano, Museo Textil de Oaxaca, Oaxaca, Mexico (hmeneses@mto.org.mx)

Applications of electron backscatter diffraction in archaeology

Shirley Northover and Peter Northover

ABSTRACT Electron backscattered diffraction (EBSD) has been used to study samples of archaeological gold, silver, copper and bronze prepared using simple metallographic etches. EBSD maps of orientation and local misorientation revealed the metals' microstructures and deformation substructures, and from this, combined in some cases with information on the crystallographic texture (also determined by EBSD), the methods of manufacture and working of the artefacts were determined. EBSD was also used to examine discontinuous precipitation at grain boundaries in a silver alloy at high resolution.

KEYWORDS electron backscatter diffraction, deformation, texture, face-centred cubic metals, sample preparation

Introduction

Electron backscattered diffraction (EBSD) [1–3] is now a well-established technique in materials science with over 2000 papers published on its use on engineering materials during the last three years, but its use in archaeology so far has been very limited [4–7]. This may be due to unfamiliarity with the wide range of different information that can be revealed by EBSD. This paper presents experimental results from EBSD studies of gold, silver, copper and bronze objects dating from between the thirteenth century BC and the present day, which demonstrate that with careful but unsophisticated specimen preparation this technique can provide valuable information, unavailable optically, on deformation and ageing that can be used to determine the methods of an object's manufacture and its thermo-mechanical history. It also shows that EBSD can reveal details of the formation of discontinuous precipitation (DP) at grain boundaries that may shed light on DP's role in silver's susceptibility to corrosion and determine if it is possible to use the morphology of DP as an unambiguous guide to an object's age.

EBSD technique

EBSD patterns are formed in a scanning electron microscope (SEM) when an electron beam is scattered by the regular array of atoms in the surface layers of a material, and are characteristic of the structure and local crystallographic orientation of the material under the beam. By systematically collecting and analysing these patterns, maps can be built up that reveal the distribution of phases present, show grain sizes and shapes, and give information on the deformation levels in the surface. The orientation measurements taken over an area can also be aggregated to show any crystallographic texture [8].

Backscattered electrons (BSE) have energies only slightly lower than those of the incident beam – but much higher than the energies of the secondary electrons (SE) that are commonly used for imaging in an SEM. The BSE that carry crystallographic contrast have been diffracted by the regular crystalline array of atoms in the metal grains close to the surface. They can escape from the surface if this happens within a layer that ranges from about 10 nm deep for heavy elements such as gold to about 100 nm for light materials such as aluminium. EBSD patterns are usually collected with the specimen tilted to about 70° to increase the proportion of these electrons that escape and so improve the signal-to-noise ratio. Such patterns can identify crystallographic orientations to an accuracy of better than 0.5° [8].

EBSD measures of deformation

The orientations measured at any point on the surface can be compared either with those in its immediate neighbourhood, to give a measure of the local misorientation and so show up sub-grain boundaries and local strains within individual grains [9], or with the orientations of all other points in the same grain to give a measure related to the total plastic strain experienced by that region of the specimen [10].

EBSD can measure texture and assist in differentiating between methods of working. Although other effects of deformation can often be observed by optical metallography, in some metals deformation traces may not always be etched successfully and resolution is limited both by the wavelength of light and by the effects of etching. EBSD overcomes these difficulties and also allows the understanding of such structures as those where grains are deformed but no internal deformation traces are visible, throwing light on deformation twinning in archaeological alloys.

Surface preparation

As EBSD is sensitive to deformation, the surface to be examined must be free of any strain induced by mechanical sample preparation; surface relief may distort the patterns so this too needs to be minimised. This can be achieved in various ways depending on the material being studied.

In the simplest case, a standard metallographic etch can be used to remove the deformed and smeared surface layer left on a metal after conventional grinding and diamond polishing. A variation on this is to use colloidal silica (which combines a chemical and a gentle physical attack) as a final polish with or without an etch. For very soft materials colloidal alumina may be used. A second approach, particularly useful for the softer metals, is to use a chemical or electrolytic polish but the ideal recipes for some archaeological alloy systems, such as As-Cu-Sb, have yet to be identified. A third method is to use ion milling and cleaning techniques but these require expensive equipment that may not always be readily accessible.

Representation of crystallographic information for face-centred cubic materials

Crystallographic planes and directions are conventionally described by Miller indices [8] and represented in space using a stereographic projection [8]. Gold, silver, copper and bronze all have a face-centred cubic structure and due to the very high symmetry of this structure all possible orientations in the crystal can be represented by those forming 1/48th of a complete sphere, which appears on the stereographic projection as the area inside a curvilinear triangle bounded by **[001]**, **[101]** and **[111]** directions. (Those not familiar with the use of Miller indices to describe crystal planes and directions need only remember that in the cubic materials a direction **[uvw]** is normal to a **(uvw)** plane, **{111}** planes[1] are the close-packed planes, the **<101>** directions are the close-packed directions and **[100]**, **[010]** and **[001]**are the x, y and z axes respectively of the cubic structure.) Whereas measurements made in the specimen frame of reference (equivalent to the stage position in the microscope) need to be displayed for all possible orientations, any representation in the crystal's frame of reference need only ever consider 1/48th of the whole sphere. EBSD data on local crystallographic orientation are frequently displayed using a colour coding for each point based on mixing red, blue and green in proportion to the orientation's proximity to **<100>**,**<111>** and **<110>** respectively.

Metalworking and crystallographic texture

Different metalworking processes such as hammering, rolling, or drawing and annealing can lead to alignment of particular crystallographic planes or directions, causing what is known as texture.[2] The details of this process are complex and only a very simple treatment will be offered here; the interested reader should consult Kocks et al. [11].

Working face-centred cubic metals usually produces plastic deformation by generating and moving dislocations and this changes the local orientation of the metal crystals. Dislocations in metals generally move more easily on close-packed planes and in close-packed directions and as deformation proceeds, grain orientations change so that such favoured planes and directions are closer to the direction of the main applied stresses. During hammering and rolling a metal, the constraints on the metal's shape while it is deforming are different in each case so the two processes produce different textures. Drawing a wire produces what is known as a fibre texture.

Table 1 Rolling and fibre textures in face-centred cubic metals.

Metal	Rolling	Fibre
Cu	(110) [1–12]	[111]
Cu	(123) [1–21]	[100]
Cu –30% Zn	(110) [–112]	[111]
Au	(123) [1–21]	[111] [100]
Au –10 at.% Cu	(110) [–112]	[111] [100]
Ag	(110) [1–12]	[100]

Table 1 shows which crystallographic planes and directions are preferentially aligned by rolling or drawing a variety of different face-centred cubic alloys [12].

In general for face-centred cubic metals, **<110>** tend to align along the direction of compression and **<111>** tend to align along the tensile direction. Annealing, which recrystallises the structure, does not randomise the metal's texture but leads to complex textures retaining and modifying particular elements of the deformation texture.

Experimental

Samples examined

Gold

A sample (area 2 mm × 3 mm) from an Achaemenid gold bowl was taken from close to where it had been soldered to a silver-11% copper alloy sculpture of a winged lion (private collection). The gold alloy contained 5.5 wt% copper and 4.5 wt% silver.

Copper

(1) Samples from two copper bolts from British warships were provided by the Hampshire and Wight Trust for Maritime Archaeology: one from *HMS Pomone*, built in 1805 that sank in 1811, and one from *HMS Impregnable*, built in 1786, that sank in 1799. The sample from *HMS Pomone* (99 wt% Cu, remainder Ag, As, Bi, mostly as oxides) was of irregular shape and the sample orientation with respect to the bolt's long axis not obvious from the geometry of the sample. A section of the bolt from *HMS Impregnable* had been turned down to make a tensile testing specimen and samples were cut from this so that it was possible to examine both longitudinal and transverse sections.

(2) Four copper pennies from the 1797 issue struck at Matthew Boulton's Soho Mint, with a composition similar to that of the copper bolts (authors' own collection).

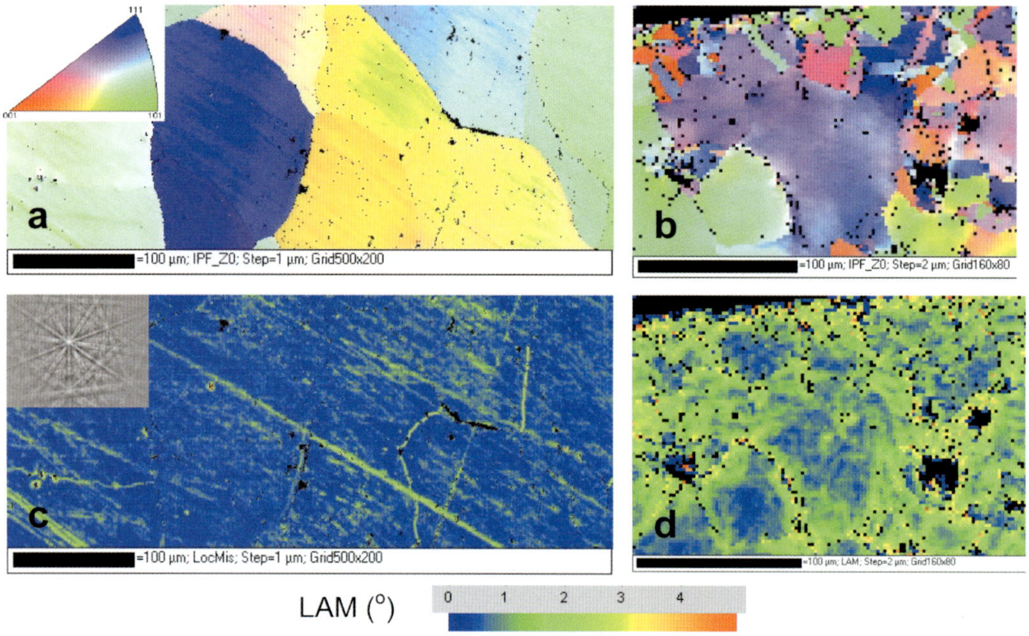

Figure 1 Orientation and local misorientation maps from Achaemenid gold bowl: (a) orientation map typical of most of the specimen with inset key to colour coding of surface normal orientations in (a) and (b); (b) orientation map from the area near the surface of the bowl close to the soldered joint; (c) map of local angular misorientation (LAM) of the area shown in (a) with inset of typical EBSD pattern from this area; (d) local angular misorientation map of the area shown in (b).

Bronze

A bronze sickle W. Hungary: BzD-Ha A1 (thirteenth century BC) – Cu, 5.4 wt% Sn, 0.82 wt% As, 0.35 wt% Sb, 0.31 wt% Ni, 0.59 wt% Pb, 0.32 wt% S (authors' own collection).

Silver

(1) Two silver Timurid coins (fifteenth century) – Ag, 3.95 wt% Cu, 0.03 wt% Bi, 0.87 wt% Pb, 0.65 wt% Au.
(2) A Samanid silver coin (ninth–tenth century) – Ag, 7.7 wt% Cu, 0.85 wt% Bi, 1.01 wt% Pb, 0.04 wt% Au.
(3) A modern Ag – 5 wt% Cu sample homogenised for 1 hour at 800° C then furnace cooled (FC).
(4) A modern Ag – 5wt% Cu sample homogenised for 20 mins at 700°C then held at 200°C for 72 hours and quenched.

The silver coins (1) and (2) come from a collection provided by a donor for joint research between L. Treadwell of the Oriental Institute, University of Oxford, and the authors.

EBSD

All EBSD in this study was carried out on a Zeiss Supra 55 Variable Pressure FEG SEM operating at 20 kV. The patterns were collected using a Nordlys II Camera with a charge-coupled device area 38 × 28 mm and analysed using Oxford Instruments Channel5 software.

Surface preparation

All the metals examined were initially polished to a 1 µm diamond finish using conventional metallographic techniques and then swab etched. Etching times were generally a few seconds except for the gold, which required several minutes. The etches used were as follows:

- Gold: aqua regia: 30 ml HCl, 20 ml HNO_3; the optimum proportions of the two acids varies depending on the silver content of the gold.
- Copper: ammoniacal hydrogen peroxide, typically 25 ml NH_4OH, 25 ml H_2O, 25 ml 3% H_2O_2.
- Bronze: 2 g $FeCl_3$, 25 ml H_2O, 5 ml HCl, 60 ml ethanol.
- Silver: ammoniacal hydrogen peroxide as for copper.

To obtain good EBSD patterns from the silver coins they then needed a light evaporated C coating (to prevent charging of the oxide layer on the surface). For purposes of comparison one copper bolt was also heavily electropolished using Struers electrolyte D2 at 24 V and 17°C.

Results and discussion

Gold

As seen in Figure 1a, over almost all the specimen, grains were very large with no twins and little orientation variation within the grains indicating a cast and homogenised structure with no history of cold work. As seen in the inset of Figure 1c, over most of the sample very sharp patterns were obtained that again indicate very low deformation levels and this is confirmed in the local angular misorientation (LAM) map [9] which, apart from remnant traces of polishing scratches from preparing the very soft metal specimen and occasional low angle boundaries pinned by inclusions, shows a LAM (taken over a 3 × 3 µm grid) of less than 0.5° over the whole specimen. There is no sign of cold work in these homogenised grains within the accuracy of EBSD.

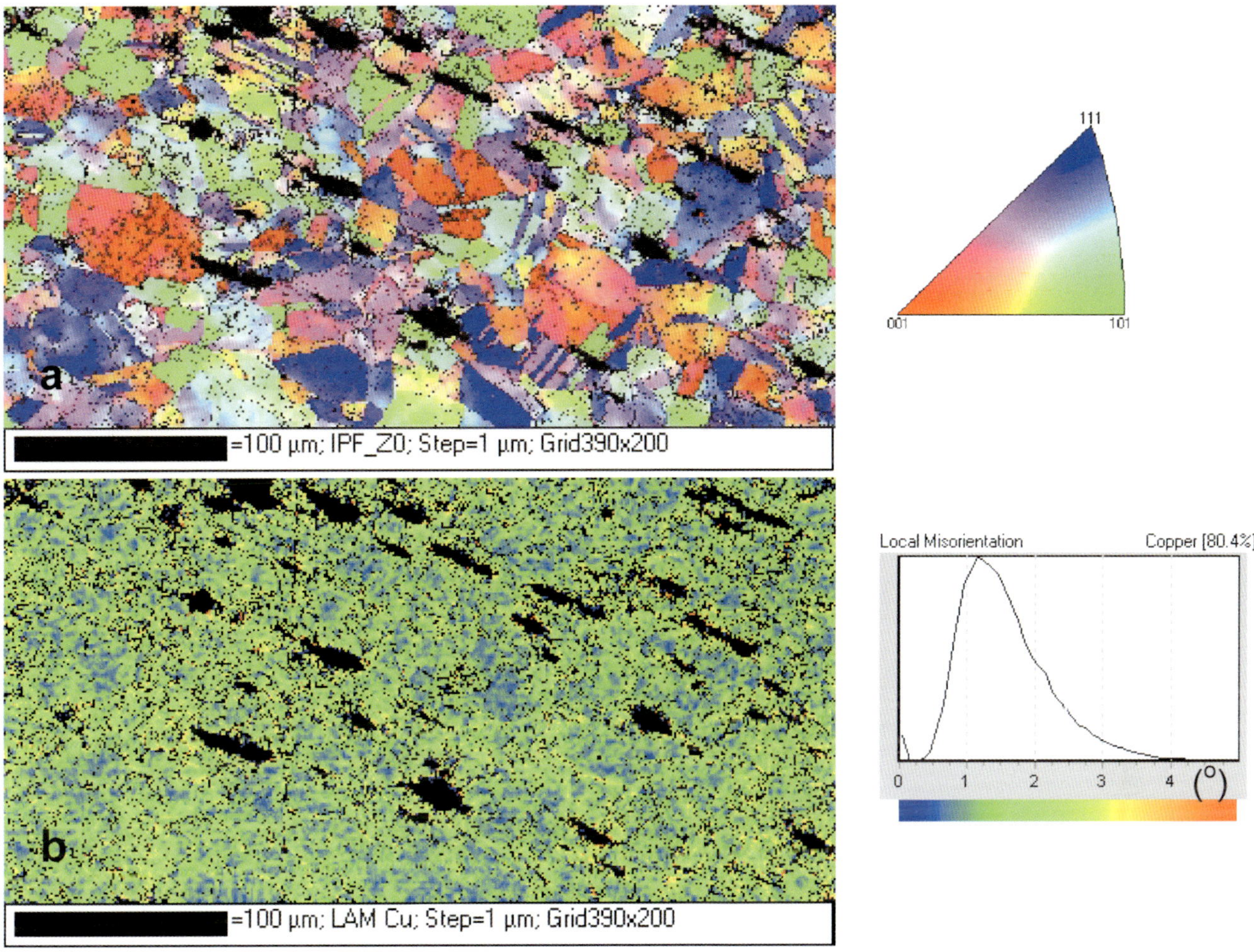

Figure 2 Copper bolt from *HMS Pomone*: (a) orientation map from close to the rolling plane and (b) local angular misorientation map from close to the rolling plane.

The orientation map and the LAM map in Figure 1b and Figure 1d both show that at one edge of the specimen, corresponding to the outer surface of the bowl close to the soldered joint, there was an area where there was an orientation spread within the grains showing that the surface had been worked; there was also a small area of recrystallised grains. The body of the material may have been homogenised by the heat of the soldering operation to fix the bowl to the winged lion's body. Any heating sufficient to homogenise the as-cast dendritic segregation in the body of the bowl would have been sufficient to cause recrystallisation at any cold-worked regions on the surface. The recrystallised grains also show a spread of orientations although when they originally formed they would have been strain free, showing that the surface had been deformed further after this heating had taken place.

Copper

Ships' bolts

Although the sample of the bolt from *HMS Pomone* was of irregular shape, the elongation of the stringers of inclusions indicated the projection of the rolling direction on the sample surface and the aspect ratios of the individual deformed particles suggested that the true rolling direction lay within a very few degrees of this plane.

In the orientation map Figure 2a, a significant number of grains are seen to have orientations close to **<110>** along the specimen normal, as might be expected for a rolled sample examined normal to the rolling plane. The variation in orientation within the grains indicates that there is considerable deformation present. Figure 2b shows that over a 3 × 3 µm grid, the LAM over most of the specimen was between one and two degrees indicating internal distortions within the grains resulting from a moderate degree of deformation.

As seen in Figure 3, the pole figures (PFs) of the bolt from *HMS Pomone* show a fibre texture with **<111>** along the rolling direction, which is surprising because **<111>** directions tend to align along the tensile directions and this is the kind of texture one might find in drawn wire rather than in rolled material. Copper ships' bolts of this period were generally rolled.

The inverse pole figure (IPF) shows (as expected from the map of Figure 2a) **<110>** aligned normal to the specimen surface and normal to the rolling direction. This would be the expected texture resulting from compression between two rollers but this **<110>** texture is weaker than the alignment of **<111>** with the rolling direction.

This bolt's strange combination of a rolled bolt with a clear tensile texture can be explained by the information from Harris [13] that 'William Collins of Birmingham took out a patent for ships' bolts on 22 October 1783 claiming to make bolts "which [...] will not be subject to the decay that has proved so fatal to those now in use". His process could be used to make either iron or copper bolts. When copper

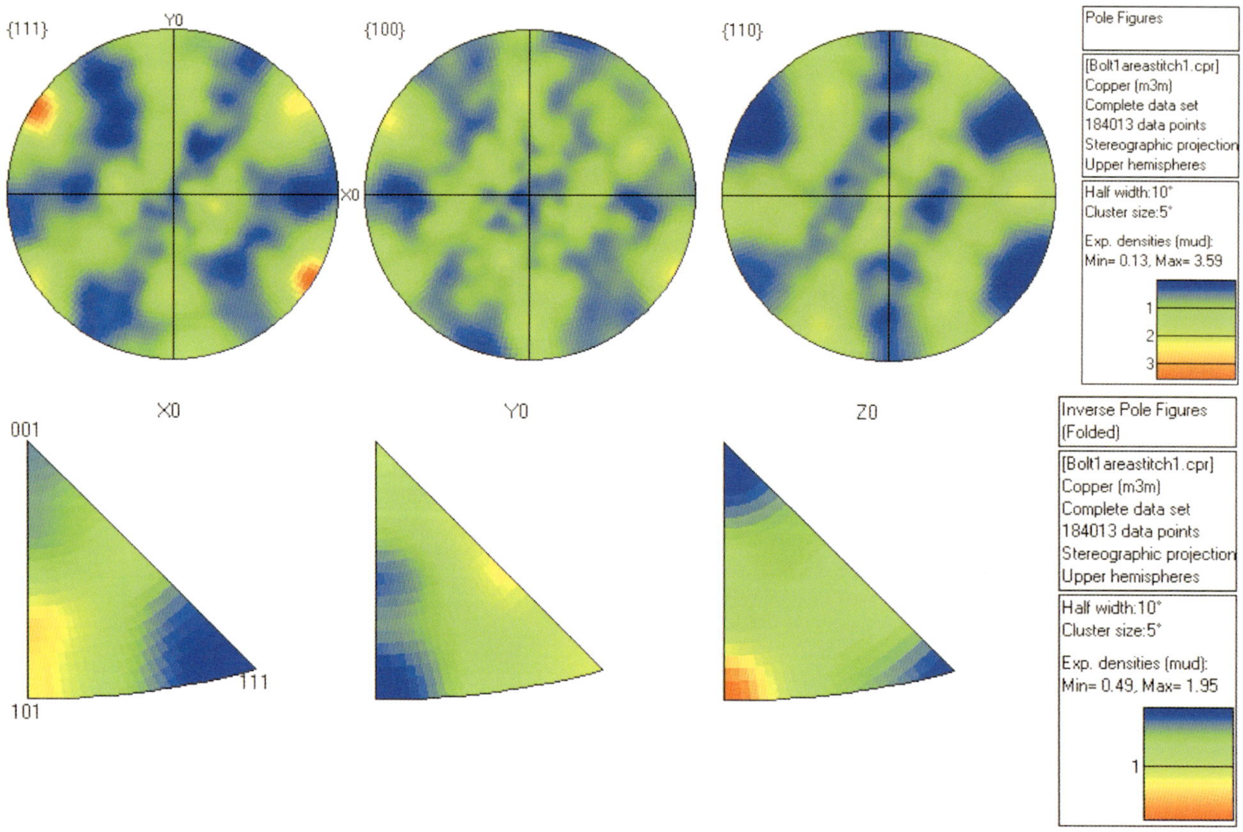

Figure 3 Results from EBSD texture measurements over approximately 2mm² from the copper bolt from *HMS Pomone*.

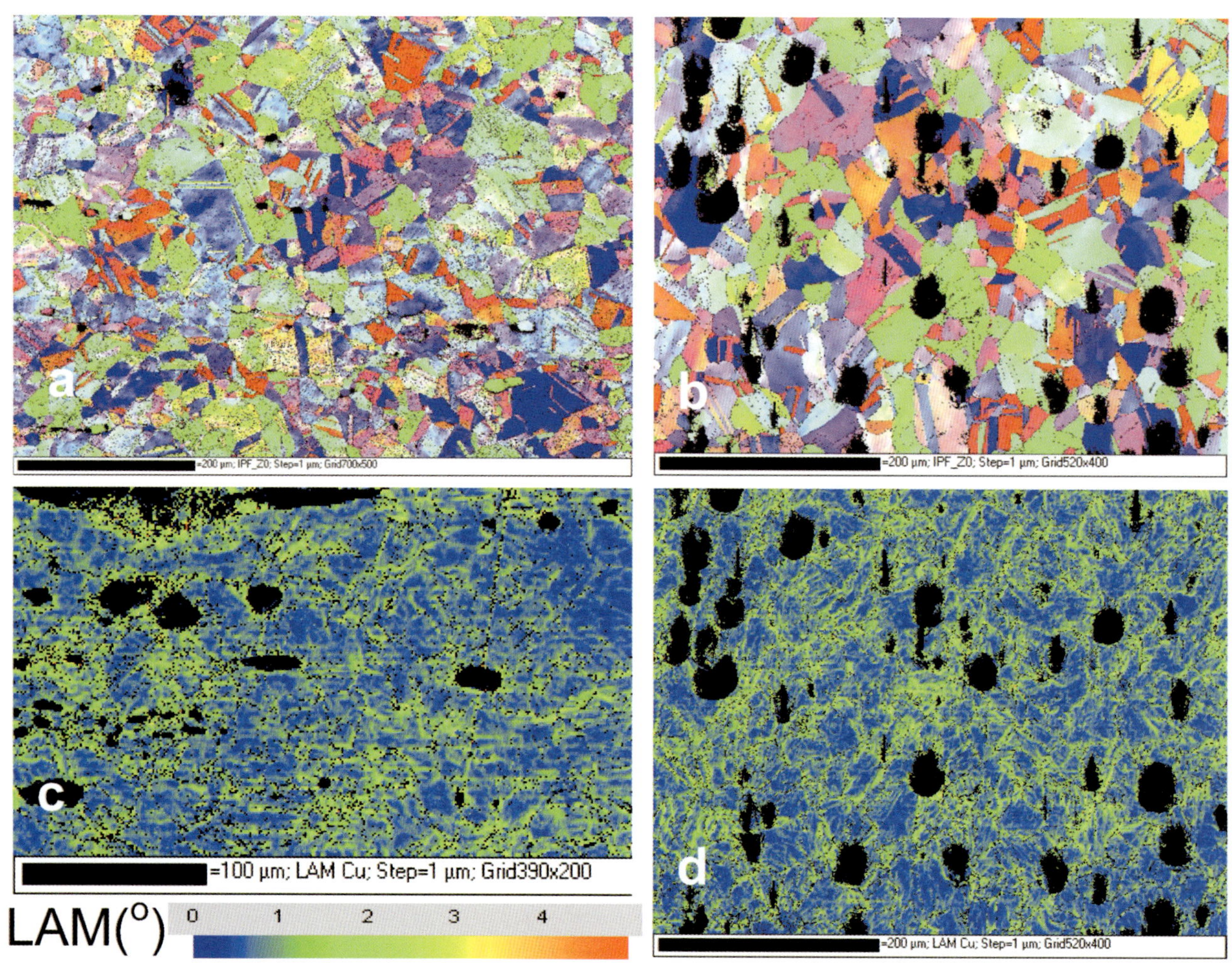

Figure 4 Copper bolt from *HMS Impregnable*: (a) orientation map of polished and etched cross-section (colour key as Fig. 2a); (b) orientation map of electropolished cross-section (colour key as Fig 2a); (c) LAM map of electropolished longitudinal section; (d) LAM map electropolished cross-section.

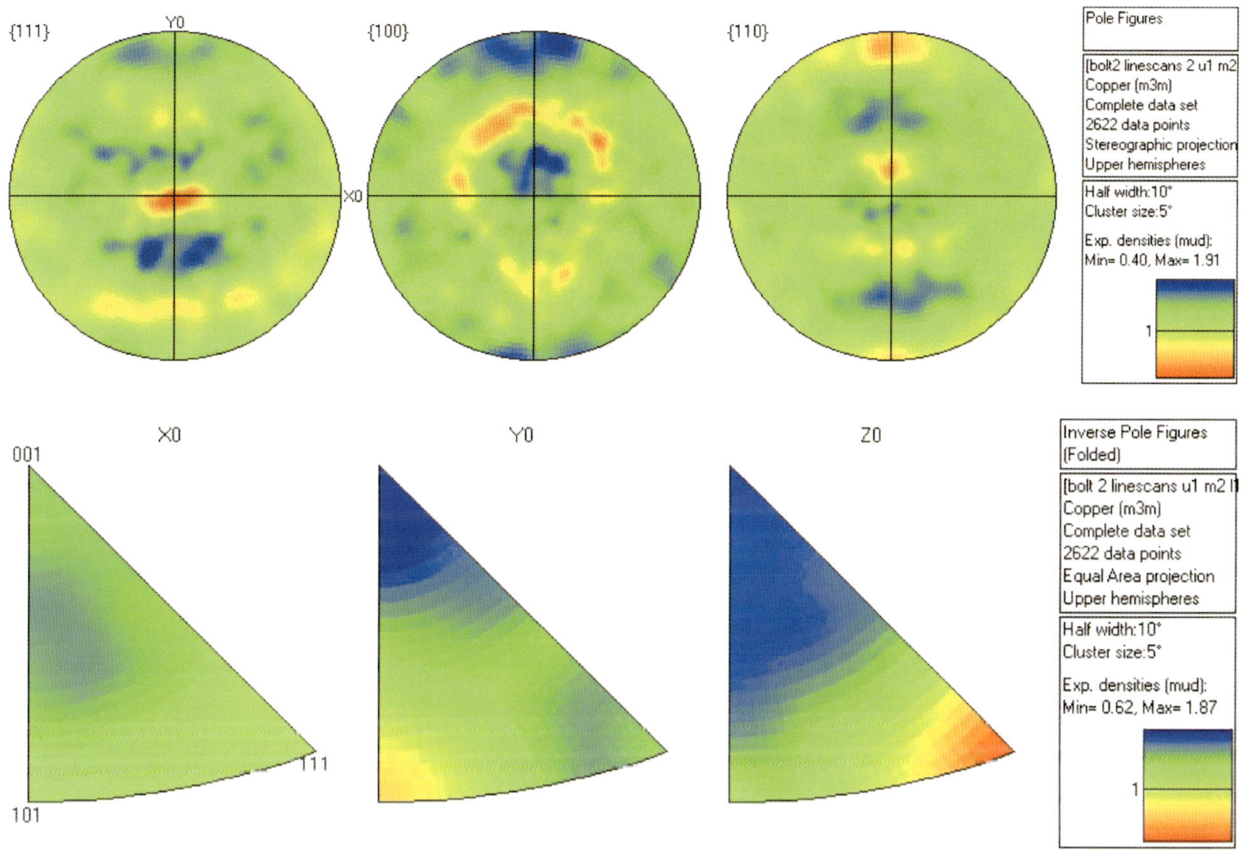

Figure 5 Results from EBSD texture measurements over approximately 10 mm^2 from the copper bolt from *HMS Impregnable*.

was to be used it was to be as pure as possible. The copper bar was to be gripped between grooved rollers and thereby pulled forcibly through steel drawplates so that small bolts would be drawn out to double their length and large ones to one and a half times.' This type of process would produce precisely the sort of texture seen here so we can be confident in saying that the bolt from *HMS Pomone* was manufactured using William Collins's process.

As seen in Figure 4a, an orientation map from the bolt from *HMS Impregnable* again showed an orientation variation within grains indicating cold work. In order to be absolutely sure that this cold work could not have been introduced during specimen preparation a sample was also prepared by heavy electropolishing and as seen in Figure 4b, there were exactly the same levels of orientation difference (as indicated by the extent of colour variations on the IPF map) within the grains in both samples. This is confirmed by the same background colour in the LAM maps of Figure 4c and Figure 4d. The holes in the electropolished samples are caused by preferential attack of the electropolish at the interface between the copper and the inclusions.

The quality of the EBSD patterns obtained with the two different types of specimen preparation was compared quantitatively in terms of both the sharpness of their lines and their contrast. Although the patterns were definitely sharper for the electropolished specimens, the contrast was slightly higher for the etched samples. Since the accuracy of orientation determination from EBSD patterns depends on both their sharpness and their level of contrast this implies that the accuracy of the orientation information from the metallographically prepared samples was comparable to that from the electropolished samples.

The LAM map of Figure 4c from the longitudinal section of the bolt from *HMS Impregnable* shows a lower degree of deformation retained in the grains than in the bolt from *HMS Pomone*; most of the LAM map is blue rather than green in Figure 2b. Deformation is concentrated in bands parallel to the rolling direction.

In Figure 4d the LAM map of the cross-section of the bolt from *HMS Impregnable* also shows much lower general levels of deformation than that from the bolt from *HMS Pomone*; in the *Impregnable* bolt the deformation is seen to be localised into a cell structure with pancake-shaped sub-grains aligned to the direction of elongation of the inclusions but with no particular other alignment. The texture of the bolt from *HMS Impregnable* measured over 9.72 mm^2 of the longitudinal section (Figure 5).

It is immediately obvious that the texture of the bolt from *HMS Impregnable* is very different from that of *HMS Pomone*. The more complex texture of the *Impregnable* bolt suggests that recrystallisation plays a much stronger part than in the first bolt and this is borne out by the much lower deformation levels retained in the grains. In the *Impregnable*'s bolt, **<111>** rather than **<110>** is oriented perpendicular to the rolling direction.

In the late eighteenth century various processes were being used to make bolts [13] and further work is necessary to determine exactly how *HMS Impregnable*'s bolts were made, but what is very clear is that the two bolts studied were made in different ways.

1797 Boulton pennies

Examination was carried out of the etched cross-sections of four of Matthew Boulton's 1797 pennies, which optically and in the SE image of the SEM appeared fully annealed.

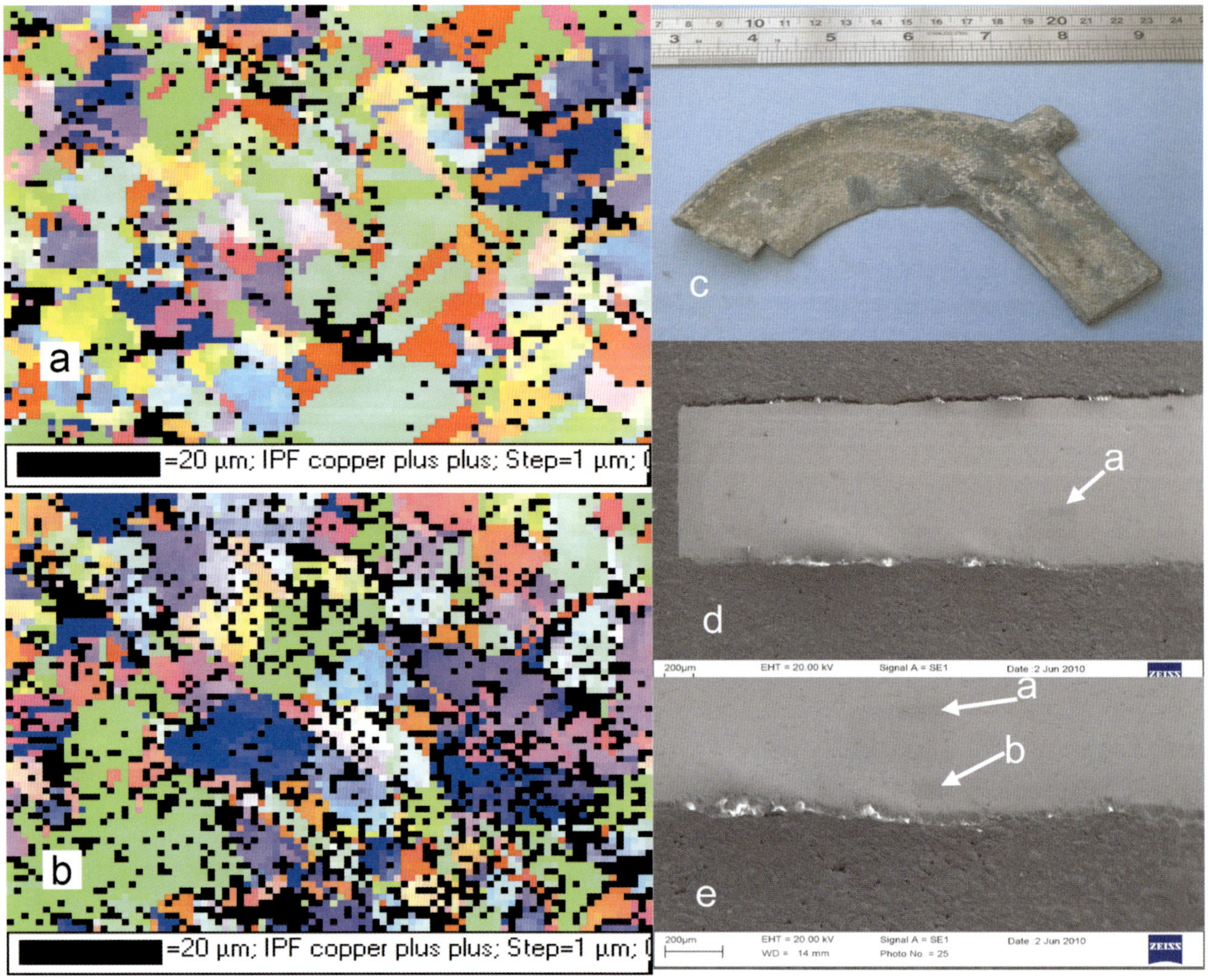

Figure 6 EBSD maps from near the edge of the blade of a bronze sickle: (a) orientation map from about 400 µm from the edge; (b) orientation map from about 100 µm from the edge; (c) position from which the sample was taken from the sickle blade; (d) position from which the map was taken relative to the broad end of the sample; (e) relative positions of (a) and (b).

Orientation maps from all four showed deformation within the grains resulting from cold striking. For two of the pennies more points were indexed if match units with slightly different lattice parameters for the copper were included although the measured orientations were unchanged: this might suggest variations in the lattice parameter caused by compositional inhomogeneities in the metal of these two coins. Optical metallography had revealed that the original as-cast segregation of arsenic had not been fully homogenised.

Bronze

The results from the bronze sickle illustrate how EBSD can be used to compare quantitatively the extent to which different areas of an object have been worked since its final anneal by comparing the deformation levels at two different places in a tapering sample taken from the sickle's blade as shown in Figure 6.

About 400 µm from the edge, the grain average misorientation (GAM) [10] over area a was 1.4° while GAM measured over area b about 100 µm from the edge was 2.0°. The higher deformation level closer to the blade edge is reflected in the higher proportion of unindexed points and the greater variation in orientation contrast within individual grains. This method could be used to map out quantitatively the amount of working different parts of the object had experienced. EBSD measurements of texture over a wider area close to the edge showed no very strong texture but a weak preferred orientation **<110>** parallel to Y, i.e. parallel to the thickness of the blade. This is consistent with hammering.

Silver

Figure 7 shows orientation maps from one Samanid and two Timurid coins. As with the Boulton pennies, with optical microscopy (OM) the etched structure looks annealed but EBSD orientation maps show deformation within the grains resulting from cold striking and this can be quantified by measuring the mean local misorientations.

One important question pertaining to silver that EBSD can address is that of discontinuous precipitation (coupled growth of a precipitated copper-rich phase and

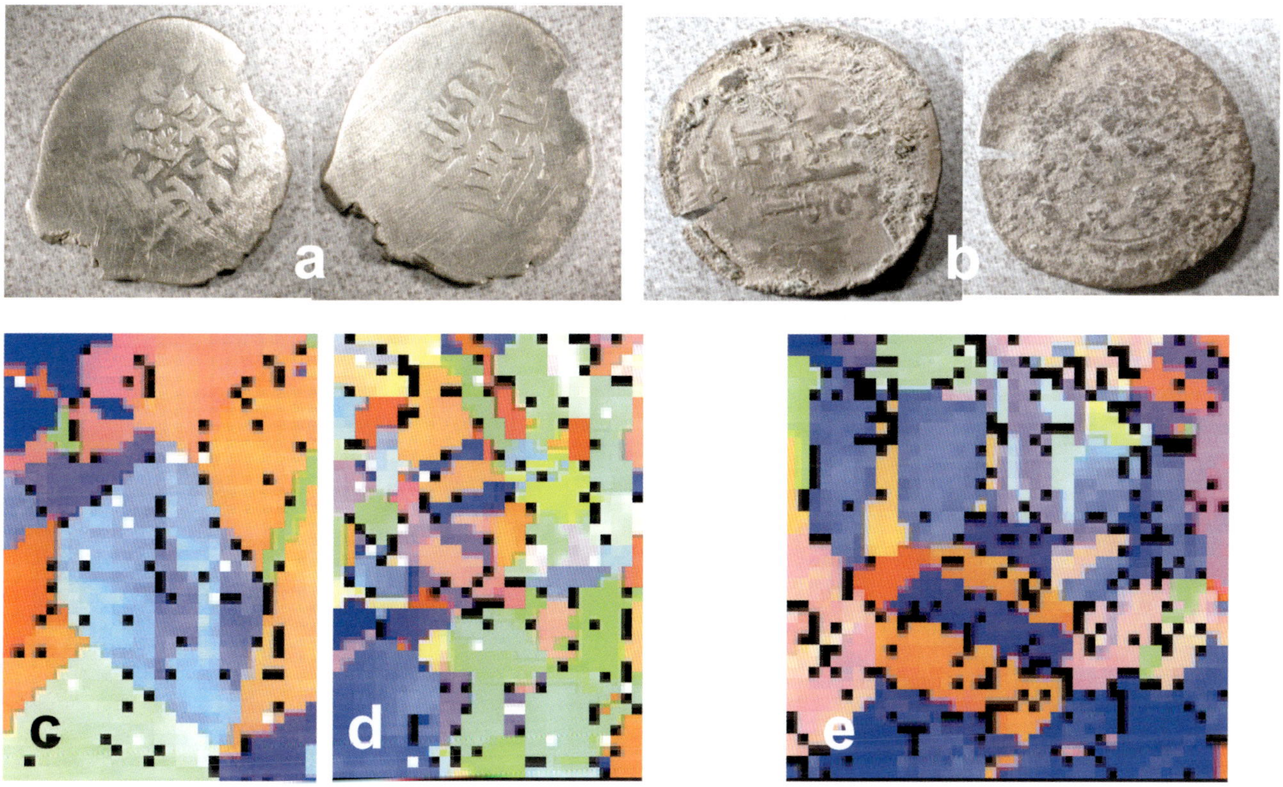

Figure 7 Orientation maps of silver Timurid and Samanid coins: (a) Timurid coins; (b) Samanid coin; (c) orientation map from a Timurid coin; (d) orientation map from the second Timurid coin; (e) orientation map from the Samanid coin.

solute-depleted regions of the matrix with a discontinuous change in composition at the reaction interface), and its validity as a marker for age. It has long been recognised that grain boundary changes can occur over archaeological time at ambient temperature in the form of discontinuous precipitation and diffusion-induced grain boundary migration. In the 1970s, Schweizer and Meyers [14] first discussed using the extent of changes identified as discontinuous precipitation of Cu at Ag grain boundaries as an indicator of age. This was based on the assumption that this type of transformation occurs only at low temperatures over very extended periods of time, but since then there has been very little characterisation of the structures other than by OM until Wanhill and his colleagues [4] investigated the Gundestrup cauldron using EBSD and concluded that remanent cold deformation rather than discontinuous precipitation was the main cause of corrosion. Energy dispersive X-ray microanalysis (EDX) of apparent cellular precipitation in a Byzantine silver paten carried out by Wanhill and the present authors [4] showed silver-rich/copper-depleted regions close to a triple point separated by an interface from the matrix that confirmed the nature of the reaction as discontinuous rather than being grain boundary migration.

In 1998, Dye [15] conducted a series of heat treatments on modern Britannia silver to explore optically the conditions of time and temperature at which different morphologies of grain boundary modifications occur. As determined on an optical scale, she succeeded in reproducing the morphologies of grain boundaries in archaeological silver that shows little stored cold work and her samples provided an excellent model system for characterising these structures more fully. EBSD allows us to investigate the crystallography as well as the morphology and some early results are presented in Figure 8.

Figure 8a shows a boundary in a sample homogenised and then held at 200°C for 72 hours where several discrete nodules of discontinuous precipitation have formed along the same boundary XY and their growth direction depends on which of the two twin variants of grain A adjoins the boundary. This shows that here the nucleation and growth of the colonies were influenced more by the misorientation across the boundary than by its plane (which simplifies study by EBSD where boundary plane determination requires serial sectioning whereas the grain boundary misorientation is directly available). In both cases the orientation of the colony is the same as that of one of the grains. In one set of colonies the boundary has migrated into grain A and in the other into grain B. (The original position of the boundary is visible in the band contrast image of Figure 8b.) In this example, the band contrast image does not show the presence of lamellae but in others the interfaces between Cu-rich and silver-rich lamellae inside the colonies are visible in the band contrast image because the change in lattice parameter there reduces the pattern quality. Usually lamellae are visible in the SE image of the etched colonies but in some instances boundaries may have migrated without formation of cellular colonies.

The complexity involved in understanding these changes at grain boundaries is well illustrated by the triple junction in an homogenised sample, held for 1 hour at 800°C + FC, shown in Figure 8c. In the SE image (Figure 8d), the structure of the cellular colonies looks very similar to that seen optically with the precipitate lamellae clearly visible but the orientation map Figure 8a shows that in this case, two apparently different colonies on each side of the boundary have both formed with the same orientation as grain Q. There are also small volumes that have grown in the same orientation as grain P. In this case, the LAM map (Figure 8e) confirms that no low angle b

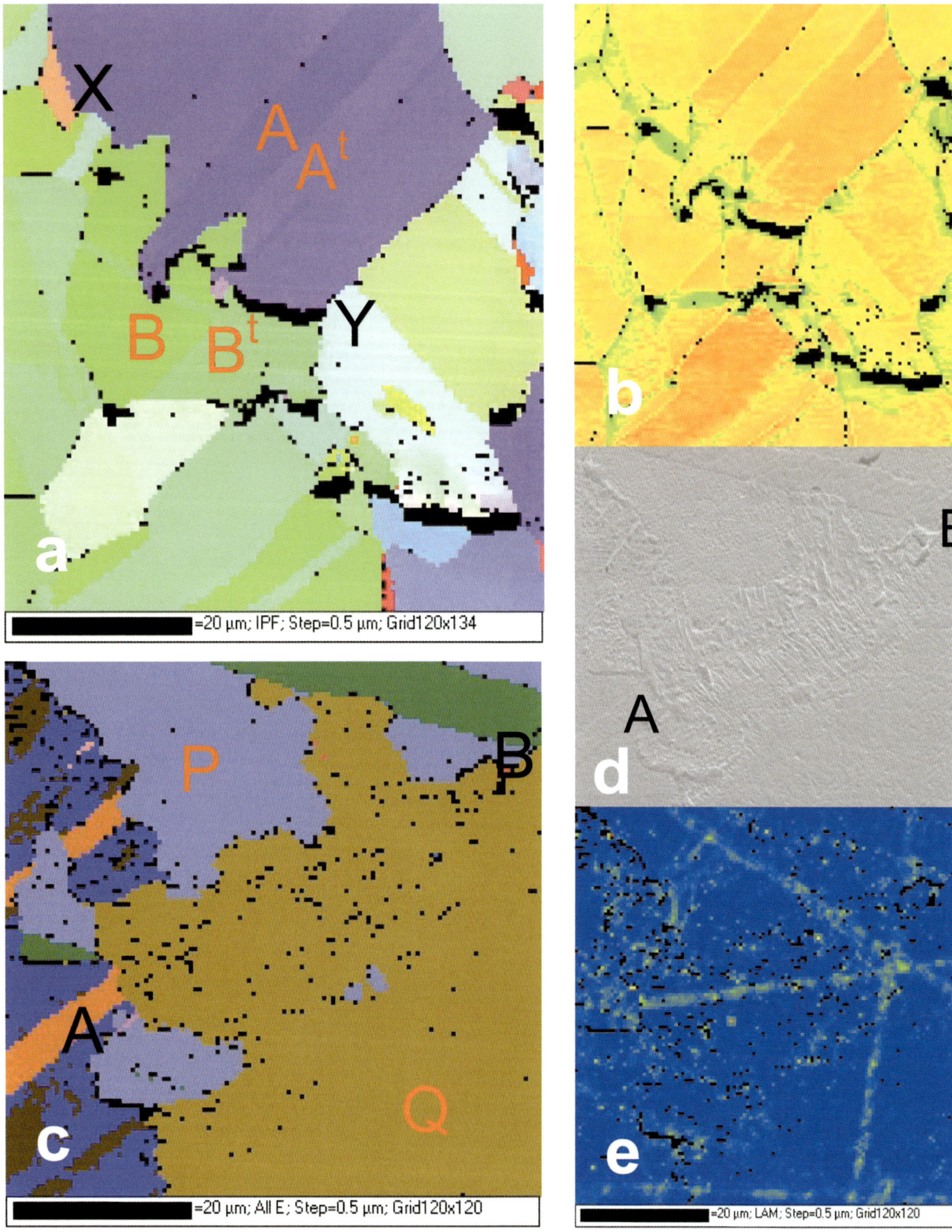

Figure 8 EBSD maps of modern Britannia silver heat treated to replicate grain boundary morphologies seen in ancient silver: (a) orientation map from the homogenised sample held at 200°C for 72 hours; (b) band contrast map of the area shown in (a); (c) orientation map around a grain boundary triple point in the homogenised sample held for 1 hour at 800°C + FC; (d) secondary electron image of the area in (c); (e) local angular misorientation map of the area in (c).

has been left at the original boundary. EBSD gives us access to the crystallographic information needed to identify the characteristics of grain boundaries at which particular types of changes occur and to help interpret these complex structures, however much research remains to be carried out.

Conclusions

EBSD can reveal a metal object's structure, crystallographic texture and distribution of plastic strain and can therefore help to indicate the methods of manufacture and shaping through their effects on the metal's micro- and macro-textures. EBSD can also be used to investigate grain boundary character and monitor changes at grain boundaries that affect the object's vulnerability to corrosion and so alert conservators to possible problems. Discontinuous precipitation at grain boundaries can be characterised at high resolution with EBSD offering the prospect of establishing conclusively the implications of discontinuous precipitation in determining an object's age and authenticity. A wide range of metals of interest can be prepared for EBSD using simple polishing and etching techniques that do not require expensive specialised equipment.

Acknowledgements

We would like to thank Gordon Imlach for his many hours of assistance in operating the SEM and to acknowledge the Open University, Faculty of Mathematics, Computing and Technology and the University of Oxford, Department of Materials for provision of laboratory facilities.

Notes

1. In this notation **{111}** represents the set of planes symmetrically related to the **(111)** plane and <**101**> is the set of directions symmetrically related to the **[101]**.
2. Also known as preferred orientation.

References

1. Maitland, T. and Sitzman, S., 'Electron backscatter diffraction (EBSD) technique and materials characterization examples', in *Scanning Microscopy for Nanotechnology Techniques and Applications*, ed. W. Zhou and Z.L. Wang, Springer, London/New York (2007) 41–75.
2. Prior, D.J., Boyle, A.P., Brenker, F., Cheadle, M.C., Day, A., Lopez, G., Peruzzo, L., Reddy, S., Spiess, R., Timms, N.E., Trimby, P., Wheeler, J. and Zetterström, L., 'The application of electron backscatter diffraction and orientation contrast imaging in the SEM to textural problems in rocks', *American Mineralogist* **84**(11–12) (1999) 1741–1759.
3. Schwartz, A.J., Kumar, M., Adams, B.L. and Field, D.P. (eds), *Electron Backscatter Diffraction in Materials Science*, 2nd edn, Springer, London/New York (2009).
4. Wanhill, R.J.H., Hattenberg, T. and Northover, J.P., 'EBSD of corrosion, deformation and precipitation in the Gundestrup cauldron', in *Ligas Metálicas: Investigação e Conservação*, ed. A.C. Ferreiro da Silva and P. Menino Homem, Faculdade de Letras da Universidade do Porto, Oporto (2008) 47–61.
5. Lejcek, P., Jager, A., Gartnerova, V., Vanckova, J., Ded, J. and Haloda, J., 'Structure/property relationship in intergranular corrosion of archaeological silver artefacts', *Materials Science Forum* **638–642** (2010) 2852–2857.
6. Zaghis, F., Peruzzo, L., Salviulo, G. and Molin, G., 'A new approach in studying Cu-Sn artefacts: measure of crystal orientation and phase distribution by EBSD analysis', XX IUCr, Florence, Italy, 23–31 August 2005, *Acta Crystallographica Section A* **A61**(C493) (2005).
7. Peruzzo, L., Fenzi, F. and Vigato, P.A., 'Electron backscatter diffraction (EBSD): a new technique for the identification of pigments and raw materials in historic glasses and ceramics', *Archaeometry* **51**(1) (2011) 178–193.
8. Randle, V. and Engler, O., *Introduction to Texture Analysis: Macrotexture, Microtexture and Orientation Mapping*, Gordon and Breach, Amsterdam (2000).
9. Lehockey, E.M., Lin, Y. and Lepik, O.E., 'Mapping residual plastic strain in materials using electron backscatter diffraction', in *Electron Backscatter Diffraction in Materials Science*, ed. A.J. Schwartz, M. Kumar and B.L. Adams, Kluwer Academic/Plenum Publishers, New York (2000) 247–248.
10. Yoda, R., Yokomaku, T. and Tsuji, N., 'Plastic deformation and creep damage evaluations of type 316 austenitic stainless steels by EBSD', *Materials Characterization* **61**(10) (2010) 913–922.
11. Kocks, U.F., Tomé, C.N. and Wenk, H.-R., *Texture and Anisotropy*, Cambridge University Press, Cambridge (2000).
12. Taylor, A., *X-ray Metallography*, Wiley, New York (1961).
13. Harris, J.R., 'Copper and shipping in the eighteenth century', *Economic History Review* **19**(3) (1966) 550–568.
14. Schweizer, F. and Meyers, P., 'A new approach to the authenticity of ancient silver objects: the discontinuous precipitation of copper from a silver-copper alloy', in *Proceedings of the 18th International Symposium on Archaeometry and Archaeological Prospection, Bonn, 14–17 March 1978*, Rheinland-Verlag GmbH, Cologne (1979) 287–298.
15. Dye, A., *Precipitation Phenomena in Silver-copper Alloys*, unpublished M.Chem. dissertation, University of Oxford (1998).

Authors' addresses

- Shirley Northover, Materials Engineering, Department of Design, Development, Environment and Materials, Faculty of Mathematics, Computing and Technology, The Open University, Milton Keynes, UK (s.m.northover@open.ac.uk)
- Peter Northover, BegbrokeNano-OMCS, Oxford University Begbroke Science Park, Yarnton, Oxford, UK (peter.northover@materials.ox.ac.uk)

Gold usage: wear marks and/or deterioration in site conditions

Alicia Perea and Oscar García-Vuelta

ABSTRACT Gold, the most malleable and ductile of metals, wears and deforms easily under continuous friction in contact with skin, cloth or other metals. Because of this, gold alloys in the jewellery market contain added elements in different quantities to make them harder and more wear-resistant. In antiquity, gold alloys contained gold-silver-copper exclusively as the main elements. It is for this reason that ancient goldwork is extremely sensitive to material loss and deformation when used for several generations.

Research under 'Project Au' has accumulated a large amount of information on surface topographies of archaeological objects on the microscopic scale, using scanning electron microscopy (SEM). Our aim is to systematically document the effects of superficial wear during the period of use, and other phenomena related to deterioration processes from the conditions of the burial site, which are often complicated to differentiate.

The literature frequently provides descriptions of ancient gold objects including expressions such as 'wear marks', 'very used', 'signs of wear' without detailed evidence or illustrations. Taking into consideration the extreme difficulty of quantifying the degree of wear of an object and even more to calculate the wear rate in relation to time, we suggest, nevertheless, that SEM is useful in documenting these topographies for comparative studies. We put forward some case studies and evidence of the variability of this phenomenon based on gold objects from very different sites in the Iberian peninsula dated over a wide period of time.

KEYWORDS goldwork, SEM-EDX, Iberian peninsula, ancient gold alloys

Introduction

One of the jewellery industry's challenges is developing alloys resistant to the abrasion produced by continuous use whilst still maintaining appearance and caratage. Few studies have characterised and quantified the extent of wear of the gold alloys by observing the surface using scanning electron microscopy (SEM) [1, 2]. Others have identified the most common causes for loss of weight, for example, of a gold wedding ring used for a year, coming to the conclusion that sandy beaches should be avoided [2]. But at present we have no systematically obtained data concerning the effects and characteristics of wear in Au-Ag-Cu alloy, which is the most frequent raw material used in the manufacture of ancient gold objects.

Table 1 Vickers hardness (HV) of some pure metals and a bronze alloy ([20] and other sources).

	Annealed	Worked
Au	20	58
Ag	22	100
Cu	50	110
Pt	37	108
Cu+10% Sn	100	250

The *Au Repertoire* [4, 5] is a database that systematically stores SEM images of surfaces of gold objects from Iberian archaeological sites. They cover the period from the inception of metallurgy until the historical period. More than a thousand micrographs have been added to this database since it was initiated in 1993. In addition to the topographic data on working traces, tools, usage and/or deterioration during burial, the *Au Repertoire* records the elemental composition of each documented microstructure using energy dispersive X-ray analysis (EDX), with a focus on brazing techniques.

The mechanical properties of Au-Ag-Cu alloys

The mechanical properties of precious metals and the working process are the starting points that reveal the behaviour of a particular alloy. Mechanical properties include hardness, yield and tensile strength, malleability and formability. They can all be modified by heat (ageing, quenching, boiling point) and/or mechanical treatments, which are not always implicit in the manufacturing process.

Hardness [6] is what best defines the capacity of an alloy to resist wear by rubbing or friction. Table 1 shows the differences in the Vickers hardness (HV) of pure gold, silver, copper and platinum, compared with a 10% tin bronze in annealed and worked conditions. Table 2 shows the HV of some standard alloys that can be found on the market.

Table 2 Vickers hardness (HV) of some industrial alloys (gold alloy data from [21]).

Carat	Composition wt%	Annealed	Worked
22	Au+5.5 Ag+2.8 Cu	52	138
18	Au+16.0 Ag+9.0 Cu	135	210
18	Au+12.5 Ag+12.5 Cu	150	225
18	Au+9.0 Ag+16.0 Cu	160	240

Today there are 18 carat gold alloys on the market with a hardness of more than 300 HV (excluding white gold alloys), obtained by adding Co, Ti, V, B, Zr, Al or one of the platinum group metals (PGMs) and various thermal treatments, designed mainly for the watch industry [7].

Ancient gold alloys

We are well acquainted with the composition of prehistoric gold alloys used in Europe thanks to the research programme on the beginnings of metallurgy undertaken in the Würtembergischen Landesmuseum Laboratories, Stuttgart, which began in the middle of the last century [8, 9] but less so with natural gold alloys, due to the need for more sophisticated, more expensive analytical techniques [10, 11]. Since the 1980s, when non-destructive analytical methods were refined, the elemental characterisation of archaeological metal objects has become standard procedure in the field of archaeometry in European countries [12, 13].

According to these data, the variation in compositions of ancient gold alloys is enormous, but always within the Au-Ag-Cu ternary system. No intentionally added elements are detected that would have notably modified the behaviour of these alloys, at least until the Roman era, and the natural trace elements or impurities – basically tin and PGMs – barely affect their mechanical behaviour. To put it another way, the mechanism for hardening a gold-based alloy was mainly plastic deformation.

In the case of Iberian gold alloys [4, 9, 14] from the beginning of metallurgy until the Roman era, the trend is of a progressive increase in the silver content to the detriment of gold and a corresponding reduction in the copper content. In the mid-first millennium BC, the average silver content in the alloys was around 10%, with the copper content never reaching more than 5%, but major regional differences have to be taken into account. For example, in the north-west of Iberia, production is characterised by the use of gold alloys with a high silver content, which, in some cases, exceeded 25%.

Bearing in mind that copper content is the element that most affects the hardness index, we can say, statistically, that the raw material used during this stage shows HV indices of less than 200 when cold worked. In general, from the first millennium BC onwards, hardness was much lower than this because many of the objects were manufactured using casting techniques, or display some kind of brazing, such as is needed for filigree and granulation. This means the metal used was annealed, i.e. in the condition in which the alloy has the lowest possible hardness, thus giving us an indication of the low resistance of ancient gold objects to abrasion and wear.

Evidence of wear traces in the production of gold in Iberia

Evidence in the *Au Repertoire* reveals numerous examples of abrasion of the original topography, the causes of which are not always easy to determine, although in most cases it is due to wear by friction as a result of continuous use. In other cases, the causes may be due to the conditions of the burial environment itself. Finally, other possible causes of abrasion may lie in the manufacture and finishing processes. This paper investigates these three possibilities with the following case studies.

The Bishop's ring (Cadiz)

The Bishop's ring was recovered during rescue excavations carried out during the remodelling work on the Bishop's Palace in the modern city of Cadiz, in an area between the two cathedrals, which has maintained its religious character from the eighth century BC to the present. The ring was found in a monumental tomb of the temple-tomb type, dated to about the sixth century BC, built with stone ashlars on an elevated platform, with a rectangular floor plan and pitched roof. It had already been looted when it was discovered, but the remains of gold braided thread and the finger ring were recovered [15].

The ring was in very poor condition (Figure 1). The structure had completely collapsed, and it displayed an ancient repair at the joint between ring and bezel; only the solid gold prismatic bezel with two chased dolphins or swordfishes could be clearly identified. The appearance of the metal surface, smooth and shiny, retained almost none of the raised topography. The SEM study revealed a complex original construction structure, covered up with filigree and granulation work that had practically disappeared. The granules can be seen on the surface as a shadow or flat rendition of the decorative motifs, such as the rosettes on the sides of the bezel.

Microanalysis of the surface in archaeological gold objects does not reveal the original alloy due to surface gold enrichment during burial, but can give an estimate of the composition. In the case of the Bishop's ring, the surface composition is typical of purified gold, with a very low silver content that was barely detectable but not quantifiable (i.e. under 1%), and copper of around 4 wt%.

This SEM study enabled us to draw a schematic diagram showing the reconstruction of the ring. This drawing was subsequently used by a commercial firm to create replicas of the jewel, which are currently sold in the site museum.

The Bishop's ring is a clear example of usage wear, which must have extended over several generations, most probably as a hereditary token attached to a specific social role. Only two examples of rings of this kind are known in Iberia: the one discussed here from Cadiz, and a second example from tomb 5 of the orientalising necropolis of La Joya, in Huelva [16]. They both weigh exactly the same: 10.5 g. The ring from La Joya, which is simpler, has a winged gryphon on the bezel and the signs of wear are less marked in a position which would indicate that it was more often used as a pendant than as a ring. Tomb 5, with extremely rich grave goods, including a Rhodian bronze jar, can be dated to the end of the seventh century

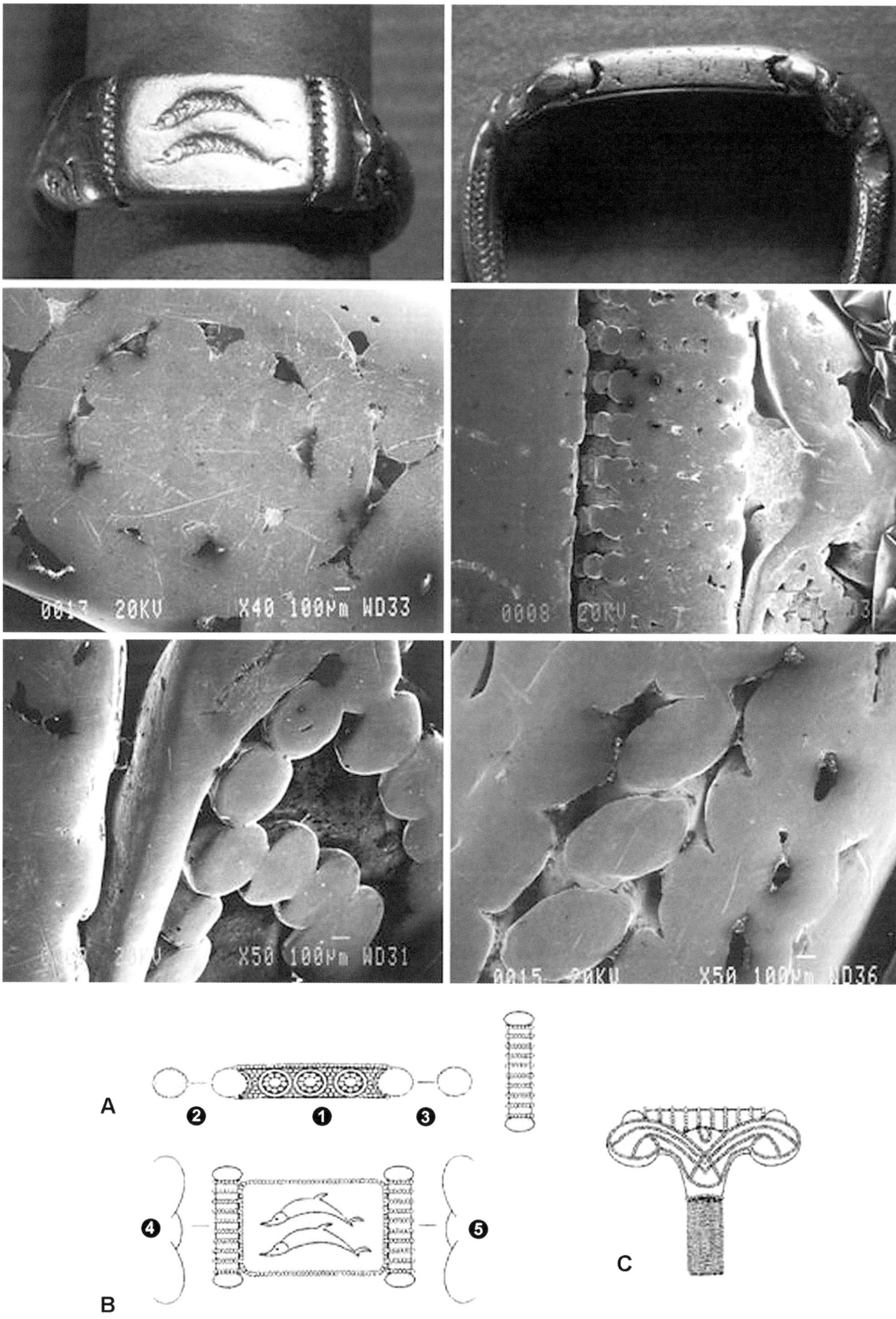

Figure 1 The Bishop's ring (top) as found in the city of Cadiz, and micrographs showing filigree and almost completely smooth granulation work (middle rows). Reconstruction of the ring (bottom).

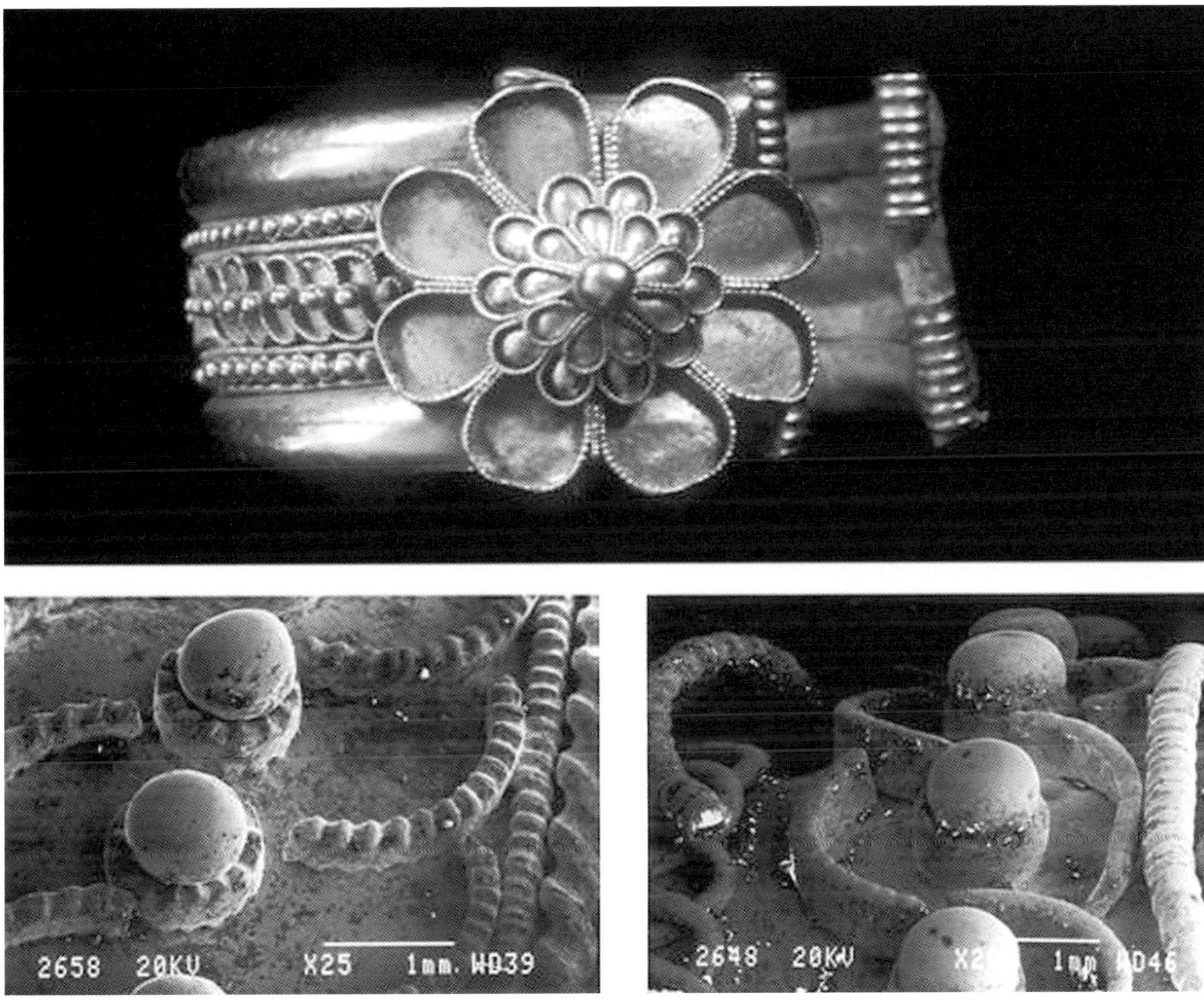

Figure 2 The type of cylindrical hoop found in some of the richest tombs of the Gaditan necropolis dated to the fourth century BC (top). Micrographs showing different rates of wear for two of these pieces of jewellery (bottom).

BC [17]. Both examples must have been personal insignia or symbols of the office held by their owners, probably in high political-religious office.

Fourth-century BC hooped earrings from the necropolis of Cadiz

The second case study is also from Cadiz. In the fourth century BC one of the ancient Phoenician colony's necropolis was located in another part of the city, Puerta de Tierra. It was excavated in the first half of the last century. This area has been much affected by changes in the coastline due to silting and erosion by the sea, so now the excavated structures can no longer be recognised.

The grave goods of the Gaditan elites included spiral and cylindrical hoops used as finger, ear or hair rings. They were mass produced, with a wide variability within the type designed to meet an economically diversified demand, which was limited to the social circle of the city itself [18]. From the technological point of view, they stand out on account of the diversity of worked wires used in relatively complex decorative friezes (Figure 2).

Some items display heavy wear traces on the surface compared with others that look as if they have just left the workshop. These wear traces are uniformly distributed and affect the whole of the surface equally. This suggests that the abrasion was not due to wear while the jewel was being worn, which is usually selective and concentrated in the most prominent parts in contact with the skin, clothing or other nearby objects.

We do not know the exact context of the tombs in which these examples appeared because the excavations of the period did not record this information. However, the whole area of the necropolis was sufficiently close to the sea to suggest that they were affected by sand and seawater over the course of time, which could have caused this type of natural abrasion.

The necropolises of Les Casetes and Poble Nou (Villajoyosa, Alicante)

Our laboratory was recently commissioned to study gold artefacts from two necropoles that came to light in rescue excavations in Villajoyosa, near Alicante. A mixed native and Phoenician population was buried in the orientalising necropolis of Les Casetes, as suggested by the grave goods, which have been dated to between the end of the seventh and the first half of the sixth century BC. Gold objects of clearly Phoenician origin were found in some of the tombs. Their typological and technical

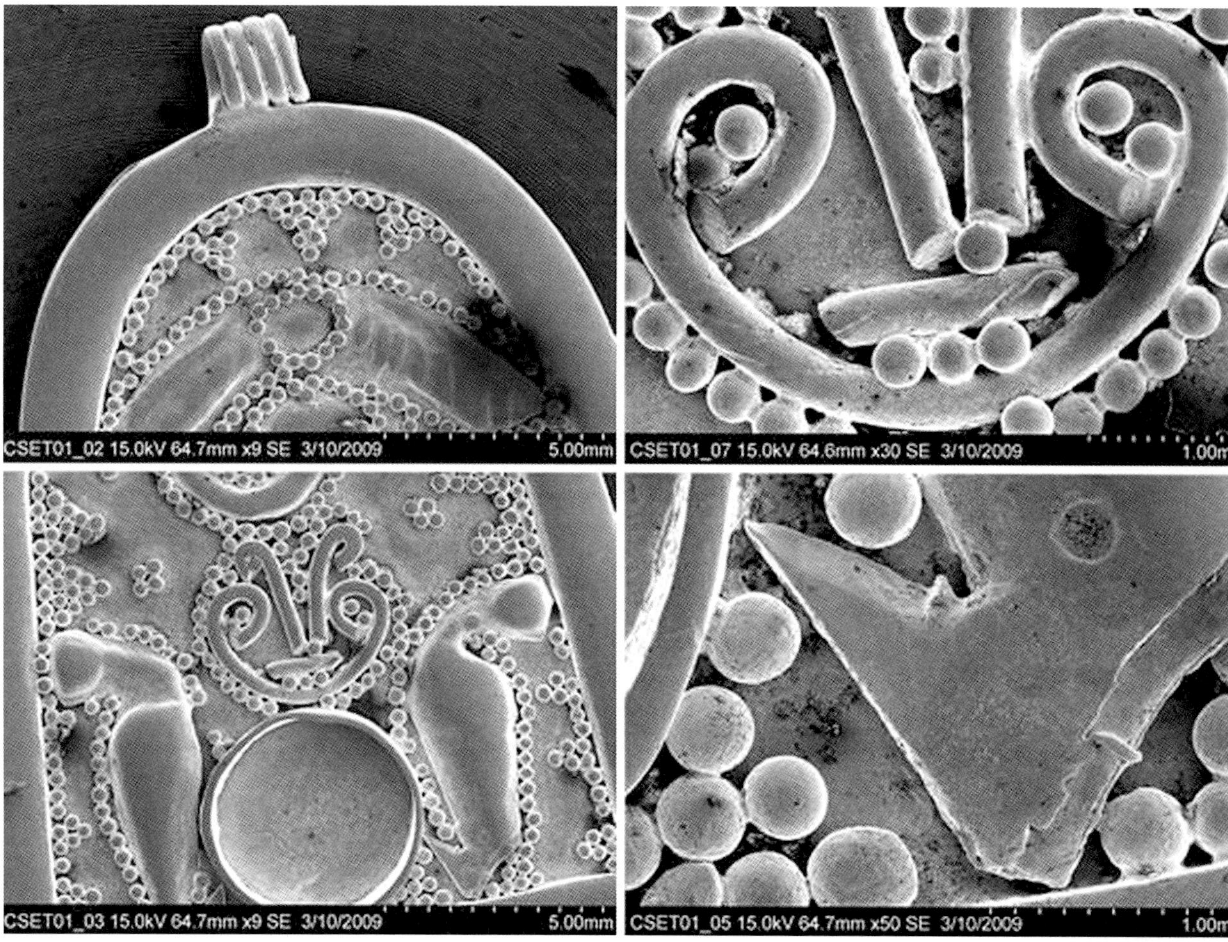

Figure 3 One of the gold pendants from Les Casetes necropolis that shows no signs of wear.

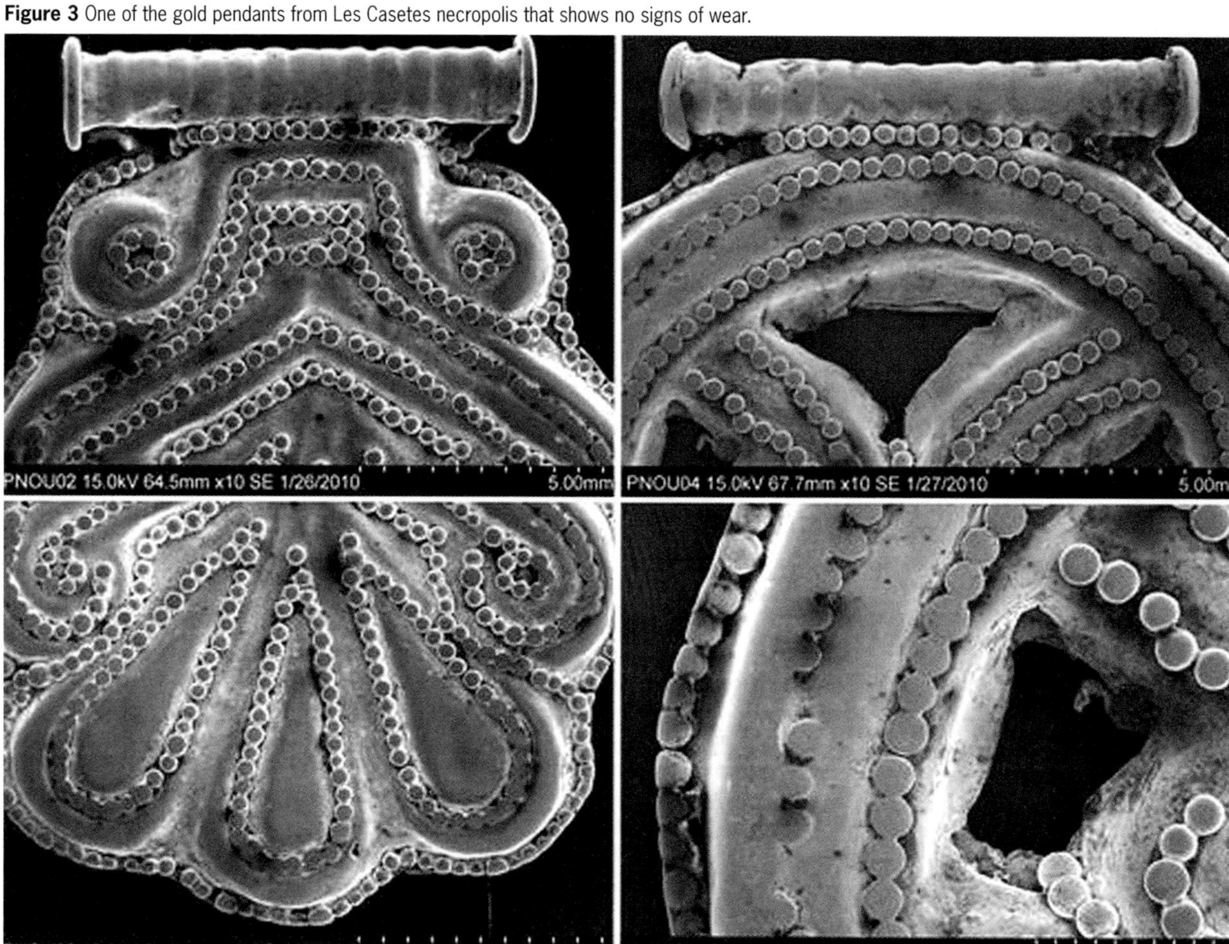

Figure 4 Two different pendants from Poble Nou necropolis showing flattened granules due to prolonged use.

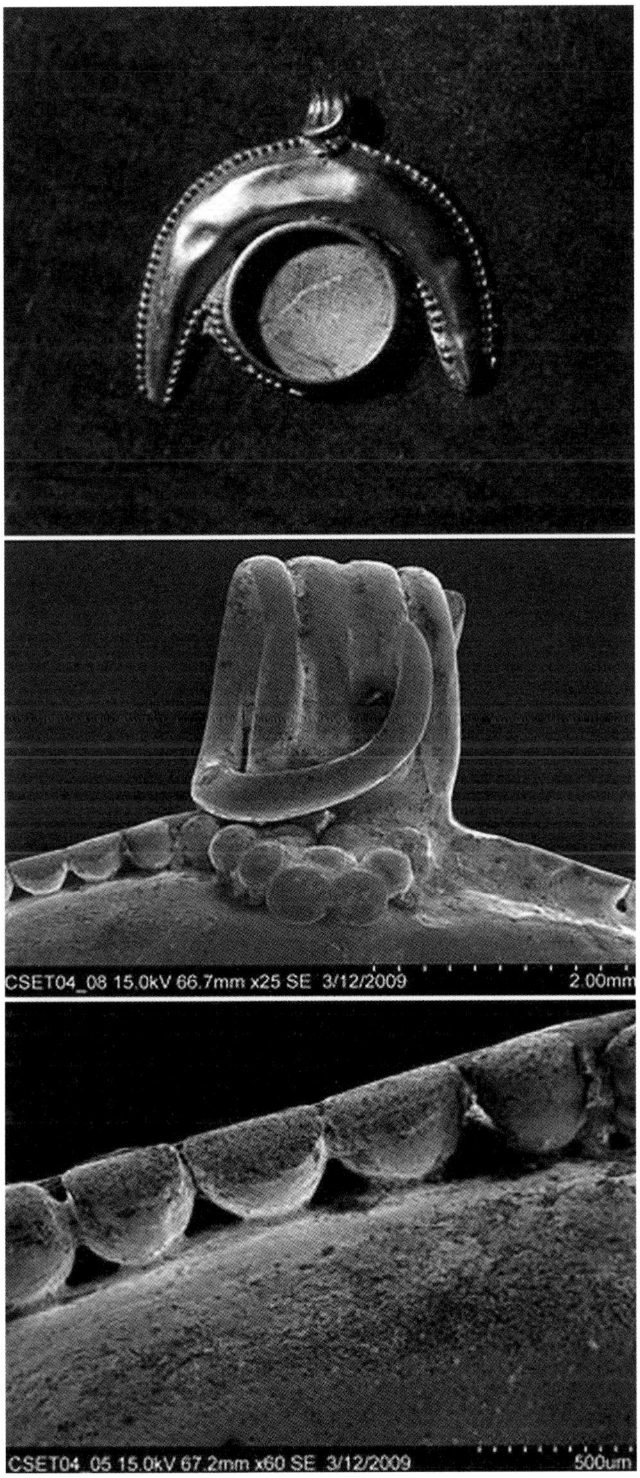

Figure 5 The astral pendant from Les Casetes showing abrasion along the granulated edge.

characteristics point to the existence of a workshop, as yet not known, on the eastern coast of Iberia [19]. The necropolis of Poble Nou was first used when the earlier one was abandoned. Entirely Iberian features were documented, although the goldwork found still had its roots in the Phoenician tradition.

Gold production at these two sites is differentiated by their wear traces. The jewellery from Les Casetes does not appear to have been used during the lifetime of its owners, at least not for long periods (Figure 3). In contrast, in Poble Nou selective wear traces, which are clear from flattened granules, indicate prolonged use (Figure 4). In both necropoles, gold objects were occasionally affected by high temperatures (visible as melting) due to the funerary rite in which the body was cremated with all the individual's belongings.

One of the pieces from Les Casetes raises a different question. The astral pendant, formed by two elements that represent the solar disc framed by the crescent moon, displays a type of abrasion that, at first sight, could be interpreted as signs of use. It consists of heavy abrasion along part of the line of granulation on the extreme edge of the pendant (Figure 5). The rest of the piece shows no sign of wear such as that seen in the examples from Poble Nou. This peculiar feature could be due to technical reasons that obliged the goldsmith to rub down the edge of the piece in order to eliminate an excess of solder or a failure during the heating process.

Conclusions

SEM is an important tool for investigating the topography of archaeological gold objects in order to identify possible wear traces and abrasions that can be produced in different ways. Since 1993, evidence of this phenomenon has been collected and the specific reasons for it are very diverse, but this study has shown that wear traces are produced in a relatively short space of time due to the characteristics of the alloys of ancient gold, thus if an object shows no clear signs of wear, it seems likely that it was not used.

On the other hand, the characteristics of these traces give different information. First, rubbing due to prolonged use, such as against the skin, clothing or other objects, for example as necklace beads. This produces wear that is uneven, selective and characteristic of the function of the pendant, ring or necklace. Secondly, the conditions of the burial site itself can produce abrasion as a result of water seepage and the movement of earth or water while the pieces are buried that can be confused with signs of use. And thirdly, a technological factor can be the cause of abrasion, such as the removal of burrs, stains and defects that occur during the manufacturing process. Differentiating between these three processes is important when interpreting the archaeological object, not just from the technological point of view, but in terms of its symbolic, political, economic and ideological significance.

Acknowledgements

This work is part of the research project 'SEM, IBA and GIS Procedures and Uses for Research in Archaeometallurgy. The pre-Columbian Gold Case Study' (HUM2006-06250/HIST), directed by A. Perea, in the frame of the programme CONSOLIDER INGENIO 2010 (CSD-TCP), funded by the Ministry of Science and Innovation, Spain.

References

1. Heidsiek, H. and Clasing, M., 'The abrasive wear of gold jewellery alloys', *Gold Bulletin* **16**(3) (1983) 76–81.
2. Wells, A., 'The wear of precious jewellery', *Wear* **112**(3–4) (1986) 363–370.
3. Steinhauser, G., 'Quantification of the abrasive wear of a gold wedding ring', *Gold Bulletin* **41**(1) (2008) 51–57.
4. Perea, A., García-Vuelta, O. and Fernández-Freire, C., *El Proyecto Au. Estudio Arqueométrico de la Producción de Oro en la Penín-*

sula Ibérica, Bibliotheca Praehistorica Hispana, CSIC, Madrid 27 (2010).
5. Perea, A., Montero, I. and García-Vuelta, O., 'Project Au and the *Au Repertoire*: a research strategy in gold metallurgy', in *Tecnología del Oro Antiguo: Europa y América / Ancient Gold Technology: America and Europe*, ed. A. Perea, I. Montero and O. García-Vuelta, CSIC, Madrid (2004) 139–146.
6. Rapson, W.S. and Groenewald, T., *Gold Usage*, Academic Press, London (1978).
7. Süss, R., van der Lingen, E., Glaner, L. and du Toit, M., '18 carat yellow gold alloys with increased hardness', *Gold Bulletin* **37**(3–4) (2004) 196–207.
8. Hartmann, A., *Prähistorische Goldfunde aus Europa I*, Studien zu den Anfängen der Metallurgie, Band 3, Berlin (1972).
9. Hartmann, A., *Prähistorische Goldfunde aus Europa II*, Studien zu den Anfängen der Metallurgie, Band 5, Berlin (1982).
10. Guerra, F., Sarthre, C.O., Gondonneau, A. and Barrandon, J.N., 'Precious metals and provenance inquiries using LA-ICP-MS', *Journal of Archaeological Science* **26**(8) (1999) 1101–1110.
11. Chapman, R.J., Leake, R.C., Warner, R., Cahill, M., Moles, N.ÇR., Shell, C. and Taylor, J.J., 'Microchemical characterisation of natural gold and artifact gold as a tool for provenancing prehistoric gold artefacts: a case study in Ireland', *Applied Geochemistry* **21** (2006) 904–918.
12. Perea, A. and Armbruster, B., 'L'archéologie de l'or en Europe', *Perspective* **1** (2008) 29–48.
13. Perea, A., 'Wands in the hand! Or Potter's powers', *Archeosciences* **33** (2010) 21–29.
14. Montero, I. and Rovira, S., 'El oro y sus aleaciones en la orfebrería prerromana', *Archivo Español de Arqueología* **64**(163–164) (1991) 7–21.
15. Perea, A., Montero, I., Cabrera, A., Feliú, M.J., Gayo, M.D., Gener, J.M. and Pajuelo, J.M., 'El ajuar de oro de la tumba fenicia del obispo. Cádiz', in *Tecnología del Oro Antiguo: Europa y América / Ancient Gold Technology: America and Europe*, ed. A. Perea, I. Montero and O. García-Vuelta, CSIC, Madrid (2004) 231–241.
16. Garrido, J.P., *Excavaciones en la Necrópolis de La Joya, Huelva (1ª y 2ª campañas)*, Excavaciones Arqueológicas en España, Madrid 71 (1970).
17. Jiménez Ávila, J., *La Toréutica Orientalizante en la Península Ibérica*, Archaeologica Hispana, Real Academia de la Historia, Madrid 16 (2002).
18. Perea, A., 'Comportamientos de mercado en la producción orfebre del taller de Cádiz', in *Intercambio y Comercio Preclásico en el Mediterráneo. Actas del I Coloquio del Centro Estudios Fenicios y Punicós, Madrid 1998*, ed. P. Fernández, C. González, and F. López, Madrid (2000) 281–292.
19. Perea, A. and García-Gandía, J.R., 'Analisis MEB e interpretación de la orfebrería fenicia de Les Casetes (Villajoyosa, Alicante)', in *Proceedings of the VIII Congreso Ibérico de Arqueometría, Teruel, 19–21 October 2009*, Seminario de Arqueología y Etnología Turolense (2010).
20. Grimwade, M., *Introduction to Precious Metals*, Newnes Technical Books, London (1985).
21. World Gold Council. *Gold Bulletin*. www.gold.org/technology/research/gold_bulletin/ (accessed 8 May 2012).

Authors' addresses

- Alicia Perea, Arqueometal Research Group, Instituto de Historia, Centro de Ciencias Humanas y Sociales (CCHS-CSIC), Madrid, Spain (alicia.perea@cchs.csic.es)
- Oscar García-Vuelta; Arqueometal Research Group, Instituto de Historia, Centro de Ciencias Humanas y Sociales (CCHS-CSIC), Madrid, Spain

A case study of ancient parchment biodeterioration using variable pressure and high vacuum scanning electron microscopy

Flavia Pinzari, Vanja Cialei and Guadalupe Piñar

ABSTRACT We present a case study of biodeterioration phenomena encountered in parchment-supported manuscripts. In this study, our understanding of the physical and chemical changes observed in parchment as a result of fungal and bacteriological attack has been addressed by means of culturing, and through the application of scanning electron microscopy with energy dispersive X-ray analysis (SEM-EDX). Two biochemical tests were performed to assess the viability of microorganisms. The luciferine-luciferase method for adenosine triphosphate (ATP) quantification was used to estimate the viability of microorganisms in parchment samples, and a test based on the measurement of fungal β (1-4)-N-acetyl-D-glucosaminidase activity was used to quantify fungal biomass. The purple stains blemishing the manuscript sheets were found to be caused by an attack of bacteria from the order Actinomycetales (presumably of the genus *Streptomyces*), although the causative agent that had provoked the stains was no longer active. Viable, but not culturable, fungal species were also present, but these could not be associated with the stains. The biodeteriorated material showed a decrease in aluminium and silicon, and an increase in sulphur and phosphorus, as documented through microanalysis.

KEYWORDS parchment, biodeterioration, scanning electron microscopy, microanalysis, variable pressure

Introduction

Parchment is a thin material made from animal skin such as sheep and goat [1], widely used as a writing support from the second century BC until the end of the Middle Ages, when it was substituted in Europe by paper. Parchment differs from leather as it was mostly treated with lime rather than tanned, therefore it is more sensitive to fluctuations in relative humidity [2, 3]. The preparation of parchment always involved a thorough cleansing of the animal skin and mechanical removal of hair by scraping. These steps were followed by desiccation using sodium or potassium chloride, adjustment of the material's pH by treatment with ammonium chloride or sulphate in combination with lime, and a wet application of potash alum. The main chemical component of parchment is collagen, a natural biopolymer with a relative molecular mass of 350 kDa [3]. Degradation of ancient parchment or vellum is a complex process that involves the chemical oxidative deterioration of amino acid chains and hydrolytic cleavage of the peptide backbone.

Microorganisms can hydrolyse collagen fibres and other proteinaceous structures of parchment; they can also act to modify its inorganic components, or produce pigments and organic acids that discolour parchment and consequently cause indirect damage [4–8]. According to Karbowska-Berent and Strzelczyk [8], *Streptomycetes* comprise the bacteria that play a major role in the deterioration of many kinds of ancient parchment documents and books. These filamentous bacteria are known to produce many types of enzymes, especially collagenases (proteases) that are capable of destroying collagen by their hydrolytic activity. They are alkaliphiles and are therefore inclined to develop on parchments (whose surfaces are prepared by rubbing with chalk). Ancient parchment also provides good conditions for the development of proteolytic fungi [4, 8].

Here we present a case study of biodeterioration phenomena encountered in parchment-supported manuscripts. In this study our understanding of the physical and chemical changes observed in parchment as a result of fungal and bacteriological attack was addressed through the application of culturing and scanning electron microscopy with energy dispersive X-ray analysis (SEM-EDX). Additionally, two biochemical tests to assess the viability of microorganisms were performed. The luciferine-luciferase method for adenosine triphosphate (ATP) quantification was applied to estimate the viability of microorganisms, and a test based on the measurement of fungal β (1-4)-N-acetyl-D-glucosaminidase activity was used to quantify fungal biomass in parchment samples.

Materials and methods

The volumes

The 'Grazie' collection is a wide-ranging archival group of volumes that dates from 1299 to 1767 and represents a register of great historical value. The entire collection has suffered serious biological damage that has sullied a number of leaves of the volumes to the extent of rendering them illegible.

The damage observed consisted of a very dense pattern of purple-coloured stains (Figure 1). Three different volumes (Nos. 3, 4 and 21) were sampled for the purposes of biological examination. Volume No. 3 was found to be particularly badly damaged; several leaves showed lacunae and missing parts, as well as fully detached fragments that were used for most of the biochemical tests. It was apparent that the microbial attack had been more intense on the flesh side than on the hair side of the parchment. The stains appeared to be distributed around the manuscript writing, but ink-coated areas were stain free, suggesting the possible presence of biostatic substances in the inks (Figure 1).

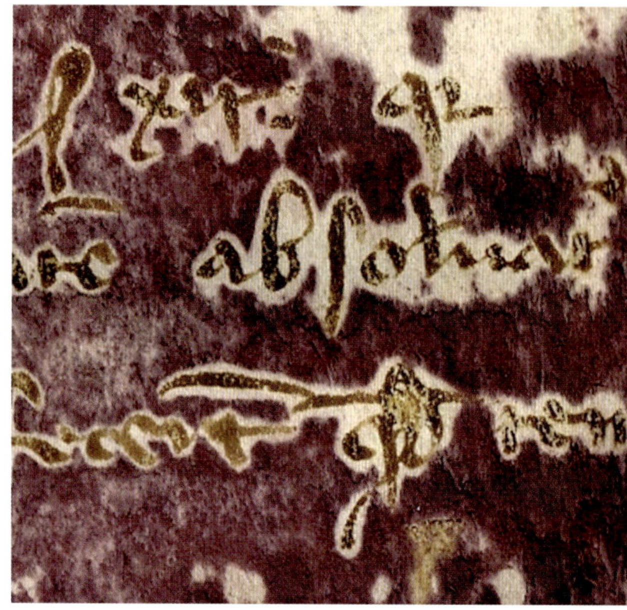

Figure 1 The dense pattern of purple stains affecting the parchment-supported manuscripts belonging to the State Archives of Venice, Italy.

Culturing methods

First, the viability of fungi and bacteria was tested using agar and broth culture media. Microbial structures sampled with cotton swabs were inoculated directly onto agar plates and the swabs were then immersed in sterile Czapek (see below) and gelatine broths.

Parchment samples in powder form taken from different pages of the volumes with the aid of sterile needles were suspended in 2 ml of sterile water and spread onto agar plates using a glass pipette (1 ml of suspension for each Petri dish). The media used to grow the inoculae were malt extract agar (MEA) and Czapek yeast agar (CYA), which were prepared according to Samson et al. [9] for fungi, and nutrient agar (5% peptone, 0.3% beef extract, 1.5% agar, pH adjusted to neutral) for bacteria.

Some of the powdery samples obtained by scratching the purple stains on the parchment surfaces with scalpels were inoculated onto M40 media [9] and nutrient agar with 10% added NaCl, with the objective of testing for the presence of xerophilic species. The growth of some filamentous fungal and bacterial species can be favoured by culturing them in media that contain high concentrations of salt or sucrose, and which feature low water activity. The culturing reagents employed were produced by DIFCO. All the inoculations were performed within the confines of a laminar flow hood so as to ensure sterility throughout the procedures. The plates were incubated at 26°C and checked daily for colony development over a 30-day period.

When fragments bearing deteriorative materials are available, a type of culturing procedure that is useful for determining whether the microbial strain is a spoiler or merely a contaminant can be performed. This procedure, commonly referred to as 'multi-point inoculum', involves washing small pieces of material (parchment, i.e. 0.5 mm in diameter) in sterile water and dividing these with the aid of sterile needles into 25 or more sub-samples that can then be inoculated directly onto solid nutritive agar and distributed according to a geometric pattern. The relevant test frame consists of a grid divided into 25 nodes that are equally spaced. The development of the same fungal or microbial organism at most of the 25 points allows for a statistically significant attribution of the damage to the organism in question. When a number of different species develop at the 25 inoculated points it is not possible to establish a strict relationship between the isolated strains and the damage observed in the heritage artefact.

SEM-EDX

Parchment samples measuring 2–3 mm in diameter were analysed using a variable pressure (VP) SEM instrument (Zeiss EVO50) fitted with a detector for backscattered electrons (BSE). Chemical characterisation of the inorganic constituents of the samples was performed by EDX. Only following an initial observation of the samples using SEM in VP mode at 20 kV were some of the samples coated in gold (using a Baltec Sputter Coater) and then subjected to further analysis in high vacuum (HV) mode. Sputtering was performed under an argon gas flow at a working distance of 50 mm at 0.05 mbar, and a current of 40 mA for 60 seconds, to create a gold film about 15 nm thick. Reference elemental intensities acquired from pure compounds (standards) are commonly used to calibrate SEM-EDX systems. In the case study presented here, conventional ZAF correction [10] integrated into the Oxford INCA 250 microanalysis package was applied to the spectrum dataset (Oxford Instruments).

ATP extraction and quantification

Some of the parchment samples affected by purple stains that were taken from different pages of volume No. 3 were analysed using the luciferin-luciferase method to determine intracellular ATP content. Parchment samples were classified according to the intensity of stains. Three categories were established, ranging from severely stained (90% of the surface) to apparently stain free, although none of the samples could be considered entirely free of microbial contaminants. The analysis required samples weighing about 100–200 mg for each extraction replica. It was possible to recover this (considerable) amount of material from the margins of pages that had been almost entirely destroyed by the biological attack. However, the control samples (showing minimal or no purple stains) were obtained from the central parts of such pages that were hidden by the binding. ATP is contained within the cells; in

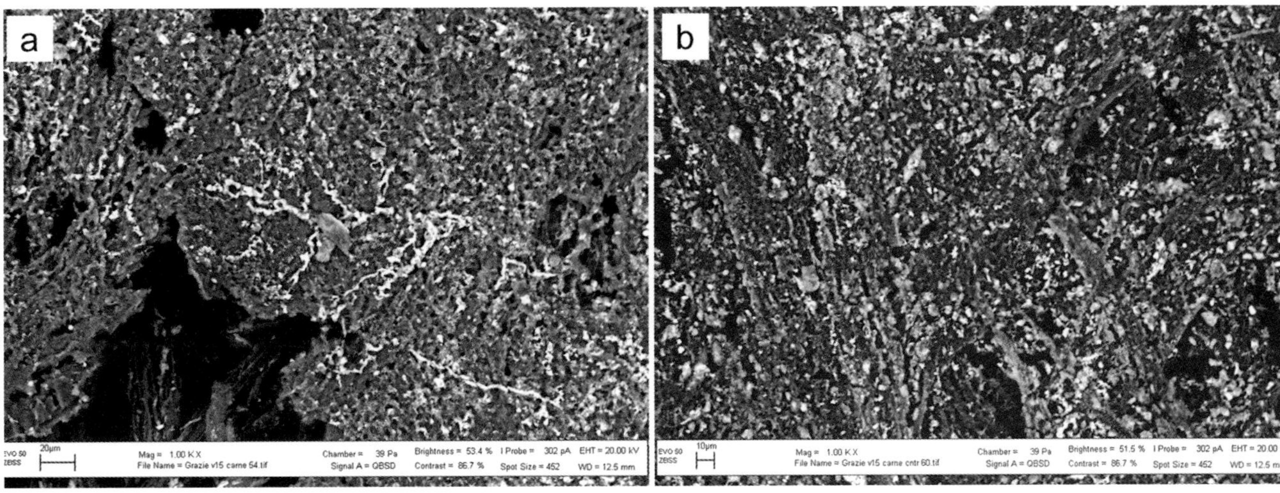

Figure 2 SEM-BSE image at low magnification showing a clear comparison of differential salts distribution in the unaltered (a) and biodeteriorated (b) areas of the parchment's surface. In the biodeteriorated parchment salts appeared to be newly crystallised into ramified, slender structures resembling fungal filaments.

order to quantify its concentration it is necessary to disrupt bacterial cells and fungal/bacterial walls. The extraction of ATP was achieved by using dimethylsulfoxide (DMSO) at 90% in TAE buffer (a buffer solution containing a mixture of tris-hydroxymethyl-aminomethane base, acetic acid and ethylene diamine tetra-acetic acid), preheating the DMSO at 100°C for 1 min, as described by Rakotonirainy et al. [11]. Quantification of ATP contained in the samples was done using the ENLITEN ATP Assay System Detection Kit (Promega) and a Turner 20/20 Luminometer (Promega). From each extract obtained (from as many parchment samples), five separate ATP measurements were performed in order to test the accuracy of the ATP concentration readings. For each test 100 μl of the extract was added to 200 μl of the luciferin-luciferase mixture (Promega). Measurements for individual ATP extracts fell within 5% accuracy bounds.

Beta-N-acetylhexosaminidase fluorescent assay

In this study the fluorometric assay method was used to evaluate fungal biomass in parchment samples. The fluorogenic 4-methylumbelliferyl (MUF) labelled substrate N-acetyl-beta-D-glucosaminide (NAG) was used to detect beta-N-acetylhexosaminidase activity in parchment samples according to Konkol et al. [12], with some modifications.

The substrate 4-methylumbelliferyl-N-acetyl-beta-D-glucosaminide (4-MUF-NAG) (Sigma-Aldrich) consists of a 4-methylumbelliferone (4-MUF) molecule linked to N-acethyl-β-D-glucosamine. The enzyme N-acetyl-β-D-glusaminidase (NAGase), which hydrolyses N-acetyl-β-D-glucosamine in the presence of 4-MUF-NAG, releases the fluorocrome 4-MUF. The 4-MUF released by hydrolysis exhibits its maximum fluorescence when irradiated at excitation wavelengths lying between 355 and 380 nm and at a pH exceeding 10. Parchment samples weighing about 50–100 mg were incubated in 100 μl of 4-MUF-NAG 5mM solution, together with 1.9 ml of 100 mM acetate buffer at pH 4.5. As a positive control, 10 μl of an *Aspergillus niger* van Tieghem spore suspension containing 100 spores/μl was incubated in 90 μl of 4-MUF-NAG 5 mM solution, together with 1.9 ml of 100 mM acetate buffer. A negative control consisting of 10 μl of distilled water incubated with 90 μl of 4-MUF-NAG 5 mM solution together with 1.9 ml of 100 mM acetate buffer was set up. After an hour of incubation at 24°C in plastic sterile vials, 100 μl of the reactant was transferred to 10 × 10 methacrylate cuvettes containing 1.9 ml of 0.1 M glycine buffer at pH 10.5. All assays were carried out in triplicate and three categories of samples were considered (ranging from severely stained to apparently stain free). Fluorescence derived from liberated 4-MUF was determined using an LS45 luminescence spectrometer set at 448 nm emission and 365 nm excitation. For the calibration of the spectrometer a standard curve was constructed using known quantities of 4-MUF sodic salt (Sigma-Aldrich Co).

Data analysis

Microanalysis data, NAGase activity and ATP content values obtained from the different parchment samples were analysed by means of one-way ANOVA, and the statistical significance of differences between samples was tested (Tukey's test, with 95% confidence interval). The statistical package included in XLStat (Addinsoft, XLStat 2009 Paris, France) was used for this purpose.

Results

Culturing methods

The incubated plates and vials displayed few fungal and/or bacterial colonies. The species present were different from those observed by means of SEM imaging. In some situations it is possible to distinguish fungal and, albeit with more difficulty due to their smaller dimensions, bacterial species by observing their structures using HV-SEM. In the specific case of certain filamentous fungi, the fruiting body shape and conidia size, shape and ornamentation can lead to positive identification at least at genus level [9]. In this case study, the fungal and bacterial species that developed *in vitro* exhibited taxonomic characteristics that enabled us to exclude the matching of them with the structures observed using HV-SEM directly on samples of biodeteriorated parchment.

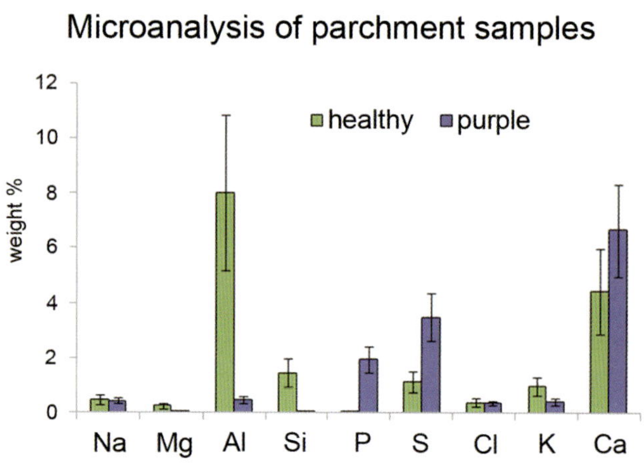

Figure 3 Summary of results of the SEM-EDX topographical analysis performed in purple-stained areas and healthy, unblemished areas of parchment samples. Measurement taking was repeated in both areas three times, and a 6 × 6 grid configuration (36 points) was used in all cases. The histogram reports average values in terms of wt% of the detectable elements, and standard deviation bars. A significant difference between unblemished and purple-stained areas was recorded only for aluminium, silicon, phosphorus and sulphur.

Furthermore, no statistically significant occurrence of individual species could be recorded on the plates, suggesting that the fungi and bacteria that grew *in vitro* were spawned by recent airborne contamination. The colonies obtained were analysed in order to identify their genera. From the qualitative point of view the bulk of genera represented on the parchment samples belonged to *Aspergillus*, *Penicillium* and *Geotrichum* and various yeast types. A few bacterial colonies were also present. No, or at least very few, colonies (chiefly belonging to the *Eurotium* and *Cladosporium* genera) were obtained from agar cultures specific for xerophilic species.

SEM-EDX

The possibility of observing uncoated specimens with a BSE detector meant that images exhibited an appreciable chemical contrast that provided much information. With VP-SEM the differences observed in patterns formed by salts on the blemished and unblemished areas of the parchment served as evidence of an active role played by microorganisms in the modification of the parchment's chemical characteristics (Figures 2a, 2b and 3). The appearance of the blemished samples indicated that the microbial attack was associated with a leaching of the mineral components of the parchment surface. A lower concentration of mineral salts in the biodeteriorated areas could readily be appreciated in the micrographs obtained using the BSE detector. Moreover, new, biogenic crystallisation of salts coinciding with fungal and microbial material was documented (Figure 2b).

Microanalysis confirmed what had been observed as a difference in the chemical contrast of the SEM-BSE images, and although EDX does not represent a quantitative analytical method due to the variability of density in the analysed material (parchment) [1], it indicated a statistically significant difference between the chemical content of damaged and undamaged areas (Figure 3). A significantly higher content of silicon and aluminium is found in the unblemished surface areas of parchment, while the P and S content is significantly higher in stained areas. Calcium appeared to be dislocated in a different way in the parchment's blemished areas, changing from a scattered distribution trend to a pattern of ramified thin structures resembling fungal filamentous structures (Figure 2b). EDX analysis of written areas indicated the presence of copper sulphate and iron sulphate. The absence of stains in ink-bearing zones and the presence of unblemished areas in proximity to calligraphic and decorative elements (Table 1) could be attributable to the biostatic effect of copper, or to the lower pH resulting from the use of iron-gall tannic inks [13].

Table 1 The results of EDX spectrometry performed on inked areas of parchment samples. Results are normalised to 100%.

Element	Wt%	Standard deviation (wt%)
C	49.27	0.86
O	36.87	0.72
Na	0.51	0.09
Al	1.17	0.07
Si	1.28	0.06
P	0.37	0.05
S	0.54	0.05
Cl	0.16	0.04
K	0.66	0.05
Ca	3.69	0.11
Fe	5.14	0.18
Cu	0.34	0.11
Total	100.00	

The observation of gold-coated samples by means of HV-SEM disclosed the presence of both bacterial and fungal cells in the damaged areas of parchment. A highly complex biodeterioration mechanism emerged from the study, with both bacterial and fungal species involved in the discoloration of parchment. Observations made using HV-SEM revealed that in most parchment samples two types of alterations were present: the formation of fissures between the bundles of fibrils and the disintegration of fibrils as a result of activity by the live microorganisms.

The formation of the fissures appeared to favour the penetration of bacterial cells and fungal hyphae into the fabric of the parchment, thereby contributing to its degradation. Bacterial cells can be distinguished from fungal conidia on account of their smaller dimensions (less than 1 μm in diameter).

Currently, the most commonly employed methods are: observation at a low temperature of frozen hydrated material (cryo-fixation, cryo-SEM, cryo-scanning), chemical fixation [3], and critical point drying (CPD) [14]. All these methods can induce shrinkage and deformation of structures, although cryofixed images offer more precise forms.

The samples used in this study were not treated prior to examination as they were already in an air-dried state, hence the dehydrated conidiophores had already shrunk and become misshapen. Taking this into account, and based on the observation of several samples, the fungal taxa present (Figure 4) resemble the genus *Aspergillus*, while the bacterial cells (mostly in chain form) can be attributed to Actinomycetes (Figure 5). Not many fungal and microbial taxa possess unique characteristics that allow them to be specifically identified by means of SEM imaging.

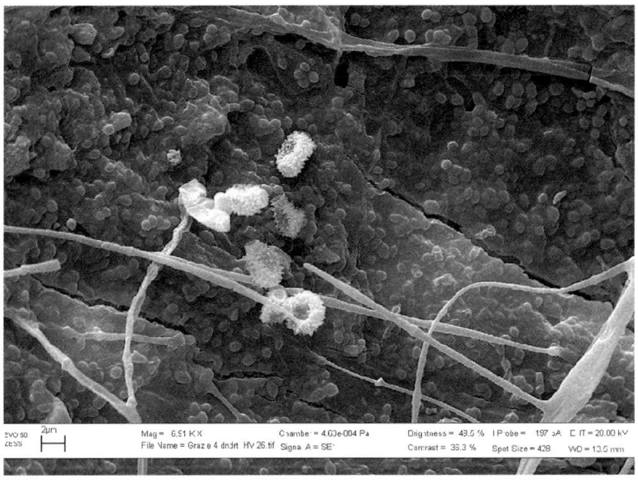

Figure 4 SEM micrograph obtained in HV mode on gold-sputtered samples of manuscript from volume No. 4, showing purple stains. SE detector, 20 kV. Fungal hyphae and conidia, as well as bacterial cells co-occur on the parchment surface.

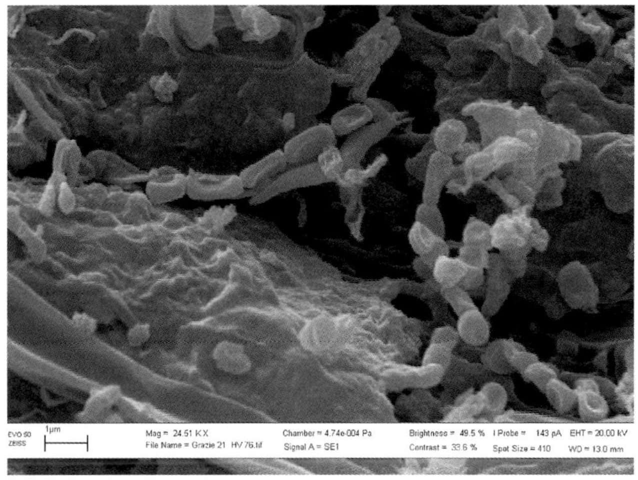

Figure 5 SEM micrograph obtained in HV mode on gold sputtered samples of manuscript from volume No. 21, showing purple stains. SE detector, 20 kV. Filamentous bacteria show typical spore chains (presumably *Streptomyces*) with individual elements measuring about 1 μm in length and less than 500 nm in width.

ATP content

ATP is responsible for coupling intracellular energy-producing and energy-requiring metabolic reactions. On the death of the cells, ATP is no longer produced and any ATP still present is quickly degraded by ATPases (enzymes that convert ATP into adenosine diphosphate ADP, and adenosine monophosphate AMP). The presence of ATP can therefore be used to characterise living organisms [11]. The results obtained by extracting and quantifying ATP from parchment samples taken from volume No. 3 are reported in Figure 6. Living organisms were present only on pages 10, 15 and 39 of the volume. All the other tested pages returned low values of extractable ATP, albeit higher than the control, but not statistically different from it. Moreover, the three categories of samples (No. 1, severely stained; No. 2, slightly stained; and No. 3, apparently stain free) exhibited no significant differences in ATP content, thus indicating that the cells present in the purple stains were no longer viable.

Beta-N-acetylhexosaminidase fluorescent assay

The test, based on a fluorometric method that uses the activity of beta-N-acetylhexosaminidase, an enzyme constitutively expressed in filamentous fungi, as a marker for fungal detection [15], was used successfully on samples of parchment. Originally developed to measure fungal biomass in soil [15], fluorescent N-acetyl-beta-D-glucosamine was adapted by Konkol et al. [12] to detect and quantify fungal biomass on certain materials found in the cultural heritage context. To the best of our knowledge, this is the first time that the method has been applied to a collagen-based material.

Figure 7 shows, in the form of a histogram, the beta-N-acetylhexosaminidase fluorescent assay results. Fungi were present only on pages 10, 15 and 39 of the volume. All the other analysed pages presented lower enzymatic activity values. Beta-N-acetylhexosaminidase activity can also be present in dead mycelia: only vapour-sterilised mycelia, in which all enzymes are denaturised, show no activity. The three categories of samples (No. 1, severely stained; No. 2, slightly stained; and No. 3, apparently stain free) showed no differences in NAG activity, thus indicating that the purple stains could not be attributed to the presence of fungal organisms. To confirm the relationship between fungal biomass and activity a linear correlation ($y = 2.4x + 39.7$) with $R^2 = 0.91$ was calculated for the ATP and NAG data.

Discussion

The purple stains affecting the leaves of the parchment-supported manuscripts belonging to the State Archives of Venice (Italy) were found to be caused by fungal and bacterial attacks. Through the use of SEM, both fungal and filamentous bacterial structures have been shown to be intimately associated with collagen fibres. Collagenolysis and deep structural damage were also documented using SEM imaging. Only in HV mode and on gold-sputtered samples was valid observation of microbial structures possible. Based on their conidia ornamentation, the fungal species present as spores and hyphae in several of the observed samples resemble the genus *Aspergillus*, while the bacterial cells (mostly in chain form) can be attributed to the order Actinomycetales (and presumably to the genus *Streptomyces*).

The VP-SEM technique, in tandem with the use of a BSE detector, disclosed the peculiar effect of the microbial attack on the salts distribution pattern of the parchment surface. The biodeteriorated zones showed a decrease in Al and Si levels, and an increase in those of S and P.

A possible interpretation of this finding is that fungal and microbial species exert a leaching effect on minerals (aluminium and silicon), and are simultaneously able to translocate and concentrate other ions such as sulphates and phosphates. This hypothesis is supported by the SEM-BSE images (Figures 2a and 2b) that show different ion distribution patterns before and after microbial damage has occurred. Alternatively, the chemical differences observed in the undamaged (integral) and damaged parchment surfaces could be due simply to the microbial erosion of the upper strata of parchment with a consequent loss of surface salts and the exposure of the underlying strata, which are of a different chemical composition, namely richer in phosphorus and sulphur.

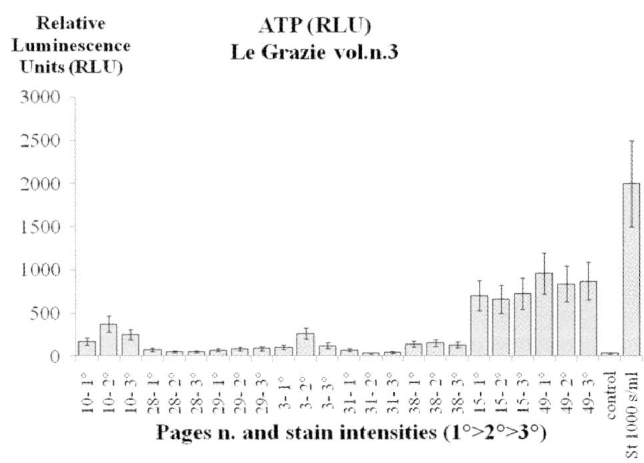

Figure 6 Bars indicate average values of ATP concentration (measured using the luciferine-luciferase luminescent assay method) in the parchment samples taken from volume No. 3. The samples, each marked on the x axis with the number of the page from which were obtained, were divided into three categories: (1) severely stained; (2) slightly stained; and (3) apparently stain free. Standard deviation is also reported.

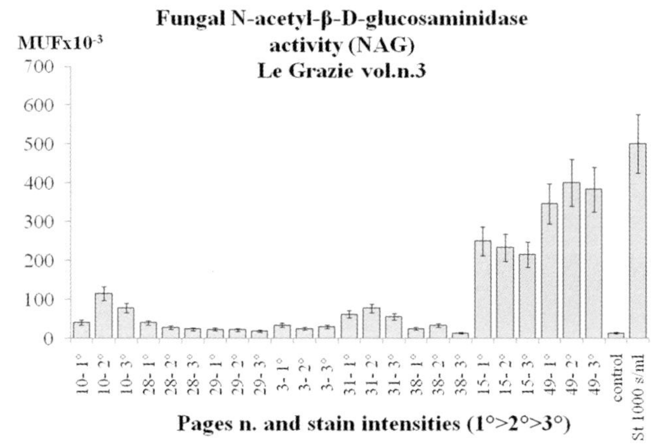

Figure 7 Bars indicate average values of beta-N-acetylhexosaminidase activity measured by means of fluorescent assay in the parchment samples obtained from volume No. 3. The samples, each marked on the x axis with the number of the page from which they were obtained, were divided into three categories: (1) severely stained; (2) slightly stained; and (3) apparently stain free. Standard deviation is also reported.

The luciferine-luciferase method for ATP quantification, used to estimate the viability of microorganisms in parchment samples, showed that the highest ATP concentration values were not significantly associated with stains, thus indicating that the agent that caused the stains was no longer active.

The measurement of fungal β (1-4)-N-acetyl-D-glucosaminidase activity, used to quantify fungal biomass in the parchment samples, detected the presence of fungal material associated only with a few pages. Purple stains were not associated with high fungal enzymatic activity. A high degree of correlation between ATP concentration and NAG activity in the samples indicated that fungi were still viable, even if they were no longer able to be cultured under the conditions used in the present study.

Different types of damage result from attacks by different microbial species, depending on the enzymes they possess and on the possibility of the co-occurrence of different species simultaneously, such as in natural successions observed on organic substrates. Evidence of the complexity of such microbial consortia in cultural heritage materials was provided by the SEM images of samples, where structures from different microbial and fungal species were visibly present within the same spatial niches (see [16] for the definition of 'spatial niche' in fungal ecology).

Conclusions

The rough identification of degradation agents is frequently unsatisfactory from the diagnostic point of view. SEM images can be of help mainly in identifying among a broad range of culturable organisms precisely which one was responsible for causing damage. In the present study, none of the small number of growing cultures could be positively coupled with what was seen with the SEM, which would indicate that the parchment biodegrading organism is no longer viable, or simply not culturable *in vitro*.

A closer microbiological study of the ancient volumes is being undertaken using molecular techniques that make it possible to investigate and monitor microorganisms based on their deoxyribonucleic acid (DNA), ribonucleic acid (RNA) and proteins. DNA-based techniques open the way to the identification of single bacterial species in sample materials without the need to culture organisms [17–19].

The combination of SEM analysis and molecular investigations has proved to be highly useful for obtaining diagnostic information relevant to the conservation of precious art objects and all manner of valuable items of cultural heritage [20–22]. SEM allows for the direct observation of causative agents, while molecular methods provide exhaustive lists of candidate species that can potentially be associated with spoiled materials, and therefore serve as a type of cross-check useful for determining which agent (or agents) actually caused degradation to occur in the heritage artefact.

Materials and suppliers

- Czapek and gelatine broths: DIFCO, Becton Dickinson, USA (www.bdbiosciences.com/)
- ENLITEN ATP Assay System Detection Kit (Cod. FF2000, Promega, USA, Madison) (www.promega.com/)
- Turner 20/20 Luminometer (Promega, USA, Madison) (www.promega.com/)
- 4-MUF-NAG, Sigma-Aldrich Co., St. Louis, MO (www.sigmaaldrich.com)
- LS45 luminescence spectrometer, Perkin-Elmer, Bucks, UK (www.perkinelmer.com/technologies/default.xhtml)
- 4-MUF sodic salt, Sigma-Aldrich Co., St. Louis, MO (www.sigmaaldrich.com)

References

1. Larsen, R., *Microanalysis of Parchment*, Archetype Publications, London (2002).
2. Reed, R., *The Nature and Making of Parchment*, The Elmete Press, London/New York (1975).
3. Florian, M.L.E., *Protein Facts: Fibrous Proteins in Cultural and Natural History Artifacts*, Archetype Publications, London (2007).
4. Kowalik, R., 'Microbiodeterioration of library materials. Part 2: Microbiodecomposition of basic organic library materials', *Restaurator* **4** (1980) 200–208.

5. Polacheck, I., 'Damage to an ancient parchment document by *Aspergillus*', *Mycopathologia* **106** (1989) 89–93.
6. Petushkova, J.P. and Koestler, R.J., 'Biodeterioration studies on parchment and leather attacked by bacteria in the Commonwealth of Socialist States', in *International Conference on Conservation and Restoration of Archival and Library Materials, Erice, 22–29 April 1996*, vol. I (1996) 195–211.
7. Rebrikova, N.L. and Dmitrieva, M.B., 'Experimental investigation of parchment manuscripts with traces of microbiological damage', in *International Conference on Conservation and Restoration of Archival and Library Materials, Erice, 22–29 April 1996*, vol. I (1996) 275–283.
8. Karbowska-Berent, J. and Strzelczyk, A.B., *The Role of Streptomycetes in the Biodeterioration of Historic Parchment*, Nicholas Copernicus University, Torun (2000).
9. Samson, R.A., Hoekstra, E.S., Frisvad, J.C. and Filtenborg, O., *Introduction to Food and Airborne Fungi*, 6th edn, Centraalbureau Voor Schimmelculture, Utrecht (2002).
10. Goldstein, J., Newbury, D., Joy, D., Lyman, C., Echlin, P., Lifshin, E., Sawyer, L. and Michael, J., *Scanning Electron Microscopy and X-Ray Microanalysis*, 3rd edn, Springer Science, Business Media Inc, New York (2003).
11. Rakotonirainy, M.S., Héraud, C. and Lavédrine, B., 'Detection of viable spores contaminant on documents and rapid control of the effectiveness of an ethylene oxide disinfection using ATP assay', *Luminescence* **18** (2003) 113–121.
12. Konkol, N.R., McNamara, C.J. and Mitchell, R., 'Fluorometric detection and estimation of fungal biomass on cultural heritage materials', *Journal of Microbiological Methods* **80**(2) (2009) 178–182.
13. Carvalho, D.N., *Forty Centuries of Ink*, Dodo Press, Gloucester (2007).
14. Florian, M.L.E., 'Fungal fox spots and others', in *On Microbes and Art*, ed. O. Ciferri, P. Tiano and G. Mastromei, Kluwer Academic/Plenum Publishers, New York (2000) 135–151.
15. Miller, M., Palojärvi, A., Rangger, A., Reeslev, M. and Kjøller, A., 'The use of fluorogenic substrates to measure fungal presence and activity in soil', *Applied Environmental Microbiology* **64** (1998) 613–617.
16. Dix, N.J. and Webster, J., *Fungal Ecology*, Chapman & Hall, Cambridge (1995).
17. Amann, R.I., Ludwig, W. and Schleifer, K-H., 'Phylogenetic identification and in situ detection of individual microbial cells without cultivation', *Microbiological Reviews* **59** (1995) 143–169.
18. Ward, D.M., Weller, R. and Bateson, M.M., '16S rRNA sequences reveal numerous uncultured microorganisms in a natural community', *Nature* **345** (1990) 63–65.
19. Piñar, G., Ramos, C., Rölleke, S., Schabereiter-Gurtner, C., Vybiral, D., Lubitz, W. and Denner, E.B.M., 'Detection of indigenous *Halobacillus* populations in damaged ancient wall paintings and building materials: molecular monitoring and cultivation', *Applied Environmental Microbiology* **67** (2001) 4891–4895.
20. Michaelsen, A., Pinzari, F., Ripka, K., Lubitz, W. and Piñar, G., 'Application of molecular techniques for identification of fungal communities colonizing paper material', *International Biodeterioration and Biodegradation* **58** (2006) 133–141.
21. Michaelsen, A., Piñar, G., Montanari, M. and Pinzari, F., 'Biodeterioration and restoration of a 16th-century book using a combination of conventional and molecular techniques: a case-study', *International Biodeterioration and Biodegradation* **63** (2009) 161–168.
22. Michaelsen, A., Piñar, G. and Pinzari, F., 'Molecular and microscopical investigation of the microflora inhabiting a deteriorated Italian manuscript dated from the 13th century', *Microbial Ecology* **60** (2010) 69–80.

Authors' addresses

- Flavia Pinzari, Istituto Centrale per il Restauro e la Conservazione del Patrimonio Archivistico e Librario, Rome, Italy (flavia.pinzari@beniculturali.it)
- Vanja Cialei, Istituto Centrale per il Restauro e la Conservazione del Patrimonio Archivistico e Librario, Rome, Italy
- Guadalupe Piñar, Institute of Applied Microbiology, Department of Biotechnology, University of Natural Resources and Life Sciences, Vienna, Austria

Arsenic and apricots: understanding the sober still lifes of Adriaen Coorte

Carol Pottasch and Kees Mensch

ABSTRACT This paper focuses on the painting technique of three still lifes by Adriaen Coorte (active 1683–1707) and the supports onto which each is painted. The mainly good state of preservation of these paintings is related to the paper painting supports and the use of starch in the isolation layer. Chemical analyses with the scanning electron microscope coupled with energy dispersive X-ray microanalysis (SEM-EDX) of cross-sections taken from the paint layers showed that the apricots are painted with an arsenic sulphide pigment and chalk/lake pigments. It is not possible to discern whether the arsenic sulphide pigment is orpiment or realgar using SEM-EDX. The SEM backscattered electron (BSE) images show that the orpiment/realgar is in the process of degradation. This can be deduced from the particles with cloudy edges and arsenic-rich deposits at interfaces (between the paint layers and around pigment particles). Coorte's painting of apricots is compared with general seventeenth-century documentary sources, which show that his painting technique is uncommon.

KEYWORDS orpiment and/or realgar, Adriaen Coorte, painting on paper, starch, magnetite

Introduction

The acquisition by the Mauritshuis of two paintings by the seventeenth-century artist Adriaen Coorte (active 1683–1707) – *Still Life with Strawberries* (1705) and *Still Life with Apricots* (1704) – led to an exhibition in 2008 (Mauritshuis, Den Haag, 23 February–8 June 2008). This was accompanied by a monograph on the artist [1]. The exhibition formed a unique opportunity for conservation treatment of these paintings and *Still Life with Two Walnuts* (1702) from the Szépmüvézeti Múzeum in Budapest, and to study Coorte's painting technique (Figures 1–3). To study the painting technique of the three paintings, cross-sections were taken. Analyses of carbon-coated paint cross-sections using a JEOL variable pressure scanning electron microscope coupled with energy dispersive X-ray microanalysis (SEM-EDX) in collaboration with the Shell Technology Centre, Amsterdam made it possible to characterise the layer build-up and the composition of the ground and paint layers.

Coorte worked in the vicinity of Middelburg [1]. His early works are relatively large pictures (on average c. 100 × 80 cm) usually depicting birds in the manner of Melchior Hondecoeter (Utrecht 1636–Amsterdam 1695). Around 1690 Coorte changed to painting small fruit and vegetable still lifes mainly depicting strawberries, apricots, asparagus, gooseberries and combinations of these. His oeuvre contains more than 60 paintings. In the early 1950s, Coorte's work started to be studied for the first time and became the subject of exhibitions in Dordrecht (Dordrechts Museum, August–November 1958), Washington, DC (The National Gallery of Art, Washington, DC, 29 June–28 September 2003) supplemented by a study of the painting technique of one work in the Rijksmuseum Amsterdam for a still life exhibition [2–4].

Painting technique

Paper support and preparatory layers

A survey of Coorte's oeuvre carried out for the 2008 exhibition showed that half of his works are painted on canvas and the other half on paper. *Still Life with Apricots* is painted on canvas, whereas *Still Life with Strawberries* and *Still Life with Two Walnuts* are painted on paper. Until now Coorte is the only Dutch artist known to use paper as a support for his paintings. The more commonly used painting supports are canvas, panel and copper. Recent research showed that the paintings on paper must have been adhered to a rigid support at a later date [5]. In the seventeenth century, paper is generally only mentioned for preparatory drawings and sketches, or, when oiled, to copy or trace. Rubens, Van Dyck, Jordaens and Rembrandt are well known for the oil sketches they made that were not intended as finished paintings.

To study the preparatory and paint layers of the still life paintings, cross-sections were taken from each. *Still Life with Strawberries* has a warm red-brown lower ground layer with a cool red-brown layer on top containing yellow, white, black and red pigment particles in both layers. EDX spectra pointed to the use of yellow ochre (Fe), bone black (P), lead white (Pb) and hematite (Fe). *Still Life with Two Walnuts* has a cool grey lower ground covered by a warmer grey-brown layer composed of lead white, bone black, hematite and yellow ochre and a very small amount of red lead. The ground of *Still Life with Apricots* has a single warm yellow earth ground containing yellow ochre, chalk, some hematite and a black iron oxide. The colour of the ground layers of the paintings were all prepared to match the subject of the painting.

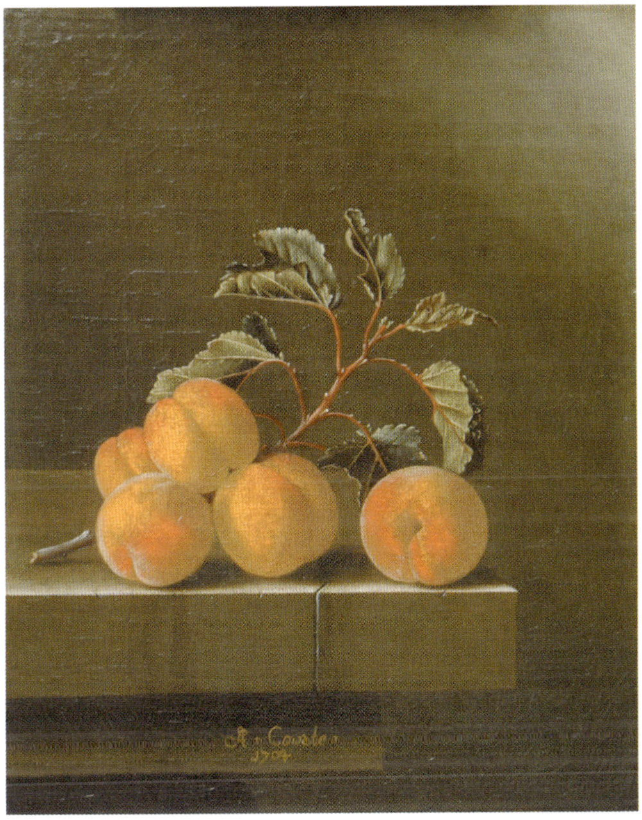

Figure 1 *Still Life with Apricots*, 1704, canvas, 30 × 23.5 cm, Mauritshuis, inv. no. 1154.

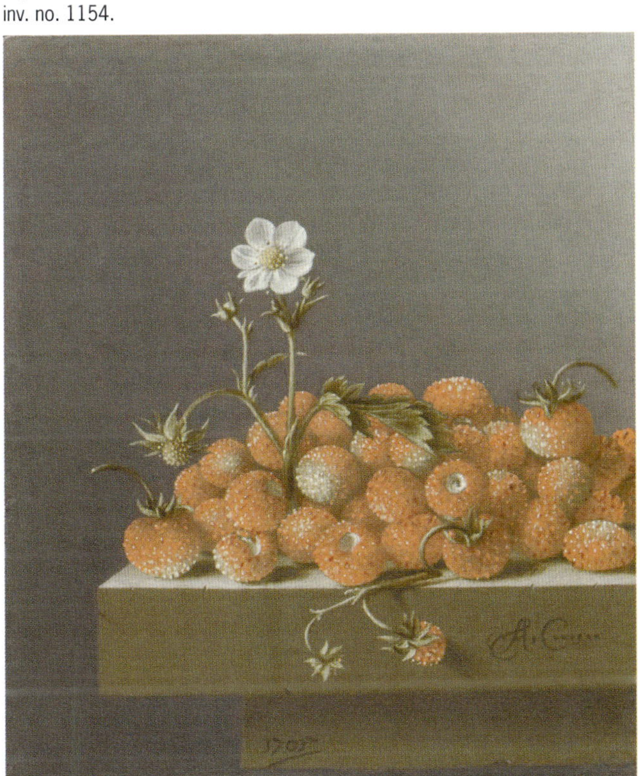

Figure 2 *Still Life with Strawberries*, 1705, paper on panel, 16.4 ×13.5 cm, Mauritshuis, inv. no. 1106.

The discovery by EDX of a conglomerate of black pigment in the yellow earth pigment mixture in the ground layer of *Still Life with Apricots*, which only contains iron and oxygen, presented an interesting aspect. The black appearance in the light microscope images and the exclusive presence of iron and oxygen suggest that it is the very fine black mineral magnetite (FeO_4) [6, 7]. Black

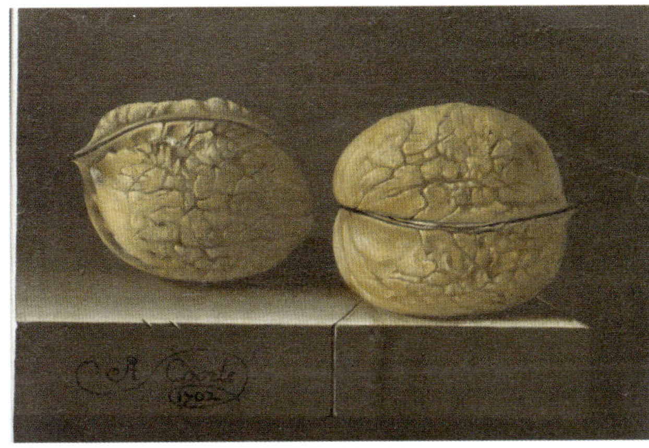

Figure 3 *Still Life with Two Walnuts*, 1702, paper on panel, 10.9 × 15.6 cm, Szépmuveszeti Múzeum, Budapest, inv. no. 75.9.

iron oxide in the form of magnetite is sometimes found together with yellow ochre. Since magnetite is known to be found in the river clay deposits of the Schelde River near Middelburg this may indicate a local source of the ochre [8].

At the bottom of cross-sections taken from all three paintings (one on canvas, two on paper) a striking feature presented itself in a brownish layer showing vague oval forms. This layer is interpreted as an isolation layer. The function of an isolation layer is to prevent the oil paint from soaking into the painting support. When examined with ultraviolet (UV) light the oval forms in the cross-sections showed a strong fluorescence suggesting the presence of starch grains (Figure 4) [9]. The SEM backscattered electron (BSE) images of this layer show large black, oval forms protruding into the ground layer. The same characteristics can be seen in all three paintings. In some cross-sections paper fibres are visible. The spectra taken from the ovals showed almost pure carbon (with traces of lead and iron from the ground) and make identification as starch plausible, but this will have to be confirmed by identification with polarised light microscopy (PLM) or Fourier transform infrared spectroscopy (FTIR). In these paintings starch appears to have been applied as an isolation layer prior to painting, a practice that is mentioned in seventeenth-century documentary sources [10, 11]. The use of starch in seventeenth-century paintings has up to now always been considered unusual, although Rembrandt's pendant portraits of *Agatha Bas* (Royal Collection, Buckingham Palace, London) and *Nicolas van Bambeek* from 1641 (Musée Royal des Beaux-Arts, Brussels), both on canvas, have recently been found to have starch-containing grounds [12].

It is remarkable that, unlike *Still Life with Apricots* where the canvas weave is pronounced, hardly any cracks are to be seen in the paint layers of *Still Life with Strawberries* and *Still Life with Two Walnuts* painted on paper supports. The hygroscopic starch may have acted as a gel that moved together with the hydroscopic paper without affecting the rigid paint layers. The starch may also have facilitated the recovery of the deformed paper support of *Still Life with Two Walnuts*. Before conservation treatment the painting was in very poor condition: the paper was deformed and partly detached from the panel support to which it was glued. After removing the paper from the panel (by paper conservator Nico Lingbeek, Haarlem) the paper support recovered well and became flat. In conclusion it can be said that the use of paper supports in combination with their preparation has played an important role in the good condition of the paintings by Coorte in general.

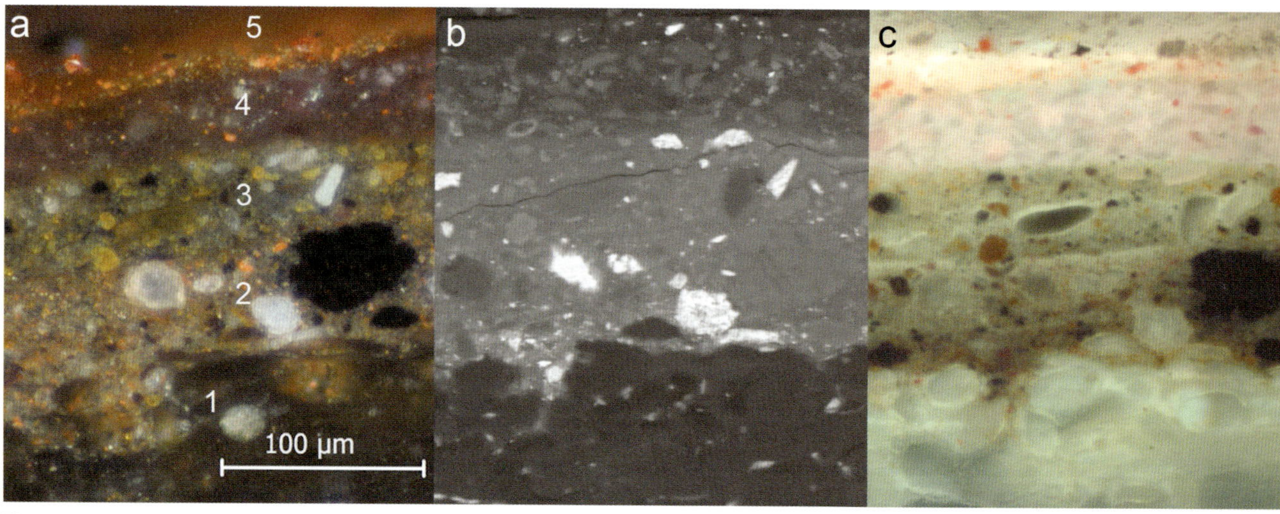

Figure 4 *Still Life with Strawberries*. Cross-section taken on the right edge from the shadow area of a strawberry: (a) bright field microscope image; (b) SEM-BSE image; (c) UV fluorescence image. Layer build-up in the cross-section: (1) the brownish forms in bright field are starch grains that appear almost black using BSE and are highly fluorescent in UV; (2) first ground layer with earth colours, bone black, chalk and lead white; (3) second ground layer with earth colours, bone black, chalk and lead white; (4) transparent red paint layer with a mixture of vermilion, hematite, smalt, red lakes – the smalt is clearly visible in the UV image; (5) similar transparent red paint layer but with more vermilion – this layer is difficult to see using BSE and has a different fluorescence.

Paint layers: the use of orpiment and /or realgar in *Still Life with Apricots*

As there is no specific contemporary documentation on Coorte's painting technique, reference will be made to general literature of the time. For the focus of this paper the research of the painting layers is only directed towards the yellow paint of *Still Life with Apricots*. This paint used in the apricots showed various degradation problems, whereas the paint layers of the other two paintings were in excellent condition. To understand how the fruit was painted, contemporary documentary sources were consulted. Two seventeenth-century sources that make a reference to painting apricots are by King [11] and Beurs [13]. King [11; fol. 43] advises a layer of massicot (in the seventeenth century generally the yellow pigment lead-tin yellow) and white, followed by glazes after a drying period. Beurs recommends painting apricots with yellow ochre and white, and glazing the shadow with umber, and 'When everything is dry use rusgeel [realgar] on the day [lit side]' [13; 38–39]. The use of realgar only as a highlight is confirmed by the Theodore de Mayerne quotation of Anthony van Dyck (1599–1641) in his treatise on painting techniques of Dutch and Flemish artists (between 1620 and 1646) [14].

Three cross-sections were taken from different areas, each relating to a different build-up of paint layers and paint degradation problems observed.[1] The orange paint in the light areas of the apricots was crumbling along fragmented cracks. The paint also appeared slightly wrinkled at the surface of these areas. Blackening of vermilion was also observed [15]. In the discoloured glazes, whitish spots were visible, giving the layers a slightly greyish appearance, on which the red highlights seemed to be floating [16].

Bright red area

The first cross-section was taken from a bright red area in the upper left apricot. It shows a build-up of three paint layers (Figure 5). SEM-EDX analysis indicated that the first paint layer on top of the ground, which is bright orange and yellow in light microscopy, is composed of orpiment/realgar (As, S), a little vermilion (Hg), chalk (Ca), hydrated alumina, and gypsum (large amount of S, Ca). The second paint layer is similar to the first but with a little more orpiment/realgar. It is made somewhat lighter with some lead white (no tin was found so lead-tin yellow can be excluded). Vermilion was mainly used in the red top layer, as well as a little orpiment/realgar and red lake. In nature, orpiment and realgar are often found together; orpiment is a semi-opaque bright yellow (As_2S_3) while realgar (AsS) is more reddish. They typically appear as elongated flakes [17]. SEM-EDX does not allow us to distinguish between the semi-opaque yellow orpiment and the reddish orange realgar as identification is based on the same elements. Nor does it shed light on the presence of other arsenic- and sulphur-containing compounds formed during the degradation process of the paint such as pararealgar or arsenolite (arsenic trioxide, As_2O_3) [18].

In this cross-section several interesting observations can be made. First, Coorte used orpiment/realgar to paint the apricots instead of massicot or yellow ochre as suggested by King [11] and Beurs [13]. Secondly, the BSE images show roundish particles instead of the usual flake-like structure of orpiment. It is also possible to observe that some particles have a cloudy outline. The presence of chalk and gypsum is noteworthy.

Middle tone: the presence of chalk and gypsum

A second cross-section sample was taken in a middle tone of the right apricot in a thin transparent orange yellow paint. This paint fluoresces red in UV light where white spots are visible under the surface. In the cross-section the paint layers appear hazy and are highly fluorescent under UV light. It is difficult to see individual particles. The first layer on top of the ground layer contains tiny roundish orange, yellow and a few red particles. It is followed by a transparent orange layer with tiny orange and white particles and the top layer is an orange crust in which orange and red particles are present. The abundance of sulphur in the EDX spectra cannot only be explained by the presence of orpiment/

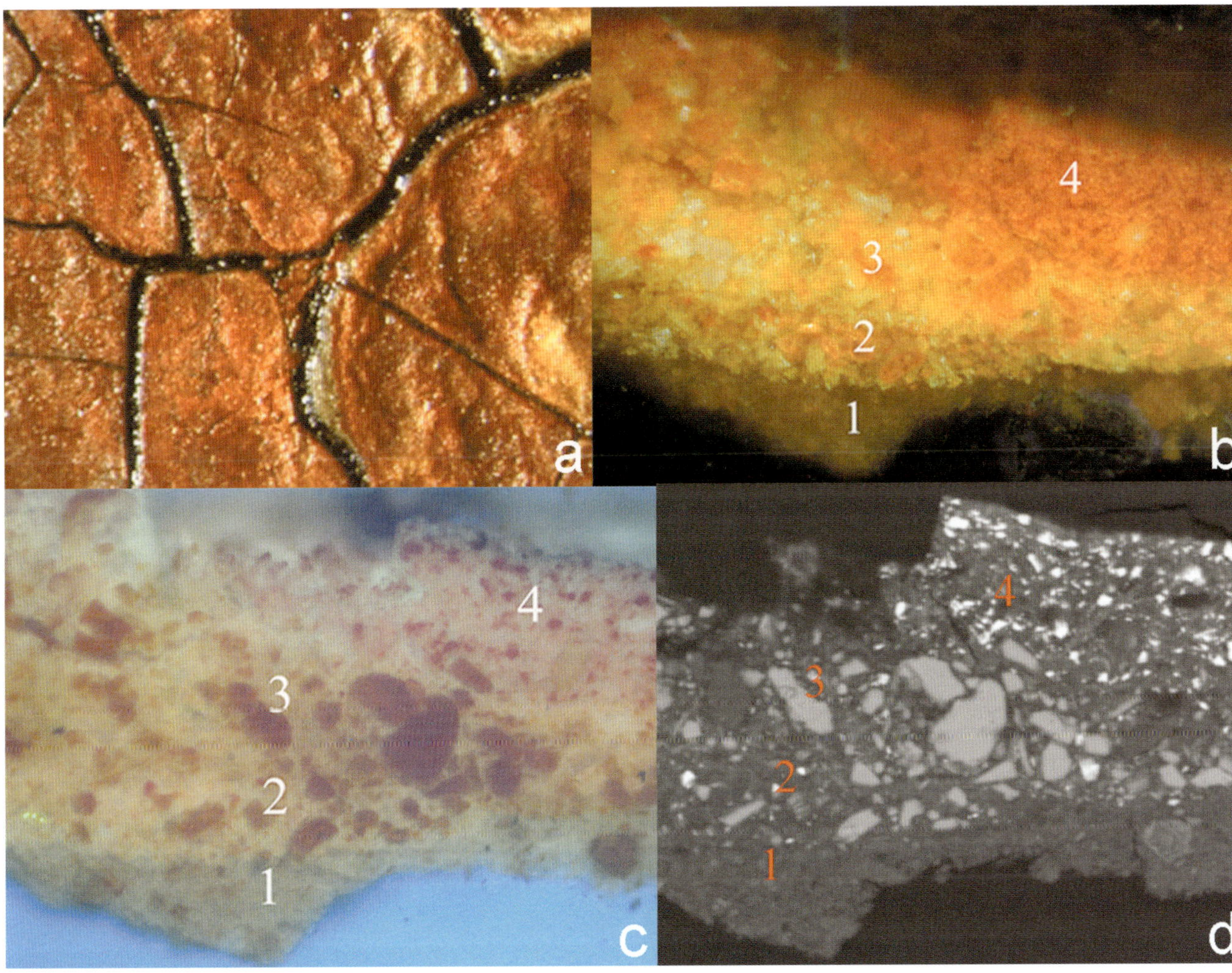

Figure 5 *Still Life with Apricots*. Cross-section taken in bright scumble of right apricot: (a) location of sample; (b) cross-section in bright field; (c) UV fluorescence; (d) BSE image. Layer build-up in the cross-section: (1) ochre-coloured ground layer; (2) and (3) yellow and orange paint layers containing orpiment and realgar; (4) paint layer containing vermilion and red lake and some orpiment or realgar. In bright field the orpiment and realgar are yellow and orange. In UV they appear orange-red. Using BSE they appear as roundish particles.

realgar and fine vermilion (bright particles in the BSE image coinciding with red coloration), and chalk. The abundance of calcium and sulphur together point to gypsum and indeed EDX point analyses of dark grey particles in the BSE images confirm the presence of gypsum particles. In the second layer orpiment/realgar, chalk and gypsum can also be identified. The pigment particles analysed in the thin top layer are vermilion, gypsum, orpiment/realgar and chalk. Again the quantity of gypsum and chalk present in the spectra is remarkable and necessitates further explanation (that will be addressed in the discussion).

Bright orange area (degraded)

The third cross-section was taken from a bright orange area of the left apricot, where the paint is very broken up and cracked. As suggested by Wallert [4], light-induced degradation of orpiment resulting in the elimination of sulphur could cause the break-up of the binding medium, thus explaining the crumbly nature of the yellow paint layers. It was noticed that these paint layers were very vulnerable to moisture. In the cross-section it was observed that the yellow, and an orange, paint layers become more and more transparent towards the surface (Figure 6). It was possible to observe under UV light that most pigment particles seem to be present in the bottom part of the layer, while the top part is highly fluorescent and contains only a few red particles. Besides orpiment/realgar and chalk a large amount of gypsum is present in the paint. The BSE image shows a thin layer visible on top of the ground layer that could not be seen in reflected light. Most pigment particles, large roundish ones and tiny flake-like ones, are present in this lower part of the thick paint layer, surrounded by tiny greyish middle tone particles. It is not clear whether these tiny plate-like particles indicate a mixture of natural and artificial orpiment. To distinguish between natural and dry processed orpiment a combination of analytical techniques would be necessary as no typical spherulitic particles are present, see [19, 20]. Keune observed [21] that the degradation of orpiment particles may result in arsenic-rich deposits at the interfaces between paint layers and around pigment particles. Several EDX spectra showed orpiment/realgar and chalk with large quantities of arsenic and calcium, but with a proportionally low sulphur peak in this lower part. Closer to the surface the relatively light particles are slightly larger and holes are also observed. Here many orpiment/realgar particles have a cloudy appearance, and much more sulphur is present. In BSE images orpiment/realgar particles appear bright, whilst gypsum appears light grey and chalk, dark grey. Spot measurements indicate that in large particles some orpiment/realgar is

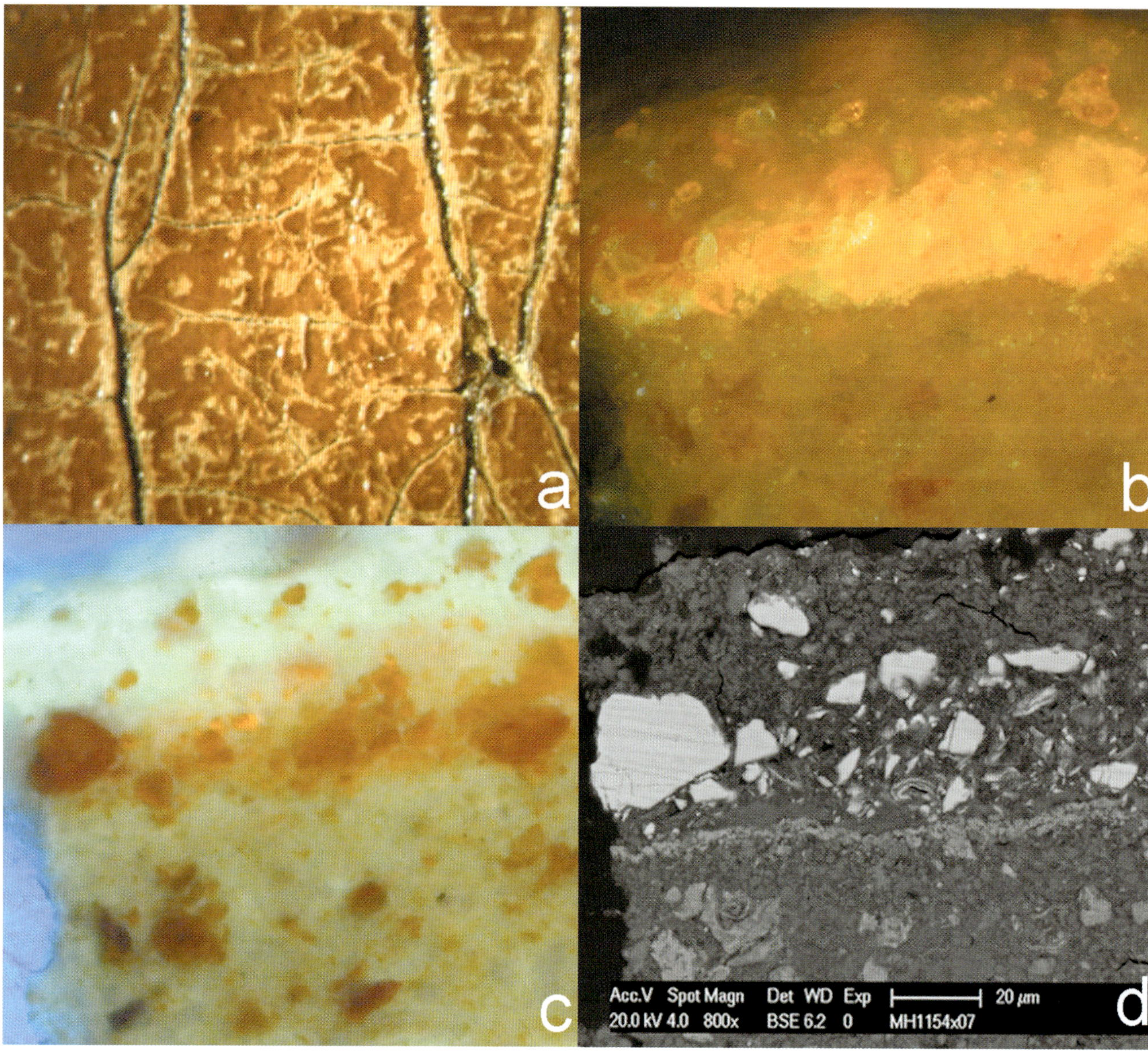

Figure 6 *Still Life with Apricots*. Cross-section taken in crumbling paint crack: (a) location of micro-sample taken in left apricot; (b) bright field image; (c) UV image; (d) BSE image (made by Katrien Keune). This detail of a cross-section shows a single paint layer on top of an ochre-coloured ground layer. The paint layer contains both yellow and orange particles. It is not possible to distinguish which arsenic sulphur compound is present. Particles of the orange realgar have settled on the ground. The paint is more transparent towards the surface. The realgar has degraded leaving only a few larger particles. Using BSE the cloudy appearance of some of the particles may show the degradation in progress. In the top part of the layer more gypsum is visible (source unknown at present).

present with peaks only for arsenic and sulphur; other measurements only showed calcium peaks for chalk, or sulphur and calcium peaks for gypsum.

Discussion

Degradation of orpiment over time

As expected from the UV light fluorescence observations, EDX measurements show less arsenic and more sulphur and calcium near the surface. It seems as if the degraded arsenic has sunk into the layer and is deposited on the interface with the ground. Light changes the pigment by eliminating sulphur from its crystal system in the form of corrosive gases, thereby changing orpiment into white arsenolite, which is highly unstable to light and humidity, and realgar into the yellow pararealgar [18]. The sulphur makes its way to the surface, while the arsenic, possibly arsenolite, migrates through the structure by water transport [21]. This phenomenon can possibly be seen occurring in the BSE image where a particle of orpiment/realgar is losing its consistency, showing a cloudy edge at the top (Figure 6d). The holes in the paint layer could be caused by partly degraded particles that are not well attached in the paint matrix and have fallen out during polishing of the cross-section or by migration of material. The roundish forms could also indicate the dissolution of the orpiment/realgar. It is notable that traces of arsenic were found in the spectra of nearly all the measurements, irrespective of layer. Even though no orpiment/realgar has been added intentionally in the ground layer, EDX analysis of areas in the ground layer of cross-section (MH1154x07) showed traces of arsenic. Keune has suggested [21] that humid conditions may trigger the arsenic to migrate through the layer. Further study is necessary to show this.[2]

In the cross-section from the bright orange area it is suggested that the top layer containing vermilion acts as protection for the orpiment/realgar middle layer, as it is much less degraded.

The presence of chalk and gypsum

The presence of chalk in Coorte's paint could either be explained by its use as a substrate for now faded yellow lake, or as an additive to adjust the colour and working properties of the paint. A medieval English documentary source [22] mentions that calcined eggs or oyster shells could be mixed with orpiment 'for making the pale yellow of orpiment paler still', which makes it possible that the shade of the pigment was still adjusted with chalk at the end of the seventeenth century. Mixing in chalk could facilitate working properties using less oil to paint [23]. Chalk is a likely component in the paint used for the apricots.

The origin of gypsum in the paint is unclear. One might speculate that it is associated with the organic lake, or depending on source, orpiment could be mined together with gypsum, quartz and many other clays, or that gypsum was added shortly after mining [24]. Small quantities of silica were found in the paint for the apricots. In the Oranjezaal (1648–1652) orpiment particles embedded in glass, quartz and gypsum were found by van Loon in yellow highlights made by the interior painter Christiaan van Couwenberg (1604–1667) [16]. This was identified as filler material to prevent sagging of the slow-drying paint. However, Coorte did not use the paint for a highlight and therefore the addition of extra gypsum is not needed. Also the presence of gypsum as a substrate for a lake is unlikely, as evidence has been found for the more common substrates of chalk and aluminium hydroxyl [25]. At present there is no explanation for the presence of gypsum particles, although the reactivity of orpiment/realgar in the presence of chalk could have resulted in the formation of gypsum.

The use of orpiment in other paintings

Instead of using lead-tin yellow (massicot) or yellow ochre, Coorte seems to have used orpiment, whereby he obtained a golden glow with a sparkle for his apricots. As commonly known towards the end of the seventeenth century, lead-tin yellow was becoming less available, while the use of orpiment was becoming more widespread. Orpiment and realgar were already used in these times as many studies have showed. An extensive study on orpiment was published recently, providing an overview of natural and artificial orpiment, its use and degradation [26]. Since the tenth and eleventh centuries, several sources mention that orpiment and realgar should not be mixed with verdigris and copper green, and also warn against the use of red lead, lead white and vermilion [24]. Coorte appears to have been careful not to get the orpiment/realgar close to copper-based pigments such as azurite, as can be seen in the leaves painted above the apricots where Coorte used lead-tin yellow for the highlights (observation with the stereomicroscope), but for the apricots he mixed in vermilion and lead white.

In a late fifteenth-century German source, the *Liber illuministarum*, a 'brilliant and magnificent gilding' is described using a mixture of orpiment, chalk, roasted vermilion and hematite [27]. Orpiment was used in a number of early sixteenth-century Dutch paintings and commonly in seventeenth-century interior painting and naval painting to imitate gold [16, 20]. A number of still life painters from the last part of the seventeenth century preceded Coorte in the use of orpiment, such as Jan Davidsz de Heem (1606–1683/84), Abraham van Beyeren (1620/21–1690) for oranges and lemons and Abraham Mignon (1640–1679) for a marigold, Martinus Nellies (active 1763–1694) for quince, and in highlights for oranges by Willem Kalf [4]. In the Mauritshuis it was used for marigolds by Daniel Seghers (1590–1645) and Rachel Ruysch (1664–1750).[3]

EDX analysis of *Still Life with Strawberries* and *Still Life with Two Walnuts* showed that Coorte also used orpiment or realgar in these paintings for layers that are not typically orange. A cross-section of bright red paint of a strawberry shows a red layer with a mixture of vermilion, presumably realgar, quartz and traces of red lead (Pb) and hematite. Also in a light area of the walnuts, Coorte mixed in a small amount of orpiment which shows that he used it instead of lead-tin yellow to provide a golden glow to the painting. Contrary to seventeenth-century beliefs expressed by Beurs [13] and De Mayerne [14] and in many other sources, the paints mixed with orpiment and realgar have retained their colour.

Conclusions

Coorte's still lifes have been little studied, but we can show that the studied paintings display an unusual painting technique that manifests itself in different ways. The paper support used in half of his oeuvre, followed by an isolation layer of starch is seldom encountered. Examination of cross-sections taken from the apricots showed that Coorte did not use the expected massicot or ochre, but used yellow paint mixtures containing orpiment/realgar, red lakes, chalk, red (vermilion) and white (lead white) in different quantities. The degradation of the orpiment/realgar by light leads to degradation of the binding medium and makes the paint crumbly. The presence of gypsum in the paint may have derived from the reaction of the free sulphur in the layer with the calcium or chalk. Although this research shows new observations about Coorte's painting technique, these results are preliminary and more paintings need to be studied.

Acknowledgements

The SEM-EDX measurements were made by Kees Mensch, microscopist and Ralph Haswell, head of the Electron Microscopy Department, Shell Technology Centre, Amsterdam. The Mauritshuis is grateful to Rob Bouwman for facilitating this collaboration. We are grateful to Katrien Keune, Annelies van Loon and Petria Noble for their valuable discussions and contributions to the text. Carol Pottasch would also like to thank Erik Veerman for his support and help in editing the text.

Notes

1. Examination of two other paintings with apricots with the stereomicroscope (sign. A. Coorte, after 1696, paper, coll. Pieter C.W.M. Dreesmann together with peaches; sign. and dated A. Coorte 1698, paper, on loan to Coll. Mauritshuis/Rijksmuseum) showed that they were painted with the same technique and exhibit the same degradation problems in the yellow paint layers of apricots.

2. Carol Pottasch would like to thank Katrien Keune and Annelies van Loon for discussing the material in this paragraph in depth.
3. Mauritshuis: Daniel Seghers (1590–1645) and Rachel Ruysch (1664–1750) used orpiment in respectively *Garland with Flowers*, 1645, MH 256, and *Vase with Flowers*, 1700 MH 151.

References

1. Buvelot, Q., *The Still lifes of Adriaen Coorte*, Den Haag (2008).
2. Bol, L., *Adriaen Coorte*, Dordrecht (1958).
3. Wheelock Jr, A.K., *Small Wonders: Dutch Still Lifes by Adriaen Coorte*, National Gallery of Art, Washington DC (2003).
4. Wallert, A. (ed.), *Still Lifes: Techniques and Style. The Examination of Paintings from the Rijksmuseum*, Rijksmuseum, Amsterdam (1999) 92–94.
5. Bijl, M., Lingbeek, N., Pottasch, C. and Noble, P., 'Technical research', in *The Still lifes of Adriaen Coorte*, ed. Q. Buvelot, Royal Picture Gallery Mauritshuis, Den Haag, Waanders Publishers, Zwolle (2008) 127.
6. Eastaugh, N., Walsh, V., Chaplin, T. and Siddal, R., *The Pigment Compendium: A Dictionary of Historic Pigments*, Elsevier Butterworth-Heinemann, Oxford (2004) 248.
7. Eastaugh, N., Walsh, V., Chaplin, T. and Siddal, R., *The Pigment Compendium: Optical Microscopy of Historical Pigments*, Elsevier Butterworth-Heinemann, Oxford (2004) 254–255.
8. Groen, K.M., 'Earth matters', *Art Matters* **3** (2005) 138–154.
9. Keith, L., Roy, A. and Morrison, R., 'Leonardo da Vinci's *Virgin of the Rocks*: technique and the context of restoration', in *Studying Old Master Paintings: Technology and Practice*, ed. M. Spring, Archetype Publications, London (2011) 72–79.
10. Witlox, M. and Carlyle, L., '"A perfect ground is the very soul of the art" (Kingston 1835): ground recipes for oil paintings, 1600–1900', *Preprints of ICOM-CC 14th Triennial Meeting, The Hague, 12–16 September 2005*, vol 1, ed. I. Verger, James & James, London (2005) 519–528.
11. Daniel King, *Secrets in the Noble Arts of Miniatura or Limming*, British Library MS Additional 12.461, (c.1653–57).
12. Sanyova, J., Saverwyns, S. and Fremout, W., 'A surprising ground layer in Rembrandt's *Portrait of Nicolaes van Bambeeck*', poster presentation, Book of Abstracts, *Studying Old Master Paintings Technology and Practice*, The National Gallery 30th Anniversary Conference, London, 16–18 September (2009) 130.
13. Willem Beurs, *De grote Wereld in het klein geschildert*, Zwolle (1692).
14. Mayerne, T. Turquet de (1620–1646) *Pictoria, sculptoria, tintctoria et...* or, the 'De Mayerne manuscript', London, British Library MS Sloane 2052, in J.A. van de Graaf, *Het De Mayerne manuscript als bron voor de schildertechniek van de barok*, The Rijksuniversiteit Utrecht, Utrecht (1958) 50–51.
15. Spring, M. and Grout, R., 'The blackening of vermilion: an analytical study of the process in painting', *National Gallery Technical Bulletin* **23** (2002) 50–61.
16. Loon, A. van, *Color Changes and Chemical Teactivity in Seventeenth-Century Oil Paintings*, PhD thesis, University of Amsterdam, Amsterdam (2007) 72–73.
17. West FitzHugh, E., 'Orpiment and realgar', in *Artists' Pigments: A Handbook of their History and Characteristics*, vol. 3, ed. E. West FitzHugh, National Gallery of Art, Washington, DC (1997) 47–80.
18. Dubois, H., Meloni, S., Metz, E., Schoonhoven, B. and Wallert, A., 'Breakdown processes of orpiment- and realgar-based sixteenth century paints', in *Deterioration of Artist's Paints: Effect and Analysis. A Joint Meeting of ICOM-CC Working Groups Paintings 1 & 2 and the Paintings Section, UKIC, British Museum, London, 10–11 September 2001*, extended abstracts of presentation (2001) 75–79.
19. Grundmann, G. and Rötter, C., 'Artificial orpiment': microscopic, diffractometric and chemical characteristics of synthesis products in comparison to natural orpiment', in *Auripigment/Orpiment. Studien zu dem Mineral und den künstlichen Produkten/Studies on the Mineral and the Artificial Products*, Materialien aus dem Institut fur Baugeschichte, Kunstgeschichte und Restaurierung mit Architekturmuseum der Technischen Universitat Munchen, ed. W. Koenigs, N. Huse, E. Emmerling and W. Nerdinger, Siegl Fachbuchhandlung, München (2007) 103–140.
20. Richter, M., Grundmann, G., Loon, A. van, Keune, K., Boersma, A. and Rapp, K.A., 'The occurrence of artificial orpiment (*dry process*) in northern European painting and polychromy and evidence in historical sources', in *Auripigment/Orpiment. Studien zu dem Mineral und den künstlichen Produkten/Studies on the Mineral and the Artificial Products*, Materialien aus dem Institut fur Baugeschichte, Kunstgeschichte und Restaurierung mit Architekturmuseum der Technischen Universitat Munchen, ed. W. Koenigs, N. Huse, E. Emmerling and W. Nerdinger, Siegl Fachbuchhandlung, München (2007) 167–187.
21. Keune, K., 'Arsenic as a marker for water-linked transport processes in paintings', in *Preprints of ICOM-CC 16th Triennial Meeting, Lisbon 19–23 September 2011* (2011).
22. Thompson, D.V., *The Materials and Techniques of Medieval Painting*, Dover Publications, New York (1956) 97.
23. Gettens, R.J. and Stout, G.L., *Painting Materials*, Dover Publications, New York (1966) 104.
24. Rötter, C., 'Natürliches und künstliches Auripigment-unter Einbeziehung von Realgar', in *Auripigment/Orpiment. Studien zu dem Mineral und den künstlichen Produkten/Studies on the Mineral and the Artificial Products*, Materialien aus dem Institut fur Baugeschichte, Kunstgeschichte und Restaurierung mit Architekturmuseum der Technischen Universitat Munchen, ed. W. Koenigs, N. Huse, E. Emmerling and W. Nerdinger, Siegl Fachbuchhandlung, München (2007) 7–102 .
25. Hermens, E. and Wallert, A., 'The Pekstok Papers: lake pigments, prisons and paint-mills', in *Looking through Paintings: The Study of Painting Techniques in Support of Art Historical Research*, ed. E. Hermens, Archetype Publications, London (1998) 269–294 .
26. Koenigs, W., Huse, N., Emmerling, E. and Nerdinger, W. (eds) *Auripigment/Orpiment. Studien zu dem Mineral und den künstlichen Produkten/Studies on the Mineral and the Artificial Products*, Materialien aus dem Institut fur Baugeschichte, Kunstgeschichte und Restaurierung mit Architekturmuseum der Technischen Universitat Munchen, Siegl Fachbuchhandlung, München (2007) .
27. Richter, M., Grundmann, G., Loon, A. van, Keune, K., Boersma, A. and Rapp, K. A., 'The occurrence of artificial orpiment (*dry process*) in northern European painting and polychromy and evidence in historical sources', in *Auripigment/Orpiment. Studien zu dem Mineral und den künstlichen Produkten/Studies on the Mineral and the Artificial Products*, Materialien aus dem Institut fur Baugeschichte, Kunstgeschichte und Restaurierung mit Architekturmuseum der Technischen Universitat Munchen, ed. W. Koenigs, N. Huse, E. Emmerling and W. Nerdinger, Siegl Fachbuchhandlung, München (2007) 167–187 (178 n. 44).

Authors' addresses

- Carol Pottasch, Royal Picture Gallery Mauritshuis, The Hague, The Netherlands (pottasch.c@mauritshuis.nl)
- Kees Mensch, Shell Technology Centre, Amsterdam, The Netherlands (sh379215-t@telfort.nl)

ESEM-EDX analyses for the characterisation and reproduction of nineteenth-century gold tesserae

Martina Raedel, Martin Sabel, Michael Bücker and Brita Unger

ABSTRACT Environmental scanning electron microscopy with energy dispersive X-ray analysis (ESEM-EDX) is a versatile working tool for the examination of gold tesserae as well as for the selection of suitable materials for their reproduction. In this paper, the ESEM-EDX analyses of original gold tesserae and their damage types are presented. These analytical results form the basis for further calculations of glass properties and experiments. These calculated values enable the determination of the relevant glass properties and allow us to choose the appropriate type of modern glass for further experiments. Subsequent experiments convert this knowledge into practice. By comparing the analytical results from the original glasses and the replicas the accuracy and precision will be shown.

KEYWORDS gold mosaics, tesserae, ESEM-EDX, glass analyses, damage mechanism

Introduction

As a popular design element, gold tesserae have been used for centuries, not just in churches. Because of their golden appearance and the elaborate process used for their production, they hold a special place among tesserae. Although there are still producers of gold tesserae today, mostly in Italy, their products are not of the same wide variety of colour and shape as seen in historical gold mosaic pictures. This became apparent during several restoration projects involving gold tesserae from the nineteenth and early twentieth century. It was difficult to fit modern tesserae, which are only available in a few shapes and are rather monochromatic, into the colourful historical original mosaic pictures.

As part of a research project on the reproduction of gold tesserae, sponsored by the German Federation of Industrial Research Associations (AiF), historical gold tessera production was recreated. Using weather-resistant glass and innovative technology, the goal of the project was to reproduce gold tesserae of the shape and colourful variety equivalent to the historical archetypes. The resulting gold tesserae were to be used not only for restoration purposes, but also as design elements in modern contexts. Based on this research project, it was intended that the versatility of the scanning electron microscope (SEM) as an analytical tool should be demonstrated.

Gold tesserae consist of a triple-layer structure, wherein the gold foil is sandwiched between two protective glass plates. More detailed information about historic gold tessera production can be found in [1, 2]. The biggest challenge in the current process was to create a weather-resistant glass composition. The melting point of the new glasses should likewise ensure sufficient contact between the layers. During the course of the project, damage mechanisms were examined and durability tests were conducted in a climate chamber with ultraviolet (UV) radiation and noxious gases.

The main purpose of this paper is to demonstrate the versatility and application range of environmental scanning electron microscopy with energy dispersive X-ray analysis (ESEM-EDX) as an analytical tool. Using original historical gold tesserae and reproduced glasses as examples, the significance of the data is discussed. The chemical composition of numerous historical and modern gold tesserae was analysed using EDX. They formed the basis for theoretical calculations of structural characteristics. As a result of the measurements and calculations, a broad basis now exists for making specific material choices for producing gold tesserae.

Even though there are many studies (for example [3, 4]) dealing with the composition of tesserae or mosaics they are seldom related to gold leaf examples. In this study, historical and modern gold tesserae were characterised in order to obtain information about their composition. In addition to structure and glass composition, special attention was paid to the quality of the interface surface bond between the glass support layer and the gold, and the protective glass layer and the gold. The study by Darque-Ceretti et al. [5] has already dealt with the bonding structures of these interfaces. The focus of their study was to compare different methods to characterise the interfaces. In our project, ESEM images of cross-sections were to provide information about the bond of the interface surfaces, air bubbles at the interface surfaces or other inclusions, as well as the profile of the gold foil.

ESEM surface images were inspected for the analysis of individual damage types, since the surface properties of the damaged protective glass layers were to be examined more closely for possible cracks and pits. The results of these analyses will be used to further investigate deterioration processes. Calculations of relevant physical characteristics, on the basis of

ESEM-EDX analyses of the examined gold tessera glasses, will contribute to the search for suitable commercially available glasses for use in reproductions.

Materials and method

Material description

At the start of the project a database was set up. The gold tesserae included in the database are unique historical specimens, taken from their original environment during restoration work, or modern samples from sample patterns. The gold tesserae originate from Germany, Austria, France and Italy. So far information on over 110 gold tesserae from approximately 1880 to the present has been entered into the database. Data captured include origin, age, appearance (including colour of the glasses and tessera size), metal coating used (silver, copper and gold) and the description of damage types as well as photographic documentation of the samples. The database is now capable of classifying a gold tessera of unknown origin and has already enabled a temporal classification as Torge demonstrated [6].

Nearly two-thirds of gold tesserae in the database were examined using ESEM-EDX. The selection of the gold tesserae analysed was made according to temporal classification, origin of the object, manufacturer and production location/country, as well as condition. Historical and modern gold tesserae were analysed, characterising each batch of samples representatively on the basis of at least one sample.

Representative gold tesserae samples from well-known companies such as Puhl & Wagner from Berlin, Germany, or Antonio Salviati from Murano, Italy, which had been productive since the 1880s, were analysed together with samples from prominent buildings such as the Marien Cathedral, Hamburg, and the Aachen Cathedral from 1880. Other samples were also analysed from art deco or neo-classical buildings from the 1900s. Some of the gold tesserae analysed were produced in Germany in the 1930s and 40s, others originate from different bequests and contain samples from various European countries from the 1950s and 60s. The gold tesserae marked as sales samples are unused therefore it is not necessary to take into account weathering processes when performing an analysis.

ESEM-EDX

ESEM-EDX permitted the dimensions and chemical composition of the supporting and protective glass matrix and metal foils of selected gold tesserae to be determined. Apart from a few exceptions, ESEM-EDX analyses were performed in the high vacuum mode of a FEI-XL-30 ESEM on carbon-coated polished sections. This required removing a few millimetres of the sample cross-section by polishing, thus the method does not qualify as non-destructive. In addition, it is important that the samples from the area to be analysed are homogeneous, flat and have been polished so that quantitative analyses can be performed. The EDX spectrometer used was equipped with an EDAX Si(Li) detector.

Accuracy and precision of the standardless quantitative analysis method was determined by comparing the results of multiple analyses on a glass sample of known composition (CRM 4006, Czech Metrology Institute). Relative expanded uncertainties were calculated using a coverage factor of $k = 2$. They yielded values of ± 2% for 70 wt% (SiO_2 concentration), ± 5% for 13 wt% (Na_2O concentration) and ± 30% for 2 wt% (Al_2O_3 concentration) for the standard glass sample.

The method's limit of detection is around 0.1 wt%. A quantitative determination of element contents below 0.1 wt% is problematic since measurement uncertainties are very high at this concentration range. The analysis of colour-bearing metallic oxides that are normally available in very low concentrations is therefore in some cases, not, or only qualitatively, possible.

The composition of the metal foils was only qualitatively determined. The lateral spatial resolution of the EDX system, the thinness of the metal foil (only 100–300 nm) as well as the fact that the foil is sandwiched between two glass layers makes quantitative determination of the composition impossible using ESEM-EDX.

The surface of the gold tesserae was tested for possible damage to the protective glass. Carbon coating was not desirable in this case so the tests were performed in the low vacuum mode of ESEM. In this mode, a water vapour pressure of about 50 Pa was produced in the sample chamber, which prevents the non-conductive sample surface from charging.

Calculation of glass characteristics

Glass characteristics were calculated using the ScienceGlass program (Version 6.5, Build 6.50.08.086, ITC, Inc., 1998–2000) to estimate relevant glass properties. The calculations were based on the composition of the glasses determined by the EDX analyses. Because of their relevance to the topic, the glass transition temperature T_g (the temperature at which a glass changes to an amorphous or glassy solid), the Littleton point (when a glass deforms visibly), the linear coefficient of thermal expansion α (which describes how much a glass will expand for each degree of temperature increase) and the temperature-dependent viscosity behaviour between room temperature and molten material were all considered. For the sake of reproducibility only characteristic values based on the ScienceGlass program Priven 2000 model were consulted. Dilatometric measurements were conducted to confirm the results of newly melted glasses, which were calculated with the ScienceGlass program. The purpose of dilatometric measurements is to determine the linear thermal expansion coefficient and its temperature dependence. It can be regarded as a combined measurement of length (extension) and temperature.

Climate tests

Sixteen historic gold tesserae were weathered over a period of six weeks in a VSK 300 type climate chamber (Heraeus Company). This climate chamber enables an exact dosage of the noxious gases. The climate programme comprised temperature-humidity

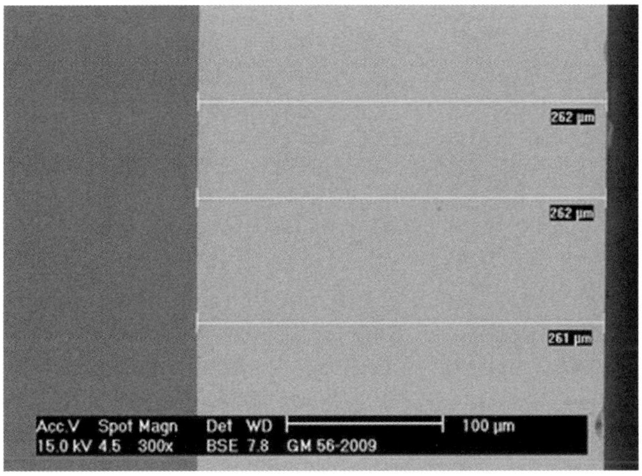

Figure 1 Typical structure of historical and modern gold tesserae (e.g. modern gold tessera from Murano, Italy, GM56), ESEM image of the surface supportive glass layer (gold layer), protective glass layer interfaces and dimensions of the protective layer.

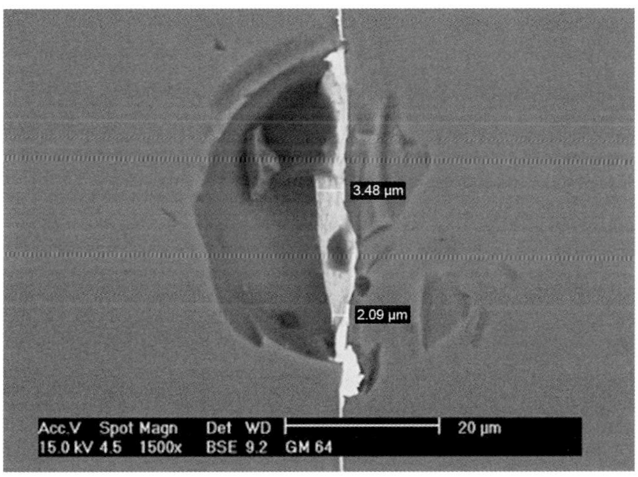

Figure 2 ESEM image of the supportive glass/gold and gold/protective glass interfaces. Note the individual hollow spaces that are clearly visible at the interface. AEG Beamtentor ('Employees' Gate', GM64), Berlin, 1896/1897.

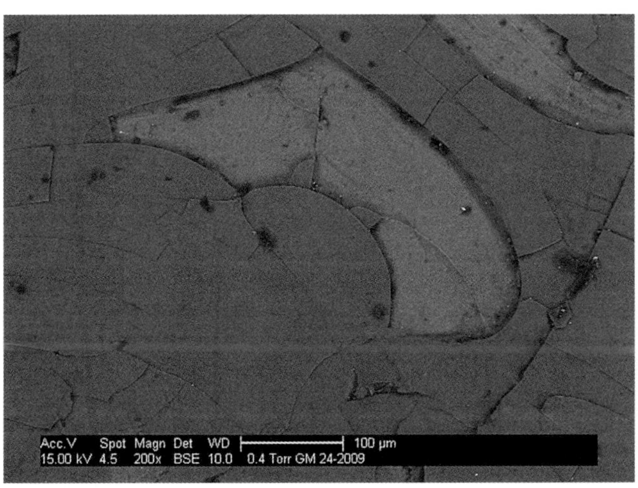

Figure 3 Damage types of a gold tessera with a crazed surface (GM24, Brahmshaus Hamburg, c. 1920): cracks and some flaking of the gel layer are visible. ESEM image at ×200 magnification.

cycles of 40°C with 50% relative humidity (RH) and 10°C with 80% RH, changing every six hours. 1 ppm sulphur dioxide (SO_2) was dosed in the first to third week, followed by 2 ppm in the fourth week and 4 ppm in the fifth week. In the final week the noxious gas concentration was 10 ppm.

Replicas

Due to the good condition of some historic gold tesserae and in order to reproduce glass similar to the traditional gold tesserae, selections of historical and modern glasses from gold tesserae were newly melted. The values obtained by EDX analysis were consulted prior to selection. The aim was to use the newly melted glasses as protective glass and combine them with commercially available resistant supporting glass. The selection of glasses to be newly melted was based on their compatibility to commercially available glass in terms of their temperature-dependent viscosity behaviour and their coefficients of thermal expansion. Another requirement was that the fusion point of supporting and protective glass should be as low as possible, i.e. under 700°C.

Results

Structure and composition of the glasses

Examination of the structure of gold tesserae with a given measured layer thickness showed that the structure remained the same throughout the investigation period from 1880 to the present – a period of 130 years. This means that, despite more modern kiln technology, the manufacturing conditions remained the same. The thickness of the protective glass layer varies from 79 to 580 µm with an average thickness of 270 µm, an example of which can be seen in Figure 1. The metal foil thickness is in the range of 140 to 500 nm with an average of approximately 270 nm.

Most of the metal foils consist of gold or gold alloys with small proportions of silver and/or copper. Nine of the metal foils examined consist essentially of silver. The technique of the application of different coloured metal foils in the production of gold tesserae has been known in Germany since the middle of the nineteenth century as reported in [2]. The thickness of the bulk glass varies between 2.5 and 5 mm.

Of the 72 gold tesserae analysed, the glass in 46 gold tesserae contains lead. This is also documented by Torge [6] who analysed tessera samples from Aachen Cathedral, classified as dating from between 1880 and 1913. The maximum lead oxide (PbO) content of the samples from the current project is 46 wt%. In about 70% of the gold tesserae tested, the composition of the supportive glass layer and protective glass layer was the same. In about 30% of samples the composition differed so significantly that, even accounting for measurement uncertainties, recognisable variations in glass composition could be seen. A thorough examination of the composition of these glasses indicated that elevated lead content has been used to reduce the melting point as shown in Table 1. Generally, protective glasses have a higher lead content, as reported by Raedel et al. [7], as do the historical gold tesserae produced by Puhl & Wagner around 1900 (supportive glass: about 5 wt% PbO, protective glass: about 15 wt% PbO) and modern tesserae of Italian origin (supportive glass: about 10 wt% PbO, protective glass: about 14 wt% PbO). The use of PbO to improve fusion temperature effects was described not only by Croveri et al. [8] for tesserae from the third to fourth century AD but is also known from various publications dealing with glass in general (for example [9, 10]).

Table 1 Selected examples of historical and modern gold tesserae with identical and different compositions. Analysis values are calculated from EDX analyses.

Origin, year (approx.)	Metal foil thickness	Material	Glass	Cover glass thickness	SiO_2	Na_2O	K_2O	CaO	MgO	Al_2O_3	Fe_2O_3	MnO	PbO	BaO
AEG Berlin, 1890 Puhl&Wagner	218	Au	Cover	480	58.3	6.4	5.6	1.6	2.8	0.3	–	–	22.8	1.8
			Bulk		56.9	11.4	1.8	10.7	0.4	0.8	–	–	15.5	2.5
Brahmshaus Hamburg 1920	400	Au+Ag	Cover	150	58.5	20.6	–	3.3	–	0.7	1.7	2.0	–	13.1
			Bulk		58.5	20.9	–	3.4	–	0.7	1.7	1.8	–	13.2
Mariendom Hamburg 1880	300	Au	Cover	350	68.6	14.4	0.3	2.1	1.1	1.0	–	–	4.0	8.6
			Bulk		66.0	13.7	0.5	1.5	0.8	4.0	–	–	4.2	9.3
Murano/ Italy 2000	250	Au	Cover	260	53.3	1.0	15.4	–	–	–	2.9	–	27.5	
			Bulk		64.6	13.2	2.3	5.8	–	0.9	1.6	–	8.8	
AEG Berlin, 1890 Puhl&Wagner	200	Au	Cover	450	68.0	17.0	5.4	6.5	0.5	1.2	–	–	1.4	–
			Bulk		67.6	18.1	1.7	2.2	–	0.8	–	–	9.6	–
Mittinger, Germany 1960	300	Au	Cover	175	63.3	15.0	4.0	10.0	0.9	0.5	–	–	6.2	–
			Bulk		62.4	15.6	4.3	8.8	1.8	0.5	–	–	6.6	–
Orsoni/ Italy 2000	200	Au	Cover	500	58.2	10.1	7.0	1.4	0.2	0.3	–	–	22.8	–
			Bulk		58.6	17.9	2.6	2.2	0.4	1.0	–	–	17.3	–
Baumhauer, legacy 1950	250	Ag+Au	Cover	380	62.4	19.7	–	3.9	0.1	0.5	–	–	13.4	–
			Bulk		55.6	19.3	–	3.0	0.2	0.7	–	–	21.2	–
France 1900	170	Au	Cover	230	71.5	20.2	0.6	5.6	–	2.0	–	–	–	–
			Bulk		69.0	19.3	0.4	5.5	–	2.0	–	–	3.8	–
Schaufelberger, legacy 1960	400	Ag	Cover	130	61.7	14.3	2.9	4.7	2.6	4.2	1.2	0.8	5.0	2.6
			Bulk		65.1	12.5	2.6	4.3	2.5	3.9	2.3	1.8	4.3	0.8

Manganese and ferrous oxides have been used as colorants in supportive and protective glasses both in historical and modern gold tesserae. Depending on the overall composition and the melting conditions (melting temperature, oven atmosphere, refining and reducing agents), an equilibrium develops between the various oxidation stages of the relevant transition metals that finally determines the colour, which is described in detail in [10] for example. Modern Italian gold tesserae additionally sometimes contain considerable volumes of copper oxides (about 1 wt%) – exclusively in the supporting glass. This is most likely the source of the bright blue and turquoise colours in the total tessera. The warm and nuance-rich colours of the historical tesserae, contrary to the modern tesserae, are probably based on the less pure source materials of that time.

Damage types

While examining potential damages with the ESEM the focus was to look at the interfaces: supportive glass/gold and gold/protective glass. For this, several gold tesserae were selected and analysed. In the polished section several hollow spaces can be detected at the supportive glass/gold and gold/protective glass interfaces. However, these hollow spaces were observed at the interfaces in several gold tesserae without having led to tangible damage types as seen in Figure 2. The metal foils of some gold tesserae show fractures or a clumping of the foil, neither of which has led to detectable damage or visual disturbance. But ESEM-EDX has limitations when examining the interfaces closely, as already reported in [11], concerning possibilities and difficulties in the evaluation of glass/gold interfaces. Specifically, the limited resolution of the EDX does not allow the observation of the migration process of the glass through the metal film.

The high levels of PbO present in some samples does not lead to damage types such as increased corrosion, as the samples tested in the climate chamber show. Other gold tesserae show a considerable amount of barium oxide (BaO), up to 16.4 wt%. Barium, like lead, is used to reduce the melting temperature of the glass and enables a better flow of the glass and bubble-free production.

The compositional characteristics more relevant to the corrosion process are high alkali contents in combination with low contents of calcium oxide (CaO). Some of the analysed gold tesserae with this composition exhibit damage typical of a crazed surface as shown in Figure 3. With the formation of cracks, a horizontal gel layer is produced and the leaching zones cover the entire protective glass. The glass network is destroyed and the horizontal gel layer chips off in slabs. Müller [12] describes these cracks on nineteenth-century glasses with high alkali content (≥ 20 wt%) and low CaO content (≤ 5 wt%). High alkali content promotes leaching, which leads to the formation of cracks in the gel layer.

The samples from the Brahmshaus, Hamburg, produced in 1920 (shown in Figure 3) have pronounced netlike crack formations with glass losses on the surface of the protective glass. The alkali present in the layers near the surface has leached out. These samples are examples of those with an alkali-rich composition.

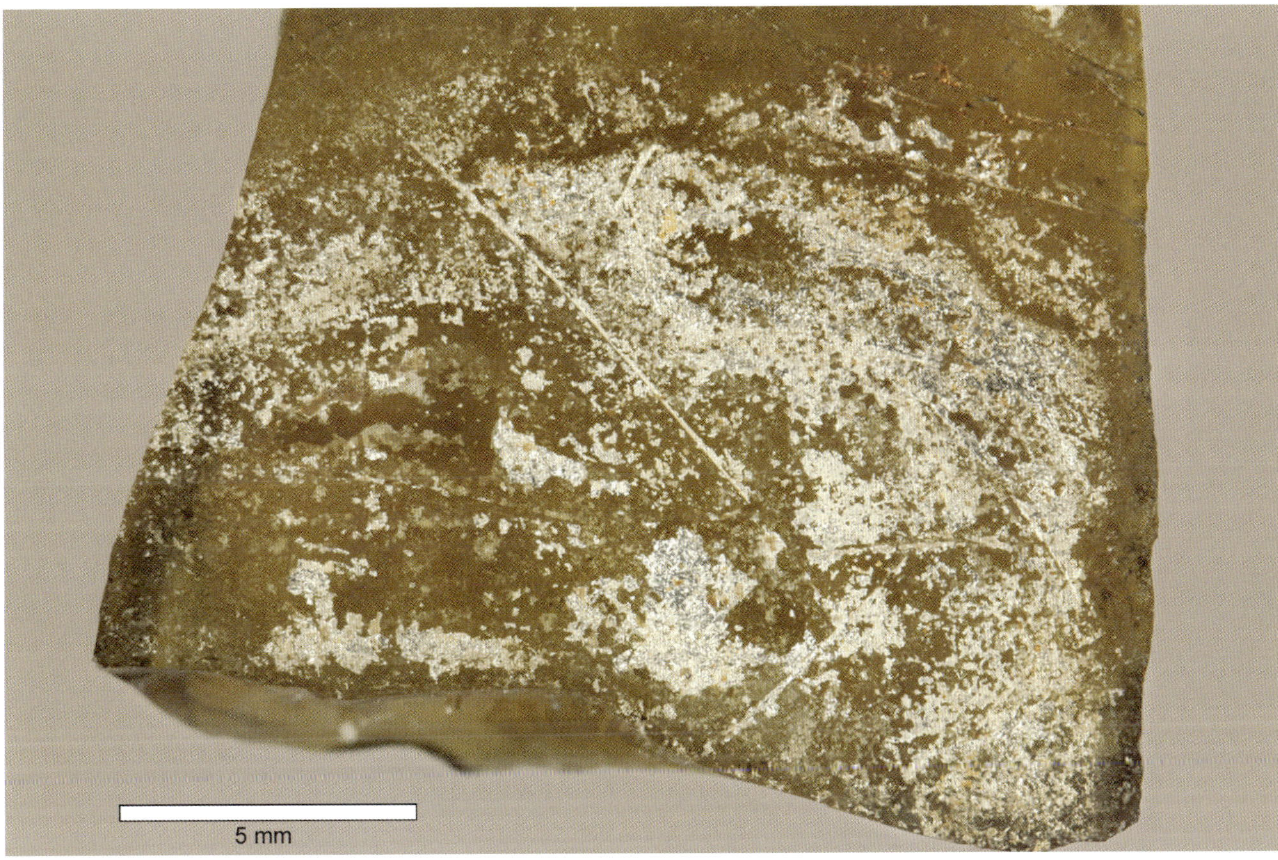

Figure 4 Corrosion products on the surface of a gold tessera after accelerated ageing (GM27, Brahmshaus Hamburg, c. 1920).

Weathering in the climate chamber

Out of the 16 weathered historical samples, three samples from the Brahmshaus in Hamburg show significant damage after weathering. The samples from around 1920 are very similar with regard to size, visual appearance and chemical composition. They originate from the same area of restoration and probably also from the same manufacturer. The supportive glass and the protective glass layers exhibit heavy corrosion deposits; the protective glass layer with the gold foil has separated from the supportive glass layer, and whitish corrosion deposits coat the entire glass body as shown in Figure 4. The damage was probably catalysed by the increased alkali content and low CaO content of these samples. The connection between glass composition and glass corrosion is discussed in several publications (for example [9, 13, 14]) and will not be discussed in further detail here.

Effects of lead and alkali contents on the viscosity of the glass

Calculating relevant glass characteristics using the ScienceGlass application, the effect of lead and alkali contents on the T_g and on the coefficient of thermal expansion α became obvious. As already known from other publications concerning historic gold tesserae (for example [15]), lead was added to glass melts to lower the fusion temperature. The elevated levels of lead are found predominantly in the protective glass layers and were probably added to achieve better fusion between these layers and the glass support.

In gold tesserae with supportive and protective glasses of the same composition, fusion is achieved without difficulty. However, in gold tesserae with varying composition, a fusion that lasts for decades can also be produced. For a more accurate examination of these samples, graphs were produced showing the viscosity-temperature gradient. They indicate that glass compositions were selected specifically in order to produce a calibrated viscosity temperature behaviour that ensures a permanent fusion of the two glasses.

Figure 5 shows two striking examples illustrating the fusibility of glasses. The examples GM18 (GM = gold mosaic/tessera) and GM99 in this figure clearly indicate that increased lead content lowers the fusing temperatures considerably. GM18 shows good fusion at approximately 730°C (log η = 6), while the lead-free GM99 tessera requires a higher temperature of approximately 825°C. While the supportive and protective glass layers of the GM99 tessera are almost identical in all characteristics, guaranteeing fusion without any problems, the linear coefficients of thermal expansion of the two glass components in the GM18 tessera differ considerably, despite their similar temperature-dependent viscosity behaviour. Differences in expansion coefficients of about 2×10^{-6} K^{-1} as in this case, would normally lead to stress cracks in the production process – however, no cracks were detected. It is therefore presumed that the protective glass had deteriorated in the environment in which it was displayed and that the alkali loss on the glass surface caused the summary degradation of the coefficient of expansion. This assumption is supported by the original location of the historical gold tessera – it came from the 1892/93 exterior mosaic image 'Muse of the Arts' from Puhl & Wagner in Berlin. This example shows how the results of analysis can also help to detect less noticeable leaching processes.

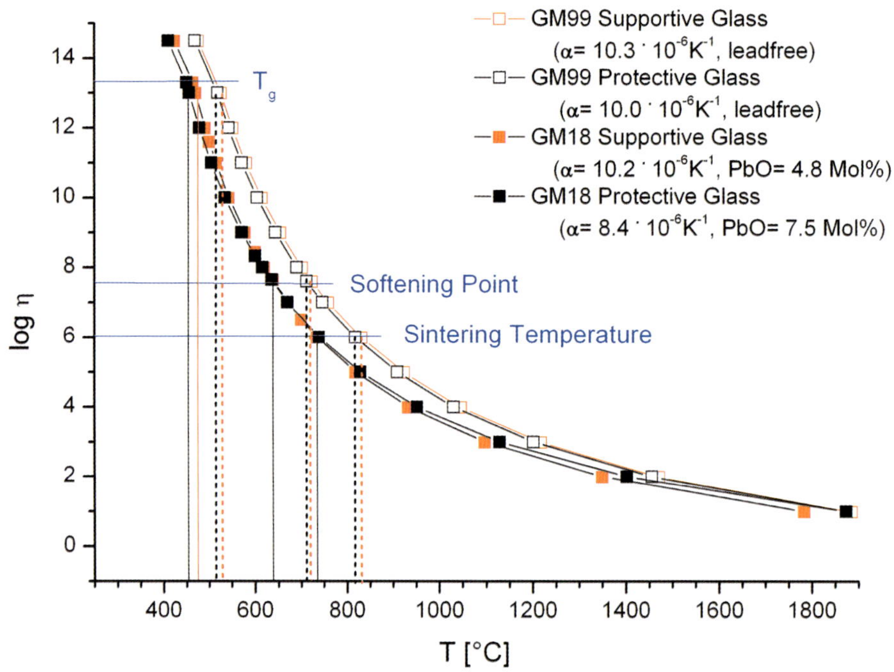

Figure 5 Temperature-dependent viscosity behaviour of the glass components of a historical (GM18, 1892, Puhl & Wagner) and a modern gold tessera (GM99, 2000, Murano, Italy), calculated from the composition using EDX analysis and the Priven 2000 model in ScienceGlass.

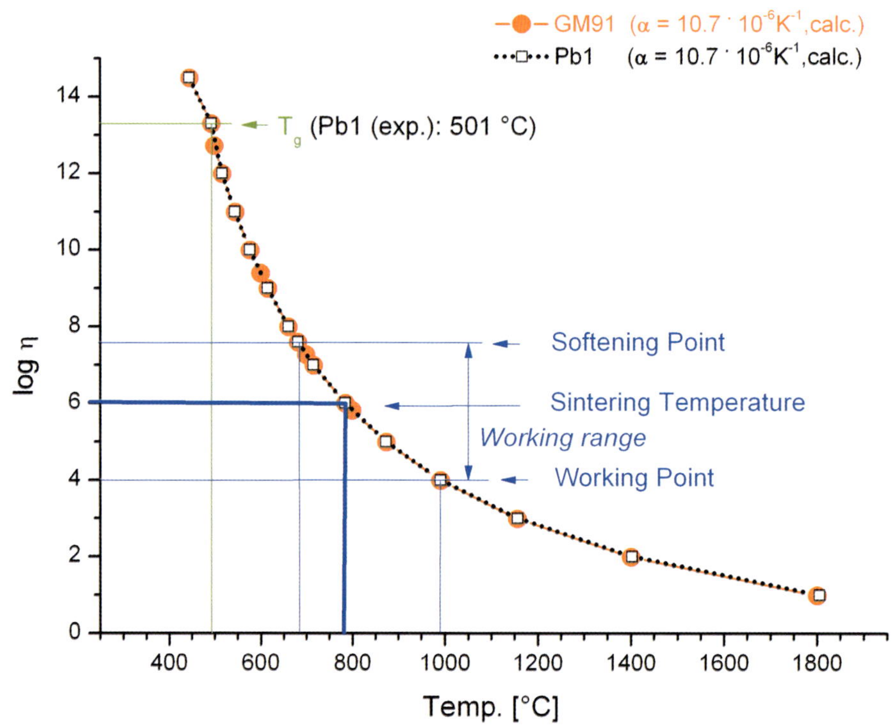

Figure 6 Temperature-dependent viscosity behaviour of supportive glass of modern gold tessera GM91 and replica glass Pb1, both: α ($T<T_g$) = 10.7 × 10^{-6} K^{-1}, calculated by Priven 2000 based on EDX analyses of the original GM91 and the replica glass; dilatometric measurement on replica glass Pb1 shows similar results: α (20–300°C) = 10.9 × 10^{-6} K^{-1}.

Replicas from newly melted glass using the analysis results

The data obtained by ESEM-EDX analysis served as a basis for the replicas. The glasses selected to produce replicas were newly melted based on the EDX analyses of the original gold tesserae recorded in the database and used as supportive and protective glass. The composition of replica glasses was supposed to be similar to those of the supportive and protective glasses and have low fusion temperatures to prevent the gold foils from being destroyed.

To be able to combine commercially available glasses as supportive glass with replicas of protective glass, low Littleton points were needed for protective glasses to enable fusion at as low a temperature as possible. Also, the glasses were expected to have a similar T_g and temperature-dependent viscosity behaviour, as shown in the example in Figure 6. Furthermore, the newly melted glasses had to be lead free or only contain small amounts of PbO to guarantee as high a chemical resistance as possible.

Dilatometric measurements were conducted on newly melted glasses calculated with the ScienceGlass program in order to reassess the data calculated by analyses. The dilatometric measurements were performed on newly melted glass to determine the T_g and coefficients of thermal expansion. These correspond to the values of the data of the original glasses calculated using Priven, despite the analysis data being affected by measurement uncertainties. Also, the curves for the temperature-dependent viscosity behaviour of the two glasses are nearly identical, as Figure 6 indicates. Thus data obtained by the ESEM-EDX analysis enabled replicas that were almost identical to the original glasses and provided a basis for further tests.

The investigation of the replicas from the newly melted glass gave promising results during the course of the subsequent experiments. The protective glasses could be combined with individual commercially available supportive glasses. However, temperatures around 700°C were necessary because of the characteristics of commercially available supportive glasses, which, as various tests indicated, impaired the golden effect of the gold leaf layer. Similar observations were described in a patent from 1997 [16]. It is probable that thinner protective glasses might reduce fusion temperature, thus achieving the optimum temperature for gold leaf of 650–680°C.

Conclusions

ESEM-EDX is a helpful tool for the analysis of glass in general and especially for gold tessera replicas. However, the limitations of this method are also evident. The versatile options of surface imaging as well as possibilities for elemental analysis in micro dimensions provide important information that aids in the understanding of the manufacturing processes and in characterising the mechanisms of damage. One major disadvantage of elemental analysis by EDX is the relatively high limits of detection of about 0.1 wt%. Alternative techniques such as inductively coupled plasma mass spectrometry (ICP-MS), which offer lower detection limits, were not used because of the comparatively high level of effort that would have been necessary for sample preparation.

Based on analytical determination of the composition of historical gold tesserae using EDX tests, it was possible to create our own model melts that enabled us to produce replicas of single tessera. These compositions were used to calculate glass-chemical characteristics that provided the basis for further materials selection for additional test series. The calculated glass-chemical characteristics of historical glasses agreed well with the experimentally determined characteristics of model glasses.

It was, however, not sufficient to rely on theoretically obtained parameters alone when selecting the glass composition. Even if the glass composition is only modified to a minor extent, empirical observations based on numerous fusion tests must be consulted in order to obtain an optimum glass combination for the tesserae as well as an optimum temperature–time programme.

The database established so far provides an overview of about 110 recorded gold tesserae from the past 130 years and from various European workshops, and facilitates the categorisation of pieces of unknown origin.

References

1. Bustacchini, G., 'Gold in mosaic art and technique', *Gold Bulletin* **6**(2) (1973) 52–56.
2. Helary, D., *Study of Gilded Layers on Vitreous Materials: Application to Gold Foil Tesserae and Gold-like Lustre Decoration on Glazed Cceramic*, PhD dissertation Ecole Mines ParisTech (2003).
3. Ricciardi, P., Colomban, P., Tournie, A., Macchiarola, M. and Ayed, N., 'A non-invasive study of Roman age mosaic glass tesserae by means of Raman spectroscopy', *Journal of Archaeological Science* **36**(11) (2009): 2551–2559.
4. Verità, M., Arena, M.S., Carruba, A.M. and Santopadre, P., 'Roman glass: art and technology in a 4th century A.D. *Opus Sectile* in Ostia (Rome)', *Journal of Cultural Heritage* **9**(Suppl. 1) (2008): 16–20.
5. Darque-Ceretti, E., Hélary D. and Aucouturier, M., 'An investigation of gold/ceramic and gold/glass interfaces', *Gold Bulletin* **35**(4) (2002): 118–129.
6. Torge, M., 'Analytische Untersuchung zur Provenienzbestimmung von Goldmosaiken im Aachener Dom', *Archäometrie und Denkmalpflege Metalla* **2** (2009) 131–133.
7. Raedel, M., Bücker, M. and Unger, B., 'Historische und moderne Goldmosaike im Vergleich', *Archäometrie und Denkmalpflege, Metalla* **3** (2010) 184–186.
8. Croveri, P., Fragala, I. and Ciliberto, E., 'Analysis of glass tesserae from the mosaics of the Villa del Casale near Piazza Armerina (Enna, Italy): chemical composition, state of preservation and production technology', *Applied Physics A: Materials Science & Processing* **100**(3) (2010) 927–935.
9. Davison, S., *Conservation and Restoration of Glass*, 2nd edn, Butterworth-Heinemann, London (2003).
10. Wedepohl, K.H., *Glas in Antike und Mittelalter: Geschichte eines Werkstoffs*, Schweizerbart, Stuttgart (2003).
11. Darque-Ceretti, E., Deram, V. and Aucouturier, M., 'Organometallic based gold decoration, formation and durability', *Surface Engineering* **24**(2) (2008) 103–109.
12. Müller, W., 'Sicherung, Konservierung und Restaurierung historischer Glasmalereien', *Forschungsbericht 217*, BAM Bundesanstalt für Materialforschung und -prüfung, Berlin. (1997).
13. Koob, S.P., *Conservation and Care of Glass Objects*, Archetype Publications, London (2006).
14. Roemich, H., 'Historic glass and its interaction with the environment', in *Conservation of Glass and Ceramics*, ed. N.H. Tennent, James & James, London (1999) 5–15.
15. Gueit, E., Darque-Ceretti, E. and Aucouturier, M., 'Glass gilding process in medieval Syria and Egypt (13th–14th century)', *Journal of Archaeological Science* **37**(7) (2010) 1742–1752.
16. Patent DE19729772B4, Bon. (describes a solution/ patent for the negative interaction between glasses with high fusing temperatures and the gold foil) (1997).

Authors' addresses

- Martina Raedel, BAM Federal Institute of Materials Research and Testing, Berlin, Germany (martina.raedel@bam.de)
- Martin Sabel, BAM Federal Institute of Materials Research and Testing, Berlin, Germany (martin.sabel@bam.de)
- Michael Bücker, BAM Federal Institute of Materials Research and Testing, Berlin, Germany (michael.buecker@bam.de)
- Brita Unger, BAM Federal Institute of Materials Research and Testing, Berlin, Germany (brita.unger@bam.de)

Quantitative energy dispersive X-ray analysis of the blue pigment smalt in the variable pressure scanning electron microscope

Marika Spring, Veronika Kugler and Stewart Bean

ABSTRACT Smalt, a blue pigment composed of cobalt-containing glass, was widely used between the sixteenth and the eighteenth century. Few quantitative analyses have been published, but they shed light on the character of the cobalt ore used to make the pigment, as well as the manufacturing process and degree of degradation. Experiments carried out to establish a reliable methodology for quantitative analysis in the variable pressure scanning electron microscope (VP-SEM) are described first. This is followed by a discussion of the results from a number of paintings, including trends in cobalt and arsenic content, and examples where quantitative analysis has enabled conclusions to be drawn about the possible original intensity of colour of degraded smalt.

KEYWORDS smalt, quantitative analysis, paintings, SEM-EDX, cobalt

Introduction

Smalt, a blue pigment composed of cobalt-containing glass, was widely used in all types of painting, and was most common between the sixteenth and the eighteenth century, although a few earlier occurrences have been reported, as well as some in works from as late as the early nineteenth century [1–5]. It was manufactured from a source of silica such as pebbles or sand, using potash as the flux, and 'zaffre' (roasted cobalt ore mixed with varying amounts of quartz or sand) to give the blue colour. It is not, however, a very stable pigment and smalt-containing paint can often deteriorate dramatically, becoming greyish or brown [6, 7].

Relatively few quantitative analyses of smalt have been published [6, 8–12], even though they shed light on the character of the cobalt ore used to make the pigment, the manufacturing process and the degree of degradation. This motivated the investigation of the composition of smalt in a large number of paint samples in the archives of the Scientific Department of the National Gallery in London (around 40). The paintings, from all over Europe, range in date from the early sixteenth century to the early nineteenth century, making it possible to consider any trends that might exist in the variation of composition over time or with geographical origin.

Quantitative analysis of smalt in paint samples using scanning electron microscopy with energy dispersive X-ray analysis (SEM-EDX) can be difficult, as the particles are generally only 10–50 μm in size and are usually mixed with other pigments. It was therefore important to establish a reliable methodology before embarking on this study. This work was part of a broader study of smalt degradation, so variable pressure (VP) conditions were chosen to avoid the need to carbon coat the samples, allowing subsequent analyses using other techniques. This constituted even more of a challenge, as scattering of the electron beam by the chamber gas needs to be taken into account. Experiments were therefore carried out, in collaboration with Carl Zeiss NTS, to determine the influence of various analytical parameters on the accuracy of the results. The first part of this paper describes this preliminary work, which is relevant not only for the analysis of smalt but also more generally when conducting EDX analysis under these conditions.

A full interpretation and description of the quantitative results from all the samples examined is beyond the scope of this short paper. Instead, the second part reports in detail a few representative examples that illustrate some initial observations on cobalt and arsenic content that emerged, based nonetheless on a preliminary consideration of the full set of results.

EDX analysis in the VP-SEM

Under high vacuum (HV) conditions in the SEM insulating samples such as paint cross-sections have to be carbon coated to make them conductive. Under VP conditions, a gas is introduced into the chamber so that there is charge compensation and the samples can be examined uncoated, in a more natural state. However, scattering by the chamber gas causes spreading of the primary electron beam, known as 'skirting', resulting in the generation of X-ray signals from areas of the specimen away from the incident beam. At higher pressures this can be a considerable distance, of the order of tens or even hundreds of microns [13, 14]. Elements from the pigments mixed with smalt can consequently appear in spot spectra of individual particles, affecting the results of quantitative analysis.

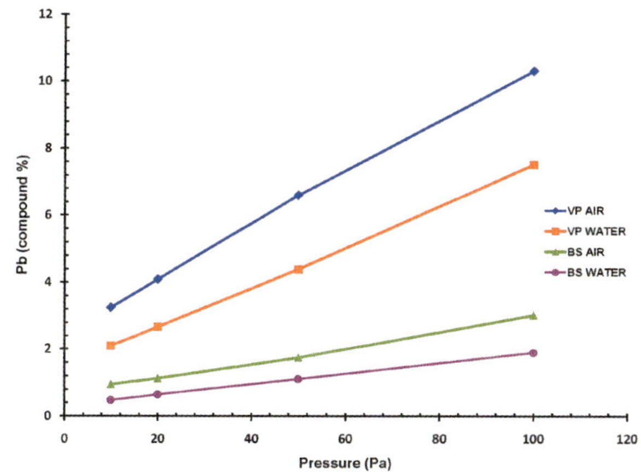

Figure 1 Lead concentration (weight % PbO) from the lead white matrix detected in EDX analysis of a smalt particle in the reference sample plotted against chamber pressure under different analytical conditions (VP = variable pressure, BS = BeamSleeve®).

The extent of the beam skirting is influenced by various factors, such as chamber pressure, chamber gas, accelerating voltage and beam gas path length (BGPL) [10, 15–17]. Their effect was investigated using a reference paint sample of smalt pigment of particle size averaging 20 µm, mixed with lead white and bound in linseed oil, embedded in polyester resin and prepared as a cross-section. The analysis was performed using a Carl Zeiss EVO MA15 VP-SEM, with an Oxford Instruments X-Max 80 mm EDX detector. The accelerating voltage was fixed at 20 kV to ensure generation of X-rays from heavy elements present in these samples; beam-skirting effects are in fact reduced at higher voltages [13]. The beam current was set at 200 pA, producing an appropriate count rate, live time was 30s, and working distance was 8.5 mm. Migration of alkali ions in the glass was not observed under these conditions.

Spectra were collected from a single smalt particle in the reference sample at several chamber pressures over the range 10–100 Pa, usually sufficient to give charge compensation on this type of sample. The measurements were made with air and with water vapour as the chamber gas. The smalt pigment itself does not contain any lead, but scattering of the beam results in X-rays generated from the lead white matrix being present in the spectrum. The amount of lead detected can therefore be considered a measure of the degree of beam skirting. The data presented in Figure 1 show that the beam skirt effect is proportional to chamber pressure, and at a particular pressure the concentration of lead detected is lower using water vapour than with air. In general, the lower the atomic number of the chamber gas, the less the scattering that occurs [18]. In addition, as water vapour is better able to compensate for charging on the sample than air, it is possible to operate at lower pressures [19].

The third variable is the BGPL. For the standard VP mode of operation the working distance is fixed at 8.5 mm because of the geometry of the EDX detector. This gives a BGPL of 16 mm. In the Zeiss SEM it is possible to reduce the BGPL by inserting a 'BeamSleeve®', a pressure limiting aperture that is mounted on the final lens assembly. Monte Carlo simulations of the electron beam trajectories at 20 keV show that with a BGPL of 16 mm (i.e. without the BeamSleeve®) around 4% of the beam is scattered, but this is reduced to around 0.5% with a BGPL of 2 mm. Repeating the analysis on the reference sample with a

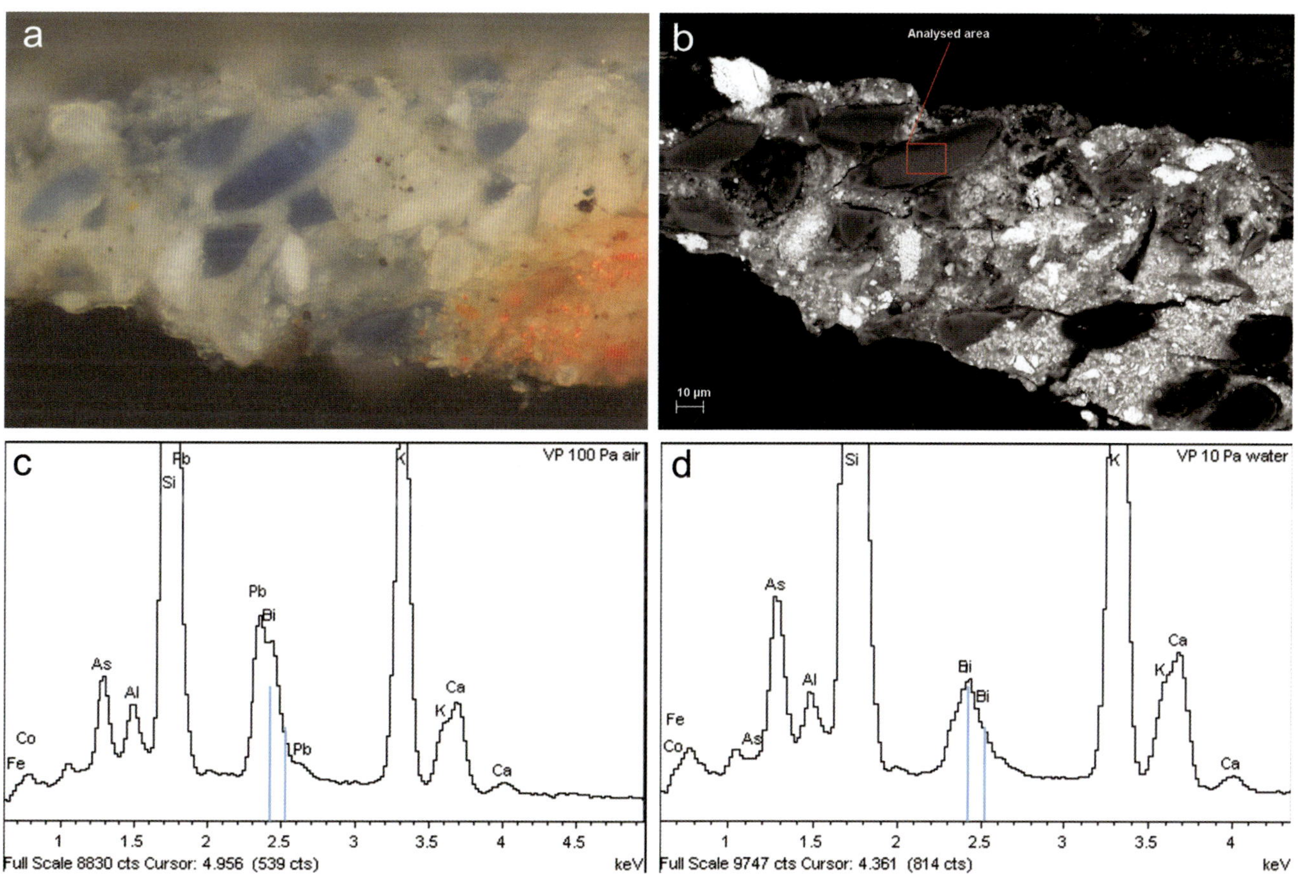

Figure 2 Paolo Veronese, *Respect* (NG 1325), about 1575. Cross-section of a sample from the sky: (a) normal light; (b) SEM-BSE image; (c) spectrum from the largest smalt particle at 100 Pa in air; (d) spectrum from the same particle at 10 Pa in water vapour (bismuth (Bi) M_{α} and M_{β} lines marked in blue).

BeamSleeve® that gave a BGPL of only 1 mm demonstrates this effect (Figure 1), since far less lead from the matrix is detected. The skirting is reduced most effectively using a BeamSleeve® with water vapour at the minimum possible pressure; 20–30 Pa was needed to prevent charging. From Figure 1 it can be seen that for this smalt particle, which is 20 μm in size, some lead is still detected at this pressure. By choosing larger particles around 30–40 μm in size it is possible to be confident that the effect of beam skirting on the analysis can be minimised.

Analysis of a sample containing smalt and lead white from the sky of Paolo Veronese's *Respect* (NG 1325) illustrates what can be gained from carefully choosing the analytical parameters. The same trends were seen when varying the conditions, evident as a decrease in the intensity of the lead peak when comparing spectra from the same smalt particle collected at 100 Pa in air and 10 Pa in water vapour. In addition, the M_α and M_β bismuth lines are close to the M_α and M_β lead lines in this region of the spectrum. In the spectrum collected at 100 Pa in air (Figure 2c), which is more greatly affected by beam skirting, the top of the peak in this region corresponds to the position of the lead M_α line and the bismuth peak is more or less obscured. In the spectrum collected at 10 Pa in water vapour, however, where the beam skirting is minimised, the peak coincides with the M_α line of bismuth, which is one of the elements associated with the cobalt ore used to make the pigment, making it easier to confirm that it is present (Figure 2d).

The beam skirting affects the spatial resolution and so is important for multiphase samples, but even on larger homogeneous samples there are issues that influence the quantitative analysis under VP conditions, including contribution of characteristic X-rays from the chamber gas, absorption of low energy X-rays from the sample by the gas (although this is only significant at higher pressures) and charging effects [10, 20]. A piece of standard Corning potash glass D several millimetres in size was analysed, so that the significance of these factors could be determined without having to consider beam skirting.

Figure 3 illustrates the effect of varying the conditions of analysis on the potassium content measured in Corning standard glass D relative to the certified value of 11.5 wt% oxide (K_2O) [21]. At higher pressures the measured value is close to the reference, but at lower pressures the difference becomes more significant due to specimen charging. The surface negative charge retards the incoming electrons so that their landing energy is less than the primary beam energy. This is more evident with the BeamSleeve® than without, probably because the shorter BGPL reduces the efficiency of charge compensation.

Charging was evident already in the backscattered electron (BSE) image, but the actual loss in the energy of the primary beam can be revealed by evaluating the Duane Hunt limit, which is the continuum energy corresponding to the energy of the incident beam as it reaches the specimen surface [10]. If this is less than the nominal beam energy then charging is occurring. An approximation can be obtained by inspecting the end region of the spectrum and estimating where the continuum counts fall to zero. Comparing the spectra from Corning standard glass D, a drop off in energy below the nominal beam energy of 20 keV can be seen in that collected at 10 Pa relative to that collected at 100 Pa, explaining the inaccuracy in the quantitative results at the lower pressure. EDX software often allows automatic calculation of the Duane Hunt limit as each spectrum is collected, with a warning given when the incident beam energy is significantly reduced; it is important to have this option enabled during quantitative analysis.

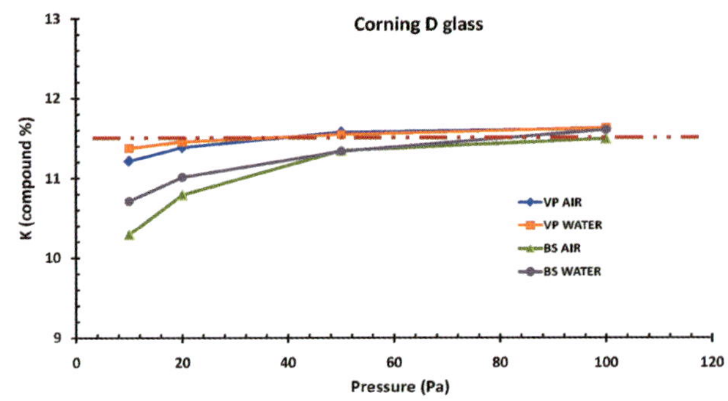

Figure 3 Potassium concentration (wt% K_2O) plotted against chamber pressure as measured on standard Corning glass D, for air and water as the chamber gas, with and without the BeamSleeve® (VP = variable pressure, BS = BeamSleeve®; dashed line represents certified value).

Considering the data for all the elements detected in the Corning standard glass D, it appears that the most accurate measurements were those made at 50 Pa in either air or water vapour. To evaluate the overall accuracy of the analysis of the reference sample using the 'standardless' calculations method in the INCA 350 software, the measurements made without the BeamSleeve® were considered, as it was only available on the SEM used for these preliminary tests and not on that which would be used for the analyses of samples from paintings. For the major elements the values are within around 5% (relative) of the certified value. This compares well with other published analyses carried out with a microbeam [22], which have pointed out that the Corning standard glasses (particularly the minor and trace elements) are not, in any case, completely homogeneous on a micron scale [21].

This level of accuracy was considered to be acceptable, taking into account other vagaries of the analyses that are inherent in this type of sample, such as variations in the results from different smalt particles within a single sample due to inhomogeneity in the composition. In addition, the thickness of the particle being analysed in a cross-section cannot be assessed, so it is possible that the interaction volume of the electron beam could extend through the particle to whatever pigments lie behind it.

Quantitative analysis of smalt pigment in paintings

Smalt in around 40 paintings ranging in date from the first decade of the sixteenth century to the first decade of the nineteenth century was analysed. The relative number of paintings in each time period is representative of the pattern of its use as a pigment, with the majority being from the seventeenth century. Paintings from a wide geographical area including Italy, Germany, England, Spain and the Netherlands were included. The preliminary assessment of the results discussed below serves primarily to indicate how quantitative analysis might be useful, as well as to identify some specific points of interest relating to the cobalt and arsenic content. These comments are based on initial consideration of the data from all

Table 1 Quantitative SEM-EDX analysis of smalt. Results from the most well-preserved smalt particle in each sample are quoted, judged on the basis of highest potassium content.

Weight % normalised	Na_2O	MgO	Al_2O_3	SiO_2	K_2O	CaO	FeO	CoO	NiO	As_2O_3	Bi_2O_3	**Ratio of atomic % Co/As**
Girolamo Romanino, *High Altarpiece, S. Alessandro, Brescia* (NG 297), about 1524. Blue of the Virgin's cloak.	11.1	1.1	0.6	67.6	4.6	3.4	2.0	3.1	0.7	4.2	1.5	**1/1**
Paolo Veronese, *The Consecration of Saint Nicholas* (NG 26), 1562. Blue of the bishop's drapery.	0.1	–	0.2	58.8	15.5	0.3	3.1	5.6	0.8	13.5	2.0	**1/1.8**
Pieter Breugel the Elder, *The Adoration of the Kings* (NG 3556), 1564. Blue of the Virgin's cloak.	0.3	0.2	0.2	59.8	15.9	3.7	6.7	7.7	0.7	3.6	1.3	**2.8/1**
Unknown artist, *Queen Elizabeth I* (NPG 2082), National Portrait Gallery, c.1575. Pattern on the silver–grey dress, now brown.	0.2	0.4	1.1	78.3	2.4	4.8	3.3	4.2	0.1	5.2	–	**1/1**
Martin de Vos (workshop), *The Crucifixion*, Finnish National Gallery, late 16th century.	0.3	0.2	0.9	59.6	13.1	1.4	2.7	8.7	3.5	4.5	5.2	**2.5/1**
Blue of the Virgin's cloak (upper row, typical particle; lower row, particle richest in Co).	0.4	0.2	0.8	52.3	14.0	2.1	2.8	14.6	3.2	6.5	3.3	**2.9/1**
Sir Anthony van Dyck, *Saint Martin Dividing his Cloak* (inv. 405878), The Royal Collection, 1618–20.	–	–	1.3	85.8	3.6	1.1	2.4	3.5	0.2	2.2	–	**2/1**
Blue of the sky, reworked area. Upper row, greyish smalt in upper layer; lower row; bluer smalt in lower layer.	–	–	1.9	70.8	10.7	2.6	1.7	3.7	2.7	5.9	–	**1/1.2**
Hendrick ter Brugghen, *The Lute Player* (NG 6347), 1624. Blue–grey ribbons on the sleeve.	0.1	0.2	1.6	68.2	15.2	1.3	3.2	3.0	0.5	5.6	1.1	**1/1.4**

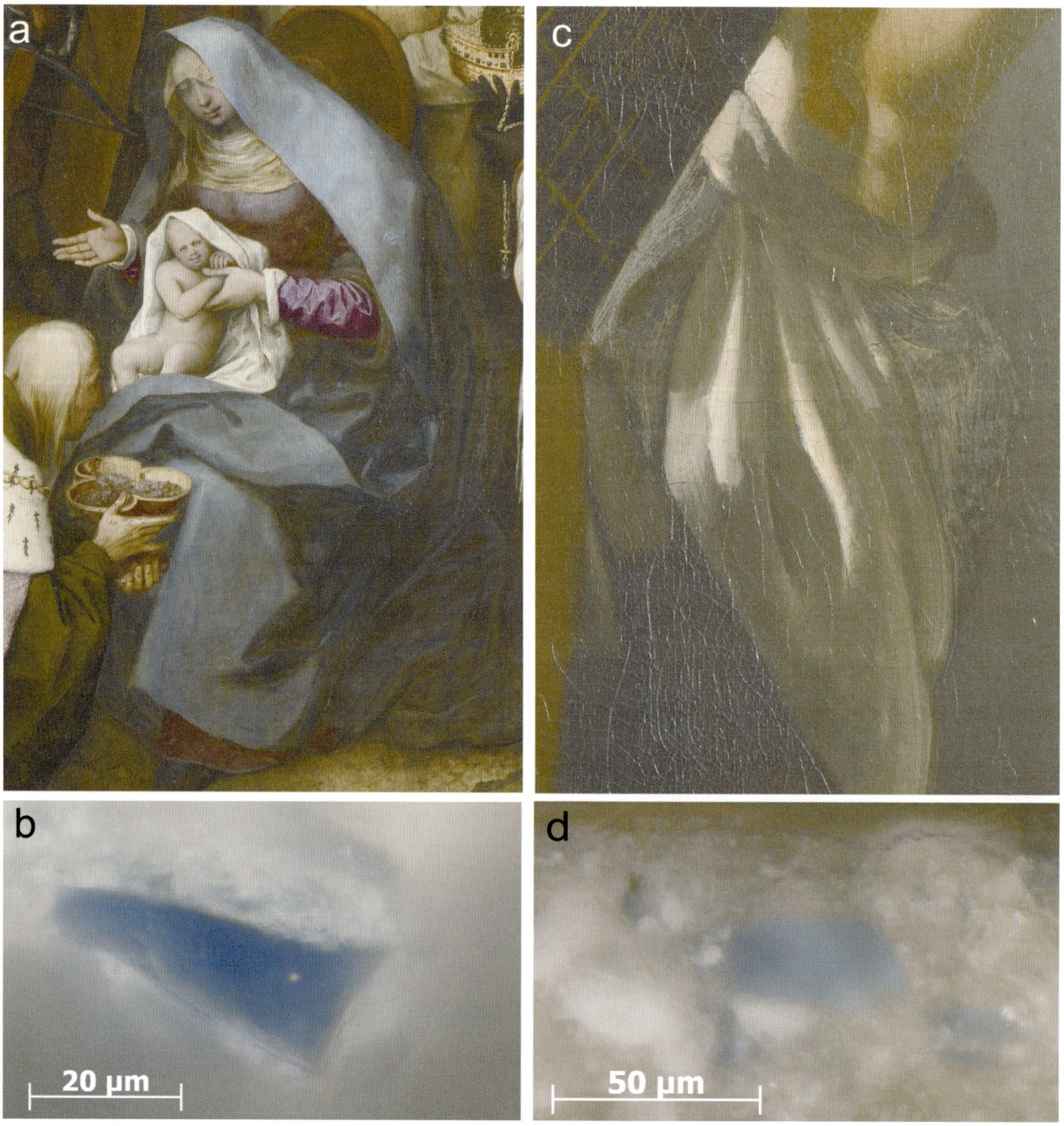

Figure 4 (a) Pieter Bruegel the Elder, *The Adoration of the Kings* (NG 3556), 1564. Detail showing the Virgin's blue cloak; (b) smalt particle in a cross-section from the Virgin's cloak shown in (a); (c) Hendrick ter Brugghen, *The Lute Player* (NG 6347), 1624. Detail showing the deteriorated smalt-containing ribbons on the sleeve; (d) cross-section of the paint of the ribbons shown in (c). Images © National Gallery, London.

the paintings analysed, but only the results from a few examples chosen to illustrate particular trends or observations are presented here (Table 1). Smalt has almost always discoloured to some extent and the loss of potassium that accompanies this change (and in extreme cases loss of cobalt) complicates interpretation of the analyses. Usually, however, one or two well-preserved particles are present in a cross-section and the results from these are sufficient indication of the original composition of the smalt.

Smalt is almost always a potash glass and differs from vessel or window glass in that it generally contains a large amount of potassium and very little other alkali or alkaline earth elements, reflected in the fact that typical recipes in historic documentary sources mention potassium-rich tartar (rather than wood, fern or marine plant ash) as the flux [2]. This can be explained by the fact that sodium-containing smalt has a less desirable duller, more purple-blue hue than that made with potassium [3, 8, 23].

It is known from historic documentary sources that smalt was sold in a number of grades. These differed in intensity of colour (and therefore in price) [24], due to variation in particle size (more finely ground smalt is paler in colour) or cobalt content [2, 3]. Smalt is far higher in cobalt content than blue vessel or window glass; in paintings it typically ranges between 3 and 7 wt% cobalt oxide (CoO). It was employed not only as an artists' pigment but for laundry blueing and tinting of paper. The lower grades of smalt were probably used for these purposes and where smalt in blue paper has been

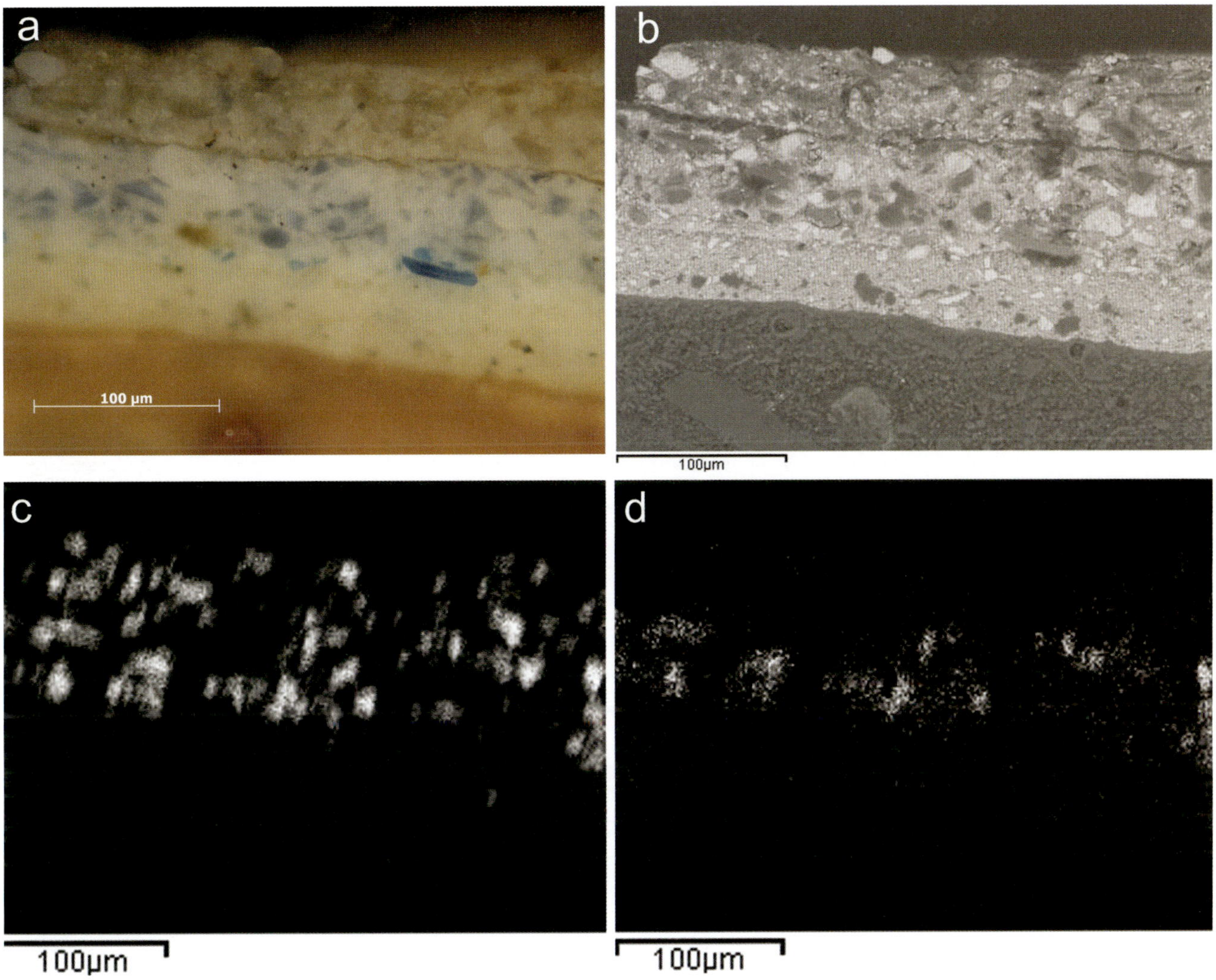

Figure 5 Sir Anthony van Dyck, *Saint Martin Dividing his Cloak*, Royal Collection. Cross-section from a reworked area of the sky showing a chalk-based ground layer, a grey priming, a thin blue layer of lead white and azurite followed by two blue layers of smalt and lead white. The uppermost layer (now rather grey) is the reworking carried out at the same time as the original addition was attached at the right edge of the painting. (a) Normal light; (b) SEM-BSE image; (c) silicon EDX map; (d) potassium EDX map.

analysed it has been found to be far lower in cobalt content, with values of around 1% being reported [3]. It seems likely that the smalt in the paintings in this study is predominantly of the higher grades.

The two paintings in which the highest levels of cobalt were detected were both from the sixteenth century; a *Crucifixion* from the workshop of Martin de Vos and Pieter Bruegel the Elder's *Adoration of the Kings* (Figures 4a and b). The smalt in both pictures is an intense blue and remarkably well preserved. In both the CoO content averaged between 8 and 9%, rising to as much as 14% in some of the particles in the painting by Martin de Vos. The smalt used for the ribbons on the sleeves of Hendrick ter Brugghen's *Lute Player* of 1624, however, contains far less CoO – only around 3% – and the effect on the intensity of the colour is evident in the weaker blue colour of the particles (Figures 4c and d). The pigment in the *Lute Player* is more deteriorated than that in the earlier paintings, but the largest and most blue particle in the sample is well preserved; it shows little evidence of the alkali leaching that is responsible for degradation as it contains 15% K_2O and can therefore be considered to be representative of the original composition of the smalt.

Looking at all 40 results, some trends in the cobalt content seem to emerge. All the examples of high cobalt smalt are in sixteenth-century paintings. In the paintings dating from after about 1600 the smalt content is rarely above 5 wt% CoO and is more often only around 3%. This might indicate that a lower proportion of zaffre was used when making the smalt; manufacturers perhaps found they could economise in this way and still achieve a marketable product. Alternatively, since several qualities of zaffre could be bought, containing different amounts of the colouring cobalt, manufacturers might have deliberately used a cheaper grade [3], or perhaps, without their knowledge, the quantity of cobalt in a particular grade might have been reduced.

This trend is interesting considering that the end of the sixteenth and beginning of the seventeenth century seems to have been a period of many changes in both the trade in cobalt ore as zaffre and the smalt-making industry. The history is complicated and patchy, depending on what documents survive or have been located in archives, but a few points can be made that may be relevant to the observations made above. Delamare states that until around 1568 (when what appears to have been the first smalt manufacturing enterprise in Saxony was set up by a Christopher Stahl) the main manufacturers of the pigment were the Dutch, although they left it to the Saxons to treat the ore and concentrated on production of the lucrative end product [3]. The sixteenth-century Paduan

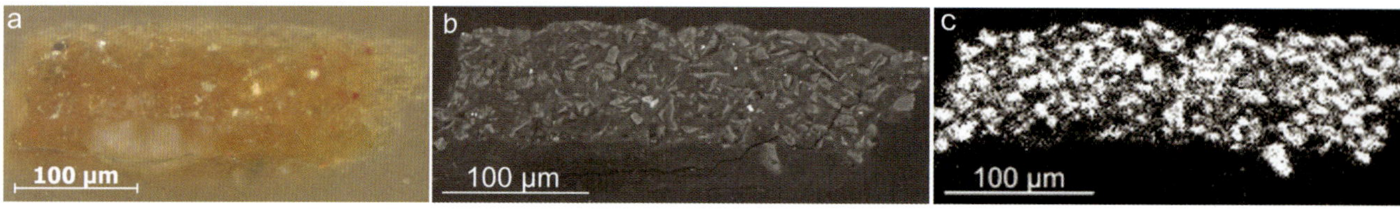

Figure 6 Unknown artist, *Queen Elizabeth I* (NPG 2082), National Portrait Gallery, c.1575: (a) cross-section from the brown lines of decoration on the silver-grey dress; (b) SEM-BSE image showing relatively small angular particles of smalt; (c) silicon EDX map locating the smalt particles.

manuscript [25] and Lomazzo's treatise of 1585 [26] both considered smalt from Flanders to be the best, and it was clearly exported all over Europe. Towards the end of the sixteenth century there was an upsurge in production in Saxony, but the Thirty Years' War (1618–1648) had an impact both on manufacturing in that region and also on export of the cobalt ore. The beginning of the seventeenth century also saw three Dutchmen being granted permission to start manufacturing smalt in England for the first time [3]. The Spanish documentary sources have recently been reviewed in relation to smalt by Véliz [27]. One of these mentions that painters complained that the quality of smalt imported from Flanders from a Master Bernard in the sixteenth century declined after he died in Germany. In addition, the relative price of smalt in Spain seems to have decreased significantly in the seventeenth century. Véliz speculates that a cheaper local supply may have become available or that the quality of the pigment was perhaps inferior to that sold in the sixteenth century, an intriguing suggestion considering the lower cobalt content observed in the seventeenth-century paintings in this study (including many Spanish works) relative to those made in the sixteenth century.

In Europe over the period that smalt was used as a pigment, skutterudite ($(Co,Fe,Ni)As_{2\text{-}3}$) and smaltite ($(Co,Fe,Ni)As_2$) appear to have been the most common and abundant ores with erythrite ($(Co,Ni)_3(AsO_4)_2{\cdot}8H_2O$) and cobaltite ($(Co,Fe)AsS$) also being used. All of them contain arsenic but with varying cobalt–arsenic ratios. The use of different ores could therefore be responsible to some extent for the variation in arsenic levels measured in the smalt in this study. Another important factor, however, was processing of the ore during preparation of zaffre, including the roasting mentioned in a number of documentary sources, which as well as converting the ores to oxides also removed the arsenic to varying degrees depending on the conditions employed [3].

Arsenic was detected in all the smalt analysed in this study, commonly at higher concentrations than cobalt. There was, however, considerable variation in the amount that was present (Table 1). The highest arsenic content, at 13–15 wt% oxide (As_2O_3), was in the smalt in Veronese's *Consecration of Saint Nicholas*. However, as the smalt in this painting is also very rich in cobalt, this simply indicates that a high proportion of ore was used in the recipe, rather than any particular characteristics of the ore or its preparation. It is more informative to consider the ratio of the atomic percentage of cobalt to arsenic. In the smalt in the *Adoration* by Bruegel, this ratio indicates almost three times as much cobalt as arsenic is present. A very similar ratio was seen in the smalt in the *Crucifixion* from the workshop of Martin de Vos. They are both far lower in arsenic content than any of the raw ores, which seems a clear indication that a roasted ore was used for the manufacture of the pigment.

In Van Dyck's *Saint Martin Dividing his Cloak*, examination of the arsenic levels made it possible to determine that two different types of smalt were present in the painting. An original canvas addition at the right side extended the painting, and, at the same time, the existing smalt and lead white sky was reworked in places with smalt of a different composition, again mixed with lead white. A cross-section from one of these reworked areas includes two layers, both containing smalt but rather different in appearance (Figure 5). That in the upper layer (the reworking) is rather grey and is low in arsenic (see Table 1), while that in the lower layer has a bluer colour and contains more arsenic than cobalt (up to twice as much in some particles).

The difference in appearance raises the question of whether the smalt in the upper layer was always paler than that of the lower layer or whether it has deteriorated. Both smalts contain around 3 wt% CoO, so this does not explain the difference in intensity of colour. EDX mapping, however, showed that the pigment in the upper layer is far more degraded than that in the lower layer (Figure 5); the silicon EDX map locates the smalt particles in both layers, but only the smalt particles in the lower layer contain enough potassium to register, indicating that in the top layer it has been leached from the pigment. This was confirmed by the quantitative spot analyses in which up to around 11 wt% K_2O was detected in the well-preserved smalt in the lower layer (although in some particles it is at lower levels and there is at least some degradation), while the degraded smalt in the upper layer consistently contains only 1–3 wt% K_2O.

Smalt is always better preserved when mixed with lead white. This usually explains differences in degree of degradation on the same painting, but here both layers contain a similar amount. It may be instead that the difference in the original composition is responsible for the variation in the state of preservation of the pigment, although it is difficult to predict or explain the stability of such a complicated glass. The large difference in arsenic content may have a bearing on its ageing behaviour, but there are also smaller yet still significant differences in the nickel and bismuth content that could also have an influence on stability.

The portrait of *Queen Elizabeth I*, known as the 'Darnley portrait', was examined during the 'Making Art in Tudor Britain' research project at the National Portrait Gallery in London. The queen wears a silver-grey dress with an embroidered pattern outlined with yellow-brown lines and filled in places with yellow and orange paint. A sample from one of the yellow-brown lines was found to contain a substantial amount of smalt mixed with some red lake (Figure 6). Initially, it was thought that the smalt may have been used to give body to the paint rather than colour as the yellow-brown lines are thick and raised, giving a texture to the surface of the dress. The quantitative analysis shows, however, that around 4 wt%

CoO is present in the pigment. It must therefore originally have had a reasonably strong blue colour and, since the smalt is mixed with red lake, the yellow-brown lines must originally have been purple or purple-blue, contrasting more strongly with the brighter yellow areas that do not contain smalt and were probably intended to depict gold embroidery. In addition, the low potassium content of all the particles indicates severe degradation, supporting the idea that the paint was not intended to be yellow-brown These conclusions, which could not have been reached without quantitative analysis, supplement the extensive historical curatorial research being carried out in parallel with the technical research. This includes consideration of Queen Elizabeth's clothes and accessories, both through portraits and the well-documented inventories that survive, in which the colour purple appears to be rare.

Conclusions

Quantitative analysis of the elemental composition of smalt in paint cross-sections can be very valuable. If conducted using EDX analysis in the VP-SEM, now routinely used for pigment analysis, a greater awareness of the influence of the analytical parameters on the results is needed relative to conventional HV-SEM. The experiments described in the first part raise some of the issues involved and show that a reliable methodology can be established, including specifically for the analysis of smalt.

Full assessment of the quantitative results from smalt in paintings collected during the study needs a more detailed consideration of historic documentary sources than is possible in this short paper, including those relating to pigments or painting practice, as well as those concerned with trade (in both the raw materials and the pigment) and the smalt-making industry. Nevertheless, some observations can already be made about which features might be informative regarding the manufacturing process. The cobalt content and the relative amount of arsenic can be of interest not only for historical reasons but also, as the example from the *Saint Martin Dividing his Cloak* by Van Dyck shows, for distinguishing between different smalts. The composition may also affect the properties of the pigment, including both appearance and stability.

It has been suggested in the past that more than one grade of smalt has been used in a single painting [28], or that it has been used as a drier or bulking agent to give body to the paint rather than for its colour [29], generally only on the basis of a pale appearance under the microscope. Quantitative analysis allows intentionally pale smalt to be distinguished from that which has deteriorated. The degree of degradation can be determined through the potassium content, whilst measurement of the cobalt content allows speculation on a firmer basis of the original strength of colour. These are the most common questions raised during examination of paintings containing smalt. The improvement in our ability to answer them offered by a more detailed knowledge of the elemental composition of the pigment, as well as a larger group of results from paintings to allow the data to be placed in context, contributes to our understanding of the original appearance of a painting in which this common pigment has been used, and therefore of the artist's intention.

Acknowledgements

We would like to thank Tarnya Cooper, Sophie Plender and Sally Marriott (National Portrait Gallery) and Libby Sheldon (University College London) for permission to include the analysis of smalt in the 'Darnley portrait'. We are also grateful to Maija Santala and Seppo Hornytzkyj (Finnish National Gallery) and Nicola Christie (Royal Collection) for their collaboration. The smalt reference sample was prepared and artificially aged by David Saunders (British Museum).

References

1. Mühlethaler, B. and Thissen, J., 'Smalt', in *Artists' Pigments: A Handbook of their History and Characteristics*, vol. 2, ed. A. Roy, National Gallery of Art, Washington DC (1993) 113–130.
2. Stege, H., 'Out of the blue? Considerations of the early use of smalt as blue pigment in European easel painting', *Zeitschrift für Kunsttechnologie und Konservierung* **18** (2004) 121–142.
3. Delamare, F., '2 – Saffre, smalt, bleu d'esmail et azur. Les tribulations du cobalt saxon', in *Bleus en poudres. De l'art à l'industrie, 5000 ans d'innovations*, Ecole des Mines de Paris, Paris (2007) 71–122.
4. Riederer, J., 'Die Smalte', *Deutsche Farben-Zeitschrift* **9** (1968) 386–395.
5. Richter, M., 'Smalt in polychromy and painting of German-speaking countries', in *Historical Polychromy: Polychrome Sculpture in Germany and Japan*, ed. M. Kühlenthal and S. Miura, Hirmer Verlag, Munich (2004) 175–203.
6. Spring, M., Higgitt, C. and Saunders, D., 'Investigation of pigment–medium interaction processes in oil paint containing degraded smalt', *National Gallery Technical Bulletin* **26** (2005) 56–70.
7. Boon, J.J., Keune, K., van der Weerd, J., Geldof, M. and Van Asperen de Boer J.R.J., 'Imaging microspectroscopic, secondary ion mass spectrometric and electron microscopic studies on discoloured and partially discoloured smalt in cross-sections of 16th century paintings', *Chimia* **55** (2001) 952–960.
8. Robinet, L., Spring, M., Pagès-Camagna, S., Vantelon, D. and Trcera, N., 'Investigation of the discoloration of smalt pigment in historic paintings by micro-X-ray absorption spectroscopy at the Co K-edge', *Analytical Chemistry* **83**(13) (2011) 5145–5152.
9. Ciliberto, E., Fragalà, I., Pennisi, G. and Spoto, G., 'Bulk and surface characterisation of early pigments: case study of a Renaissance smalt', *Science and Technology for Cultural Heritage* **3** (1994) 163–168.
10. Cavallo, G., 'The blue pigment used in Vallemaggia (Switzerland) in the half of 19th century by painters Vanoni and Pedrazzi', *Microchimica Acta* **155** (2006) 121–124.
11. Seldes, A.M., 'A note on the pigments and media in some Spanish colonial paintings from Argentina', *Studies in Conservation* **39** (1994) 272–276.
12. Santopadre, P. and Verita, M., 'A study of smalt and its conservation problems in two sixteenth-century wall paintings in Rome', *Studies in Conservation* **51** (2006) 29–40.
13. Newbury, D.E., 'X-ray microanalysis in the variable pressure (environmental) scanning electron microscope', *Journal of Research of the National Institute of Standards and Technology* **107**(6) (2002) 567–603.
14. Mathieu, C., 'The beam-gas and signal-gas interactions in the variable pressure scanning electron microscope', *Scanning Microscopy* **13**(1) (1999) 23–41.
15. Mansfield, J.F., 'X-ray microanalysis in the environmental SEM: a challenge or a contradiction?', *Mikrochimica Acta* **132** (2000) 137–143.
16. Doehne, E., 'A new correction method for high resolution energy dispersive X-ray analyses in the environmental scanning electron microscope', *Scanning* **19** (1997) 75–78.
17. Bean, S.J. and Kugler, V.M., 'The effect of the charge compensating gas and beam gas path length on X-ray analysis in the variable pressure SEM', *Microscopy and Microanalysis* **12** (Supp. S02) (2006) 1490–1491.

18. Khouchaf, L., Mathieu, C. and Kadoun, A., 'Electron microbeam changes under gaseous environment: CP-SEM case and microanalysis limits', in *Microscopy, Science, Technology, Applications and Education*, vol. 2, Formatex Microscopy Series No. 4, ed. A. Méndez-Vilas and J. Diaz, Formatex, Spain (2010) 1273–1279.
19. Stokes, D.J., 'Recent advances in electron imaging, image interpretation and applications: environmental scanning electron microscopy', *Philosophical Transactions of the Royal Society A, Mathematical, Physical and Engineering Sciences* **361** (2003) 2771–2787.
20. Le Berre, J.F., Demopoulos, G.P. and Gauvin, R., 'Charging effects on the spectral quality of X-ray microanalysis using the variable pressure scanning electron microscope', *Microscopy and Microanalysis* **13**(Supp. 2) (2007) 1478–1479.
21. Vicenzi, E.P., Eggins, S., Logan, A. and Wysoczanski, R., 'Microbeam characterization of Corning archeological reference glasses: new additions to the Smithsonian microbeam standard collection', *Journal of Research of the National Institute of Standards and Technology* **107**(6) (2002) 719–727.
22. Kuisma-Kursula, P., 'Accuracy, precision and detection limits of SEM-WDS, SEM-EDX and PIXE in the multi-elemental analysis of medieval glass', *X-ray Spectrometry* **29** (2000) 111–118.
23. Terczynska-Madej, A., Cholewa-Kowalska, K. and Laczka, M., 'The effect of silicate network modifiers on colour and electron spectra of transition metal ions', *Optical Materials* **32** (2010) 1456–1462.
24. Fabri, R., '"Eenen ramenant van verf ende pinselen": some aspects of the materials used by seventeenth-century cabinet makers in Antwerp', in *Trade in Artists' Materials: Markets and Commerce in Europe to 1700*, ed. J. Kirby, S. Nash and J. Cannon, Archetype Publications, London (2010) 366–374.
25. Merrifield, M.P., *Medieval and Renaissance Treatises on the Arts of Painting*, John Murray, London (1849) 649.
26. Lomazzo, G.P., *Trattato dell'arte della pittura, scoltura et archittetura*, Milan (1585) 191.
27. Véliz, Z., 'In quest of a useful blue in early modern Spain', in *Trade in Artists' Materials: Markets and Commerce in Europe to 1700*, ed. J. Kirby, S. Nash and J. Cannon, Archetype Publications, London (2010) 389–400.
28. Leighton, J., Reeve, A. and Burnstock, A., 'A "Winter Landscape" by Caspar David Friedrich', *National Gallery Technical Bulletin* **13** (1989) 44–60.
29. Roy, A. and Kirby, J., 'Rembrandt's palette', in *Art in the Making: Rembrandt*, ed. D. Bomford, J. Kirby, A. Roy, A. Rüger and R. White, National Gallery Company, London (2006) 35–47.

Authors' addresses

- Marika Spring, National Gallery, London, UK (marika.spring@ng-london.org.uk)
- Veronika Kugler, Carl Zeiss NTS Ltd, Cambridge, UK (kugler@nts.zeiss.com)
- Stewart Bean, Carl Zeiss NTS Ltd, Cambridge, UK (bean@nts.zeiss.com)

SEM and TEM analyses of chrome yellow and chrome orange dyestuffs used for imported and domestic cotton fabrics (*Touzan*) in Japan in the nineteenth century

Nahoko Sugioka and Masahiro Kitada

ABSTRACT The microstructures of chrome yellow and chrome orange dyestuffs in cotton fibre used for a vertically striped fabric, called *Touzan* in Japanese, have been investigated. This valuable fabric was brought to Japan from the fifteenth to the nineteenth century from Asian countries. The threads are dyed dark red, yellow, orange and blue. In the nineteenth century, conventional plant-based dyes were replaced by artificial dyes, such as chrome yellow and chrome orange (also known as chrome red), developed in Europe. These dyestuffs were traded to the Southeast Asian countries and then used for the fabrication of domestic *Touzan* in Japan. However, the scientific details of the dyes have not yet been classified because of the limited number of investigations carried out. *Touzan* specimens imported into Japan and made in Japan in the nineteenth century were used in this study; chromium (Cr) and lead (Pb) were detected on the surface and within the yellow and orange threads. The metallic compounds lead chromate ($PbCrO_4$) and dilead pentaoxochromate (Pb_2CrO_5) were detected by X-ray diffraction (XRD); these compounds being chrome yellow and chrome orange (chrome red). Fine particles of dilead pentaoxochromate (Pb_2CrO_5) were found to be concentrically distributed along the outer circumference of the fibres. Fine particles are formed in defects such as cracks by preferred precipitation. Several types of lead chromate crystals with different shapes and crystal structures (orthorhombic $PbCrO_4$ and monoclinic $PbCrO_4$) were observed. The number of defects in the fibres of the imported *Touzan* specimen was much smaller than that in the fibres of the Japanese *Touzan* specimen.

KEYWORDS microstructure, lead chromate, dilead pentaoxochromate, cotton fibres, nanoprecipitate

Introduction

The valuable fabric called *Touzan* in Japanese was brought into Japan from the fifteenth to the nineteenth century from Southeast Asian countries [1, 2]. The fabric was used for clothes worn by the upper echelons of society at that time. Even now, antique pieces are highly valued, for example, as pouches to wrap tea bowls used in the Japanese tea ceremony. In our previous investigation [3], it was found that some yellow threads of *Touzan* fabric indicated a specific dyestuff. Some of their dyestuffs were natural yellow dyes, although the others were not. Initially we focused on the yellow threads, and similar results were also obtained from orange threads. It is thought that in the nineteenth century, natural yellow and orange dyes were replaced by chrome yellow and chrome orange (chrome red) artificial dyes, which were developed in Europe [3–5]. Lead chromate ($PbCrO_4$), called chrome yellow, was developed in Europe in the early nineteenth century. This artificial compound has been widely used as a pigment in oil paints and in printing inks. The use of chrome yellow as a dyestuff was one such application. *Touzan* were also produced in Japan for a limited period in the latter half of the nineteenth century. However, the scientific details of the dyes have not yet been classified because of the limited number of investigations. The primary purpose of our study was to investigate the microstructures of chrome yellow and chrome orange (chrome red) compounds used as dyestuffs. In addition, we aimed to observe the distribution of chrome yellow and chrome orange (chrome red) compounds in cotton fibres, to obtain information about the manufacturing region of the fabric by comparing the microstructures, and to search for a part of the international historical relationship between Europe and Japan in the nineteenth century.

Materials

A piece of *Touzan* fabric (property of M. Kitada) imported into Japan in the middle of the nineteenth century was selected for scientific analysis. The specimen used is shown in Figure 1a. It is vertically striped and woven from cotton threads. The image shows an enlarged optical micrograph, where the colours of the threads can be seen to be blue, white, dark red and yellow. The structure is a plain weave, and the warp and the weft are twin threads. The dyestuff used for the yellow threads of *Touzan* fabric imported into Japan in the middle of the nineteenth century was investigated first. Then samples of orange threads of *Touzan* fabrics imported into Japan in the middle

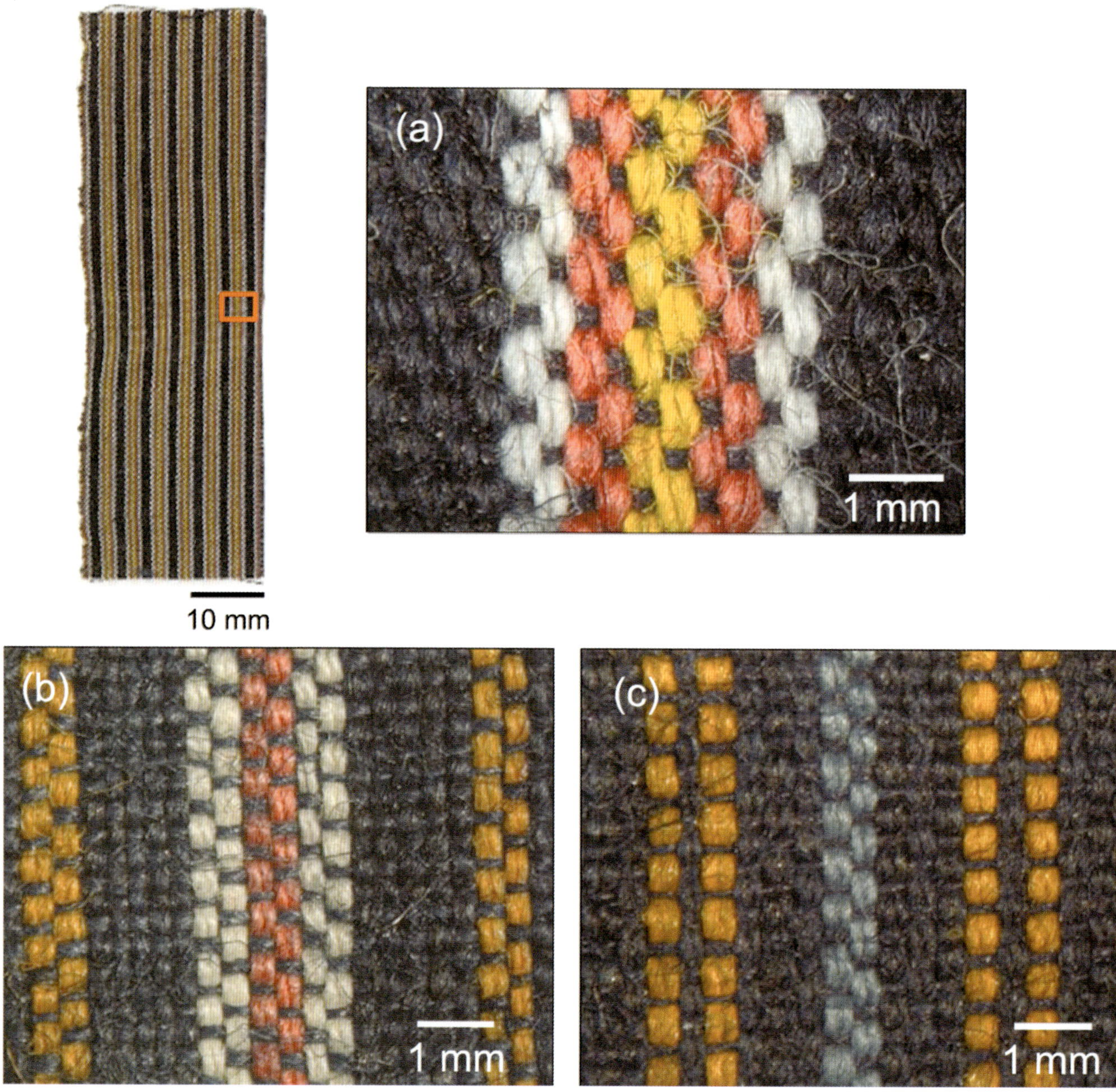

Figure 1 (a) Optical micrographs of imported *Touzan* fabric with yellow threads; (b) imported *Touzan* fabric with orange threads; (c) Japanese-made *Touzan* fabric with orange threads.

of the nineteenth century and orange threads of *Touzan* fabric made in Japan in the latter part of the nineteenth century (also the property of M. Kitada) were analysed. The threads of the imported *Touzan* fabric are thin, and twin threads are used for both the warp and weft (Figure 1b). The threads of the Japanese *Touzan* fabric are thick, and a twin thread is used for the warp whereas a single thread is used for the weft (Figure 1c). The imported *Touzan* fabric was compared with the Japanese *Touzan* fabric, focusing on the orange threads.

Methodology

Analysis proceeded as follows: colour was measured with a spectrophotometer, then the crystal structure was analysed with an X-ray diffractometer (XRD). Elemental composition was investigated by energy dispersive X-ray analysis (EDX), the macro- and microstructures were observed by X-ray computed tomography (X-ray CT), and field emission scanning electron microscopy (FE-SEM) (Hitachi S-4800). The nanostructure of the dyed fibre was observed by transmission electron microscopy (TEM) (Hitachi H-9500). A thin film for observing the nanostructures was prepared by the focused ion beam method (Hitachi FB-2200).

Results and discussion

Yellow thread of imported *Touzan* fabric

Fibre size and X-ray CT

Cotton threads are used for *Touzan* fabric, as described above. It has been suggested that the length and thickness of cotton fibres depend on the place of origin [6]; for example that cotton fibres from Egypt, India, and Turfan (China) are usually thin and long, whereas those from Japan and Southeast Asia are historically thick and short [7]. If these suggestions can be adopted, it may be possible to pinpoint

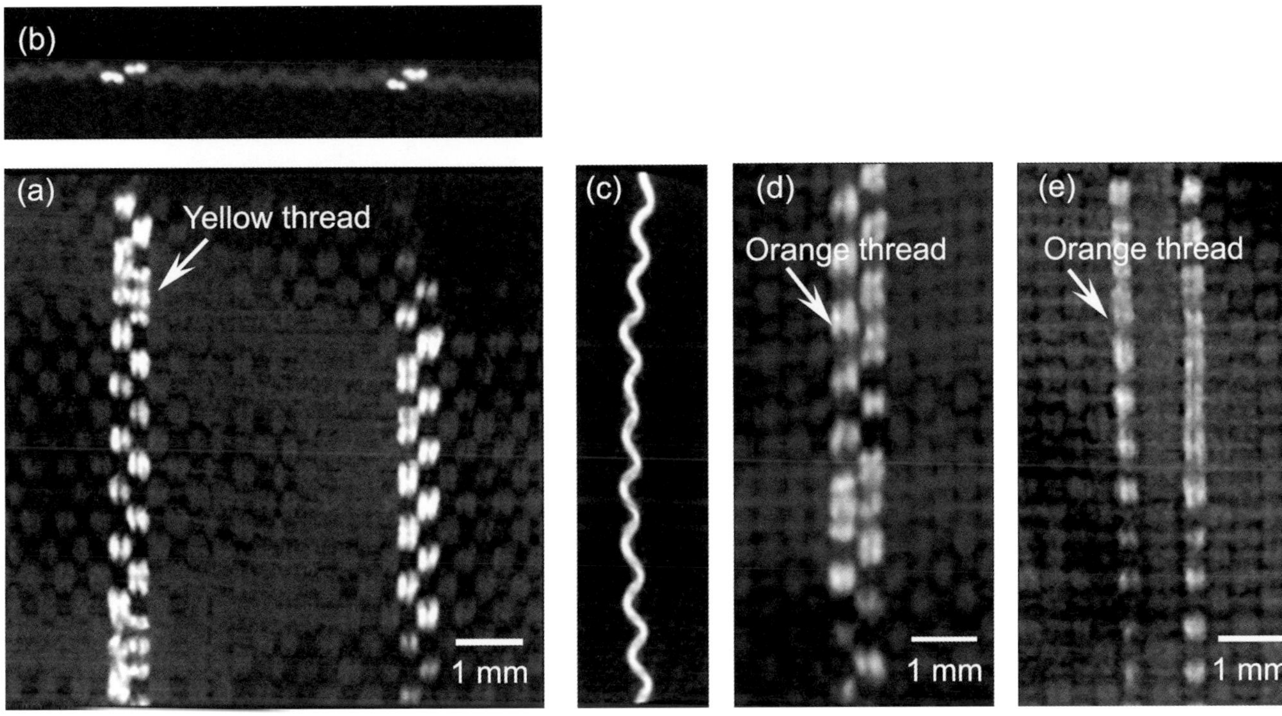

Figure 2 X-ray CT image of *Touzan* fabrics shown in Figure1: (a) cloth face of imported *Touzan* fabric with yellow threads; (b) and (c) cross-section of imported *Touzan* fabric with yellow threads; (d) cloth face of imported *Touzan* fabric with orange threads; (c) cloth face of Japanese-made *Touzan* fabric with orange threads.

the region of cultivation from the fibre length. The average fibre length of the imported *Touzan* sample used is 27.2 mm, and the longest fibre is 34.7 mm. The cotton fibre length can be grouped into five classes currently in use [8]; extra long cotton fibre (34.9 mm and above), long cotton fibre (28.6–33.3 mm), medium long cotton fibre (26.2–27.8 mm), medium cotton fibre (20.6–25.4 mm), and short cotton fibre (20.6 and below). On this basis, it falls into the categories of medium long or long cotton fibres, so we deduce that imported *Touzan* fabric with yellow threads is thought to be made from non-Japanese cotton.

X-ray CT images of imported *Touzan* fabric with yellow threads are shown in Figure 2 (a, b and c). When the fabric structure was analysed by X-ray CT, it was found that the X-ray absorption of yellow threads was very high, shown by the bright area. Figure 2a is parallel to the cloth face and Figure 2b and Figure 2c are normal to the cloth face. It clearly shows that heavy elements exist in the yellow threads (as the very bright area contains the metallic elements).

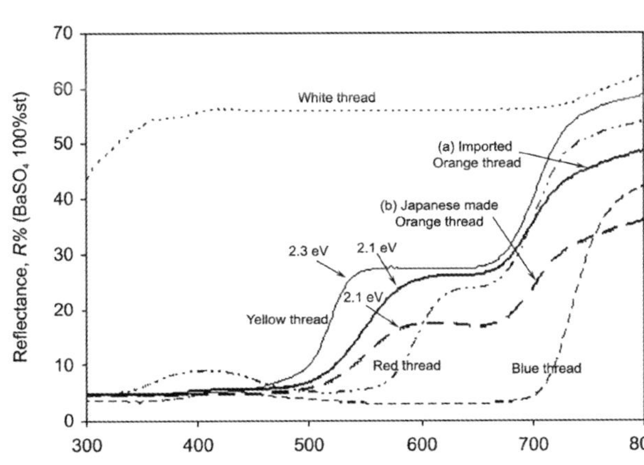

Figure 3 Reflectance spectra of white, yellow, red and blue threads from imported *Touzan* fabric; (a) imported orange thread; (b) Japanese-made orange thread.

Optical properties

The reflectance spectra of white, yellow, red and blue *Touzan* threads are shown in Figure 3. For the yellow thread, the main absorption edges are at 536 nm with an energy of 2.3 eV and at 720 nm with an energy of 1.7 eV. These absorption edges determine the colours and the thread colour is close to yellow. The band gap of 2.3 eV indicates that the yellow dyestuff material is a semiconductor compound. For the white thread, reflectance is very high over a wide area. For the dark red thread, the main absorption edges are at 610 nm and 710 nm with a broad peak near 400 nm. For the blue thread, the main peak is at 420 nm with a high absorption edge near 740 nm and reflectance is very high, above 700 nm; however, visual sensitivity in this area is very low. Therefore, the broad peak near 420 nm, corresponding to blue, is clearly recognisable by eye.

SEM, EDX and XRD

SEM backscatter electron (BSE) images of the yellow thread surface are shown in Figure 4. Many small particles were observed on the surface. These particles were not observed on thread of other colours. The enlarged image superimposed on the left of Figure 4 shows that the particles are very small, 1–2 μm in length and 0.2–0.7 μm in width. It is thought that these particles on the surface are yellow dyestuff compounds. Some elements were detected on the yellow threads by EDX. Sodium (Na), chlorine (Cl), potassium (K), and calcium (Ca) were all found in the cotton fibre by measuring the white thread that seems not to be dyed. Chrome (Cr) and lead (Pb) were not found on the cotton fibre itself. It is thought that the particles observed contain those of two elements. The average contents of the elements (excluding carbon and oxygen) on the yellow thread surface are Na 6.3 mol%, Cl 5.8 mol%, K 5.1 mol%, Ca 5.1 mol%, Cr 42.2 mol%, and Pb 35.5 mol% by EDX. The atomic ratio Pb/Cr is about 1.2. This result suggests that the dyestuff material contains equivalent amounts of Pb and Cr.

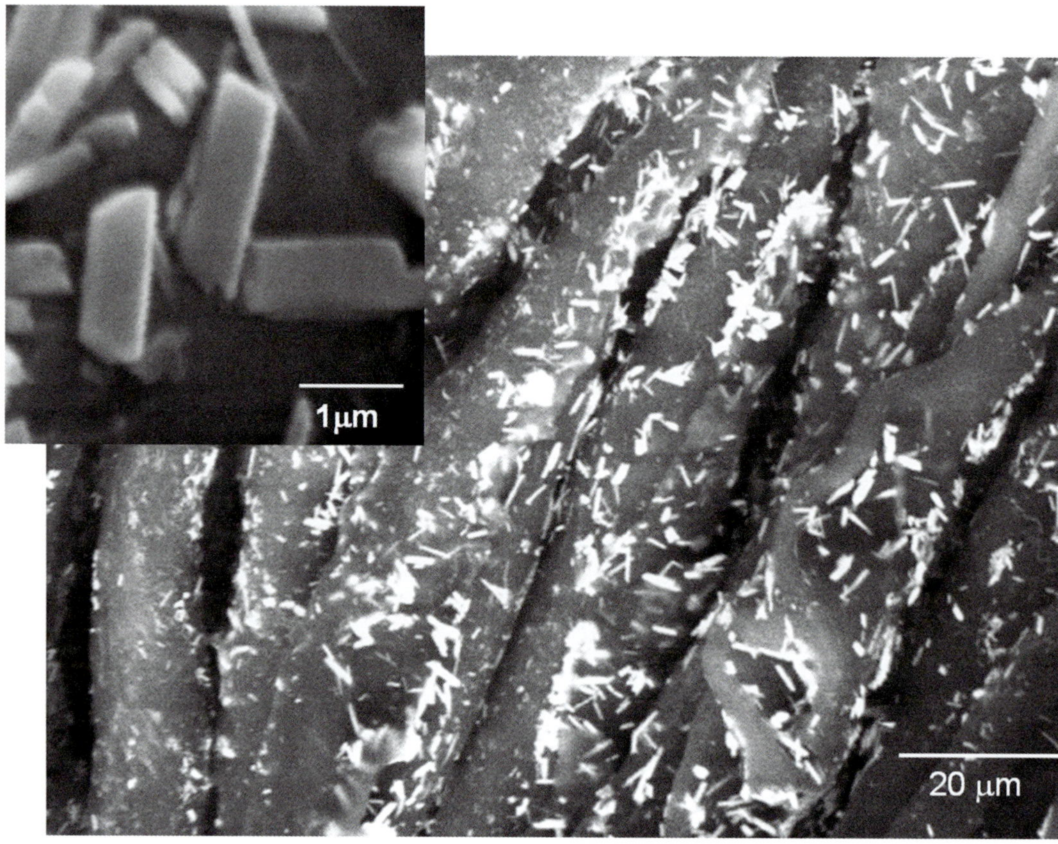

Figure 4 SEM images of precipitates on yellow thread surface of imported *Touzan* fabric. Enlarged particle image is superimposed.

The X-ray diffraction (XRD) pattern of the yellow thread is shown in Figure 5. Diffraction peaks of monoclinic lead chromate ($PbCrO_4$) [9] and those of cellulose ($(C_6H_{10}O_5)_n$) [10] and very weak peaks of orthorhombic lead chromate ($PbCrO_4$) [11] were detected in the sample of yellow thread. Therefore, the yellow thread was dyed with lead chromate ($PbCrO_4$).

An 1881 publication [12] provided the method used for dyeing and precipitating lead chromate in this study. The outline of the procedure is as follows: (1) cotton threads were immersed in a lead acetate ($Pb(CH_3COO)_2$) aqueous solution for 50 minutes and then rinsed in water. (2) The threads were immersed in a potassium dichromate ($K_2Cr_2O_7$) aqueous solution for 20 minutes and then rinsed in water. (3) The threads were then immersed in a lead acetate aqueous solution again.

The usual reaction for forming $PbCrO_4$ is:

$$2Pb(CH_3COO)_2 + K_2Cr_2O_7 + H_2O$$
$$\rightarrow 2PbCrO_4 + 2CH_3COOK + 2CH_3COOH$$

thus when potassium dichromate is added to lead acetate, lead chromate is formed as a yellow precipitate. In this reaction, monoclinic $PbCrO_4$ and orthorhombic $PbCrO_4$ compounds coexist [13]. In the early stage of this reaction, a large amount of orthorhombic $PbCrO_4$ and a small amount of monoclinic $PbCrO_4$ compounds are generated. Their compounds changed to almost monoclinic $PbCrO_4$ due to long time reaction (such as 20 minutes). Aqueous potassium acetate (CH_3COOK) and acetic acid (CH_3COOH) remain in the aqueous solution. The particles observed on *Touzan* yellow threads during SEM imaging are lead chromate ($PbCrO_4$) generated by the above reaction. Lead chromate is insoluble in water. Consequently, it is presumed that *Touzan* cotton threads are dyed yellow.

Diffusion into the fibre

To clarify the existence of the dyestuff elements in the fibre, cross-sectional concentration profiles of lead and chromium were measured by EDX. Pb and Cr are distributed uniformly throughout the fibre. Therefore, Pb and Cr elements penetrated through the fibre surface from the dyestuff solution and diffused into the fibre. Since the atomic ratio Pb/Cr is approximately 1 from the fibre edge to the middle of the fibre as seen through EDX, it is thought that lead chromate ($PbCrO_4$) precipitates in the fibre. The solubility limits of Pb and Cr in the cotton fibre are unknown. Pb and Cr that could not be dissolved in the cotton fibre were precipitated as lead chromate ($PbCrO_4$). It is thought that the concentrations reached the solubility limits therefore we believe that lead chromate ($PbCrO_4$) crystals precipitate in the fibre.

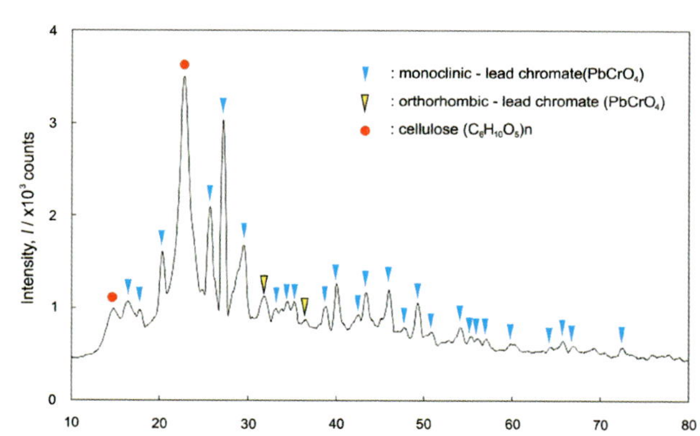

Figure 5 XRD pattern of yellow thread from imported *Touzan* fabric.

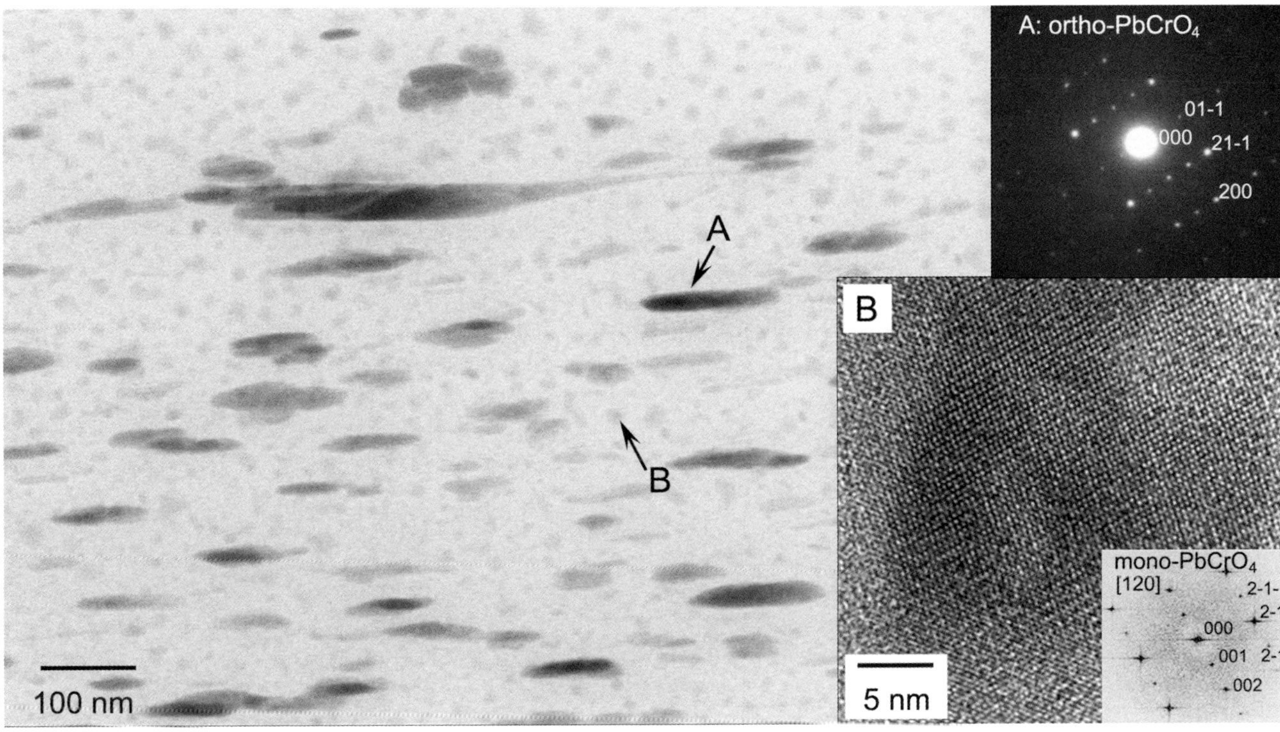

Figure 6 TEM image of yellow cotton fibre from imported *Touzan* fabric. A: large acicular precipitate; B: very small precipitate (crystals).

TEM observations

To classify the presence of dyestuff elements in the fibre, the yellow fibre of imported *Touzan* fabric was observed using TEM (Figure 6). Many nanometre (nm) sized particles can be observed in the cotton fibre. Many acicular particles (indicated by A) and very small grains (indicated by B) can also be observed. The longitudinal direction of the acicular particles is parallel to the fibre direction. The very small grains are distributed randomly. The size of the acicular particles is 50–100 nm long and 10–20 nm wide. The size of grain B is 3–5 nm. The electron diffraction pattern of the acicular particles shown in Figure 6A matches that of orthorhombic lead chromate ($PbCrO_4$) [11]. Although large amounts of monoclinic $PbCrO_4$ and small amounts of orthorhombic $PbCrO_4$ are generated on the fibre surface according to EDX, the converse result is observed inside the fibre. The reaction inside the fibre seems to be a closed system, so that the methods of forming crystals seem different to a normal system; this needs clarification through further research. The electron diffraction pattern of the very small grain was not obtained because the grain was too small. Therefore, the crystal lattice image of the very small grain shown in Figure 6B was obtained. The reproduced electron diffraction pattern by Fourier transform matches that of monoclinic lead chromate ($PbCrO_4$) [9]. Therefore, two types of crystal precipitates exist in the cotton fibre.

Orange threads of imported and Japanese-made fabrics

Fibre size and X-ray CT

The average fibre length of the imported orange thread *Touzan* specimen used is 27.5 mm, and the longest fibre is 35.2 mm. The average fibre length of the Japanese orange thread *Touzan* specimen used is 23.7 mm, and the longest fibre is 25.2 mm. As mentioned above [8], it is presumed that different sources of cotton were used for orange threads of the imported and Japanese *Touzan* specimens.

X-ray CT images of the imported and Japanese *Touzan* specimens with orange threads are shown in Figure 2d and Figure 2e. When the fabric structure was analysed by X-ray CT, it was found that the X-ray absorption of orange threads was very high, shown in the bright area. This also applies to yellow as well as orange threads of the imported and Japanese *Touzan* specimens.

Optical property

The reflectance spectra of the orange threads of the imported and Japanese *Touzan* specimens are shown in Figure 3a and Figure 3b. Absorption edges appear at 590 and 720 nm for both specimens, indicated by arrows. However, the reflectance of the Japanese *Touzan* specimen, labelled (c), is lower than that of the imported *Touzan* specimen, labelled (b). Because a low reflectance indicates a low brightness and a dark colour, it is clear that the orange threads of the Japanese *Touzan* specimen are dyed darker than those of the imported *Touzan* specimen.

Diffusion into the fibre

The distributions of Pb and Cr in the fibres were examined. Pb and Cr are almost uniformly distributed throughout the fibres in both the imported and Japanese *Touzan* specimens. Although non-uniform distributions are also observed in the Japanese *Touzan* specimen, the Pb and Cr dyes have diffused deeply into the fibres in both. The measurement results also indicate that the molar concentration ratio of Pb/Cr is about 2:1 for both the imported and Japanese *Touzan* specimens, according to EDX. XRD indicates that the spectral peaks of lead and chromium in the two specimens match those of dilead pentaoxochromate (Pb_2CrO_5). We suggest both specimens were dyed with dilead pentaoxochromate (Pb_2CrO_5) [14].

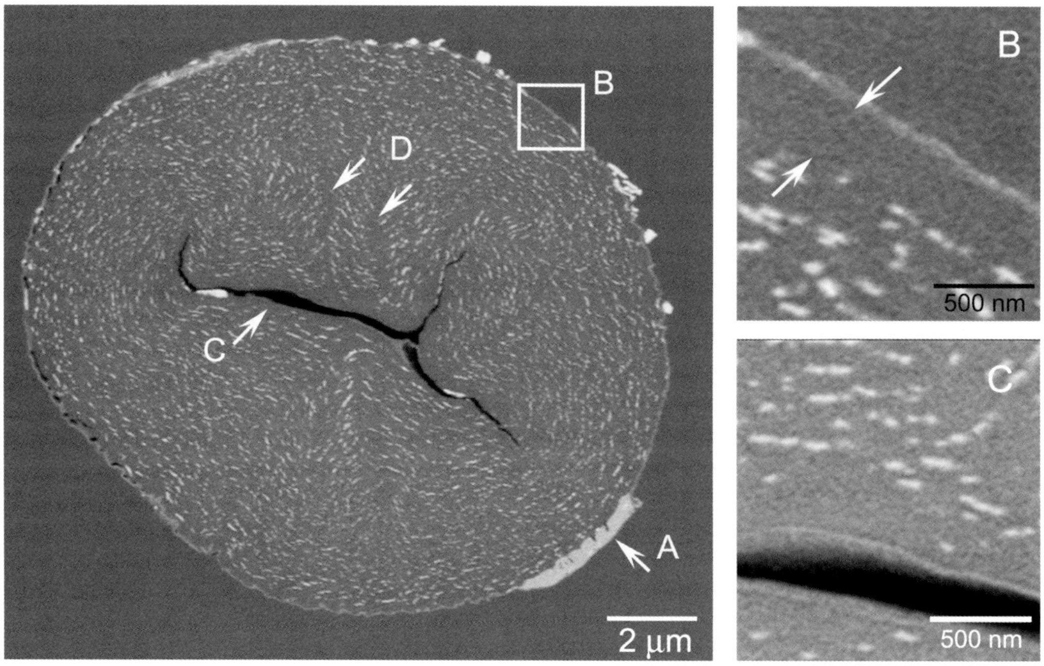

Figure 7 SEM image of a typical cross-section of an imported orange fibre.

As mentioned above for the lead chromate ($PbCrO_4$) dyeing reaction, the method of precipitating dilead pentaoxochromate (Pb_2CrO_5) was as follows. (1) Cotton threads were immersed in lead acetate ($Pb(CH_3COO)_2$) aqueous solution with calcium hydroxide ($Ca(OH)_2$) for 50 minutes and then rinsed in water. (2) The threads were next immersed in potassium chromate (K_2CrO_4) aqueous solution at 90°C for 20 minutes and then rinsed in water. The usual reaction for forming Pb_2CrO_5 is:

$$[2Pb(CH_3COO)_2 + Ca(OH)_2] + K_2CrO_4$$
$$\rightarrow PbO \cdot PbCrO_4 + 2CH_3COOK + Ca(CH_3COO)_2 + H_2O$$

thus when potassium chromate is added to lead acetate with calcium hydroxide, dilead pentaoxochromate (Pb_2CrO_5) (or $PbO{\cdot}PbCrO_4$) is formed as an orange precipitate. It is thought that Pb_2CrO_5 (or $PbO{\cdot}PbCrO_4$) and $PbCrO_4$ coexist in this reaction. Because of a reversible reaction under certain conditions, it is possible that $PbCrO_4$ may coexist. The particles observed on *Touzan* orange threads are dilead pentaoxochromate (Pb_2CrO_5) generated by the above reaction. As dilead pentaoxochromate (Pb_2CrO_5) is also insoluble in water, the orange compounds remain after washing.

Cross-sectional SEM images

Orange threads from imported fabrics

A typical cross-sectional SEM-BSE image of an orange thread fibre from the imported *Touzan* specimen is shown in Figure 7. To observe the cross-section of a fibre using FE-SEM, it is important to make a soft material, such as cotton, as flat as possible. Thus, the optimisation of the sample pre-treatment method was an important (and time-consuming) step. First, we obtained the cross-section of the specimen by mechanical polishing, but no particles were observed. Next, the surface of the specimen was flattened by ion milling; the residues of mechanical polishing were removed, and then the cross-section was observed. Particles were successfully observed by optimising the conditions of ion milling.

The dark region at the centre of the cross-sectional image represents the lumen. As indicated by arrow A, a dilead pentaoxochromate (Pb_2CrO_5) layer is observed on the surface of the fibre. Fine precipitates with a diameter of 100–500 nm are concentrically distributed around the lumen inside the cotton fibre which, in this case, has an almost circular cross-section. Although a cross-section of cotton fibre is often bean-shaped, it seems that the shape was deformed in this example as a result of dyestuff spreading inside the fibre. Figure 7B shows an enlarged image of the vicinity of the fibre surface enclosed by square B. In the vicinity of the fibre surface, a precipitation-free zone is observed. The precipitation-free zone, indicated by an arrow, is considered to be of high density. In the enlarged image of the vicinity of the lumen, indicated by arrow C, precipitation-free zones are also observed. In addition, as indicated by arrow D, a region with a smaller amount of precipitates than other regions is observed inside the fibre. Regions with a smaller amount of precipitates in this region are considered to have a high degree of crystallinity. In some regions, precipitates are irregularly distributed, probably because of the structure of the fibre. This may also reflect the growth process of cotton fibres.

Orange threads from Japanese-made fabrics

A typical SEM-BSE cross-sectional image of the fibre of a Japanese *Touzan* orange thread is shown in Figure 8. The lumen is observed at the centre of the fibre, and particles form a relatively large block of precipitate. Since the atomic ratio Pb/Cr found by EDX is approximately 2 (0.80:0.41), it is thought that dilead pentaoxochromate (Pb_2CrO_5) particles are present in this fibre. The particles are concentrically distributed around the lumen, but their distribution pattern is quite different from that of the imported *Touzan* specimen. Many particles, considered to be dilead pentaoxochromate (Pb_2CrO_5), with a length of 100–500 nm and a width of 50–80 nm are observed on the cross-section of the fibre, whereas in the vicinity of the fibre surface, particles larger than those in other regions are

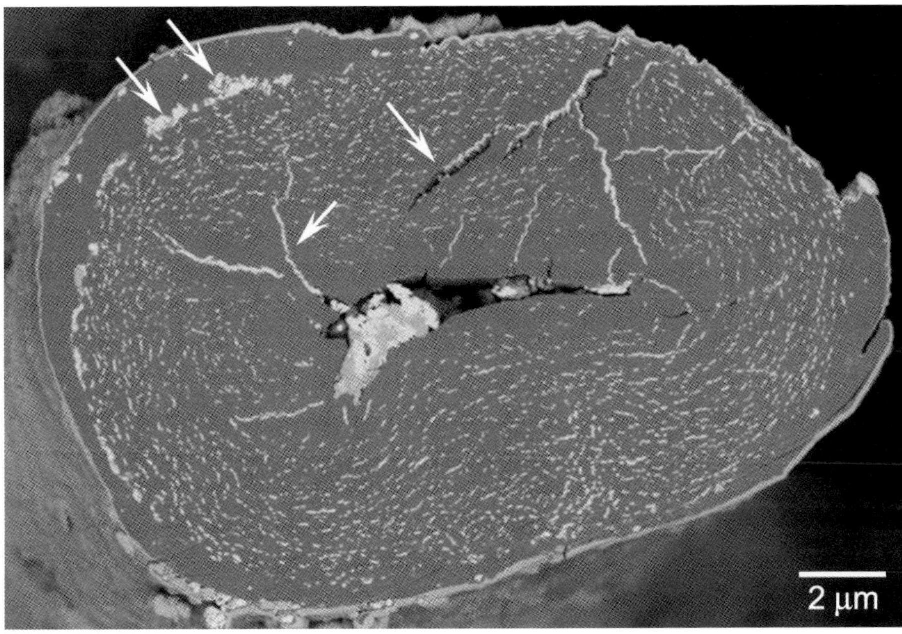

Figure 8 SEM image of a typical cross-section of a Japanese-made orange fibre. Arrows indicate preferential precipitation areas in cotton fibre.

observed. Precipitates forming zigzag lines and large cracks, as shown by the arrows are observed. Precipitation-free zones are also observed around those lines and cracks. These zones may be generated by the preferred precipitation of Pb and Cr atoms in defect regions such as cracks or by competitive crystal growth after precipitation. Precipitation-free zones are similarly observed in the vicinities of the fibre surface and the lumen. These zones are considered to be dense structure zones of cellulose, as inferred from the distribution of precipitates. Thus, precipitation-free zones are considered to have two patterns of precipitation. One is caused by preferred precipitation around cracks, and the other is due to a structural factor of the fibre observed in the vicinities of the surface and the lumen.

TEM observation

Orange threads from imported fabrics

To further examine the state of the precipitates, the specimens were observed using TEM. A TEM image of an imported *Touzan* specimen is shown in Figure 9. As indicated by arrow

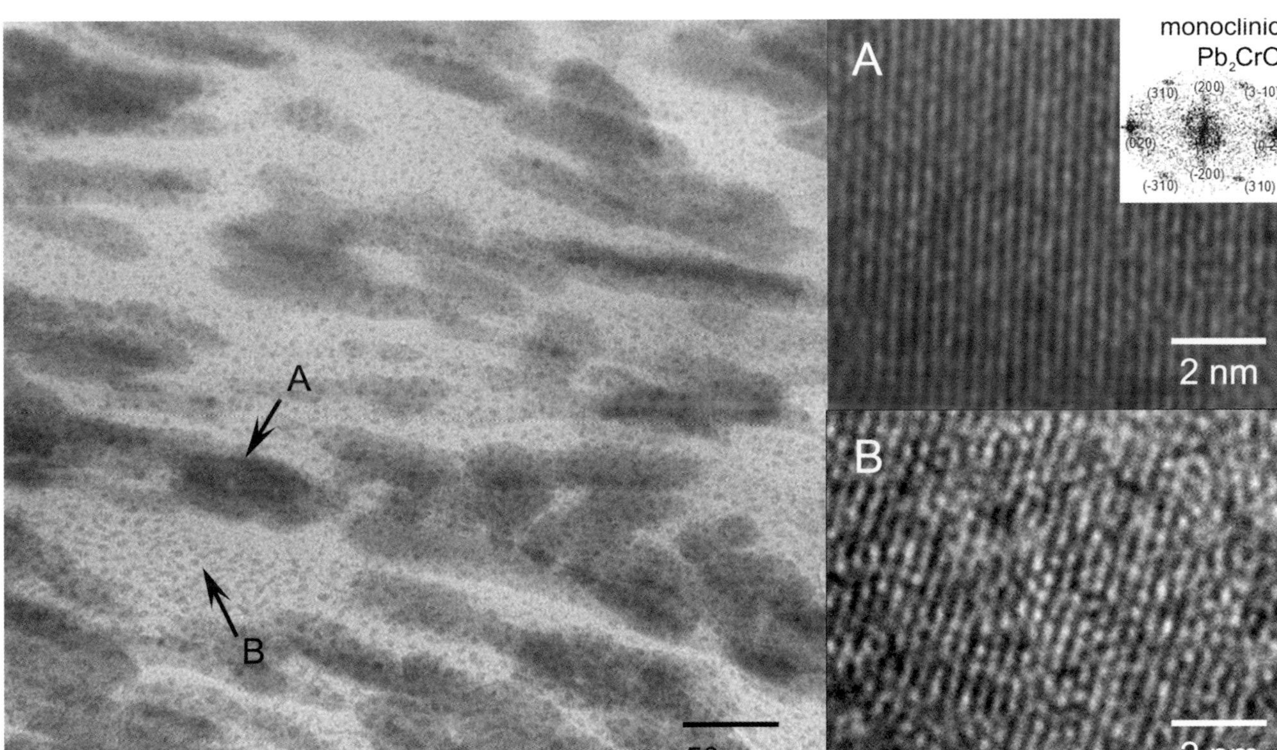

Figure 9 TEM image of imported orange fibre. Typical lattice images of precipitates and diffraction patterns. Labels A and B indicate acicular and polygonal precipitates, respectively.

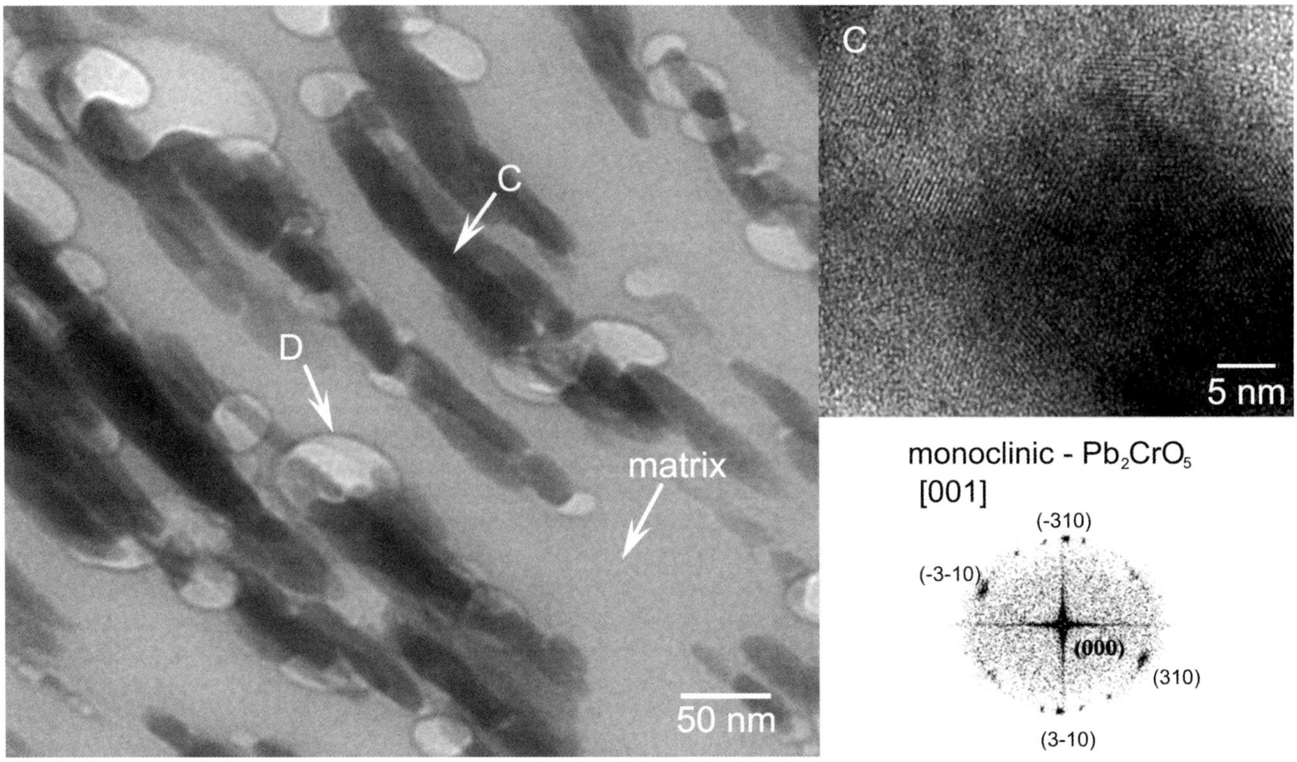

Figure 10 TEM image of Japanese-made orange fibre. Typical lattice image of precipitates and diffraction pattern. Labels C and D indicate precipitate and void, respectively.

A, acicular particles precipitate in a certain direction. The particles are 100–200 nm long and 20–50 nm wide. In addition, very fine small particles with a width of 3–5 nm are observed around the acicular particles, as indicated by arrow B. No electron diffraction pattern was obtained, so the crystal lattice images of particles A and B were obtained, and the diffraction patterns reproduced by Fourier transform were analysed. According to the analytical result, acicular particles, labelled A, were orange monoclinic dilead pentaoxochromate (Pb_2CrO_5) [14], and the very small particles, labelled B, were yellow monoclinic lead chromate ($PbCrO_4$) [9]. Thus, two types of compound coexist in the orange thread, but the colour is determined by the colour of dilead pentaoxochromate (Pb_2CrO_5), the principal component of chrome orange (chrome red).

Orange threads of Japanese-made fabrics

A TEM image of an orange thread in the Japanese *Touzan* specimen is shown in Figure 10. As in the imported *Touzan* specimen, acicular particles are observed in the fibre, as indicated by arrow C. The particles are 100–300 nm long and 20–50 nm wide, similar to those of the imported *Touzan* specimen. For the Japanese *Touzan* specimen, many small cavities are observed around the acicular particles, as indicated by arrow D. This result suggests that Japanese *Touzan* fabrics comprise cotton fibres with many defects such as cracks and cavities, visible in the SEM image (Figure 8). However, very fine small particles are not observed in the Japanese *Touzan* specimen, indicating a microstructure different from that of the imported *Touzan* specimen. Moreover, Pb and Cr were detected in the acicular particles, indicated by arrow C. Since slight amounts of Pb (0.3 mol%) and Cr (0.1 mol%) were also detected in the precipitation-free matrix zone, Pb and Cr atoms are soluble in the fibre matrix (cellulose 91%, H_2O 7.8%, pectin 0.5%, wax and fat 0.4%, mineral 0.2%) as an interstitial solid solution. So that the crystal lattice image of acicular particles, labelled C, could be obtained, the diffraction pattern reproduced by Fourier transformation was analysed. According to the analytical result, the acicular particles were found to be orange compounds, that is, monoclinic dilead pentaoxochromate (Pb_2CrO_5) [14]. Although the acicular particles appear to be large in the SEM image (Figure 8), in the TEM, the particles can be seen to consist of very fine nanometre-sized crystals.

Conclusions

Examination of the microstructures of chrome yellow and chrome orange (chrome red) dyestuffs found in cotton fibres used for a vertically striped *Touzan* fabric has resulted in a number of observations, which are summarised briefly here. Many acicular lead chromate ($PbCrO_4$) particles, 1–2 μm in length and 0.2–0.7 μm in width were present on the surface of the imported *Touzan* yellow fibre. *Touzan* yellow threads were dyed by lead chromate ($PbCrO_4$), not only on the fibre surface but also inside the fibre. Crystal precipitates with two different structures exist in the yellow fibre. Fine acicular orthorhombic lead chromate ($PbCrO_4$) crystals were observed in the yellow fibre. The length and width of the crystals were 50–100 nm and 10–20 nm respectively. Nanometre-sized (3–5 nm) monoclinic lead chromate ($PbCrO_4$) crystals were also distributed randomly in the fibre. Fine dilead pentaoxochromate (Pb_2CrO_5) particles were distributed concentrically along the outer circumference of the orange fibre. Fine particles were formed in defects such as cracks by preferred precipitation. Acicular monoclinic dilead pentaoxochromate (Pb_2CrO_5) crystals were observed in the imported orange fibre. The length and width of the crystals were measured as 100–200 nm and 20–50 nm respectively. Nanometre-sized (3–5 nm) monoclinic lead chromate ($PbCrO_4$) crystals were also distributed randomly in the imported fibre. Acicular monoclinic dilead pentaoxochromate (Pb_2CrO_5) crystals were also observed in the fibre of the Japanese-made fabric. The length and width of the crystals were 100–300 nm and 20–50 nm respectively. Many small cavities were observed around the acicular particles in the Japanese *Touzan* specimen. The number of defects in the fibre of the imported *Touzan* specimen was much smaller than those in the fibre of the Japanese *Touzan* specimen. The quality of fibres in Japanese *Touzan* fabrics appeared to be lower than that of imported *Touzan* fabrics. There seems to be a strong interaction correlation between the precipitation phenomenon and the microstructures in the cotton fibres.

Many kinds of data support the theory that artificial dyestuffs from Europe were introduced with technology to Japan by European innovations in the nineteenth century. As the chrome yellow and chrome orange (chrome red) dyestuffs are superior in durability to natural dyestuffs, it is thought that this is the reason why natural dyestuffs were replaced by these artificial dyestuffs.

Acknowledgements

This work was supported by KAKENHI, Grant-in-Aid for JSPS Fellows, Young Scientist (A) (No. 24680082) and Grants-in-Aid for Scientific Research (B) (No. 22360295). We gratefully acknowledge the financial support of the Idemitsu Foundation of Culture and Social Welfare, Mishima Kaiun Memorial Foundation, Hattori-Hokokai Foundation and Kurata Memorial Hitachi Science and Technology Foundation. We would also like to thank Hitachi High-Technologies for the nanometre-sized structure observation and Rigaku Corporation for assisting with X-ray CT measurement.

References

1. Urano, R., *Touzan*, Bunka syuppankyoku (1972).
2. Hata, H., *Touzankou*, Akane Sousyo IV, Akaneya Syobou (1932).
3. Sugioka, N. and Kitada, M., 'Metallic compound of yellow dyestuff used for *Touzan* cotton fabric imported in the late Edo period', *Journal of the Japanese Institute of Metals* **72** (2008) 863–867.
4. Okamura, K., *Tateyama Touzan*, Iseikatsu Kenkyukai (1987).
5. Uemura, R., *Minzoku to Sensyoku-bunka*, Seibunsya (1943).
6. *Sen'i-kougaku (II)*, 3rd edn, The Textile Machinery Society of Japan (1996).
7. Watanabe, K., *Menshihu no Kisochishiki*, Konan Shoin (1950).
8. *Sen'ibinran*, 2nd edn, The Society of Fibre Science and Technology, Japan, Maruzen (1994).
9. Joint Committee on Powder Diffraction Standards (JCPDS) card No. 08-0209.
10. JCPDS card No. 50-2241.
11. JCPDS card No. 38-1363.
12. *Zoutei Seiyou-Sensyokuhou*, vol. 1, Tokyo-hu Kangyoka (1881).
13. Fukusawa,T. and Iwatsuki, M. 'State analyses of precipitates in systematic qualitative analysis of lead(II) and chromium(IIII) ions', *Japan Analyst* **26**(6) (1977) 379–383.
14. JCPDS card No. 29-0768.

Authors' addresses

- Nahoko Sugioka, Graduate School of Cultural Conservation, Tokyo University of the Arts, Tokyo, Japan (sugioka.nahoko@fa.geidai.ac.jp)
- Masahiro Kitada, Graduate School of Cultural Conservation, Tokyo University of the Arts, Tokyo, Japan (kitada.masahiro@palette.plala.or.jp)

Metallurgical and chemical characterisation of Venetian silver denarii dating to the late twelfth century

Irene Calliari, Michele Asolati, Andrea Saccocci, Francesco Grazzi and Antonella Scherillo

KEYWORDS silver coins, medieval, surface enrichment, metallography

Introduction

During the tenth to twelfth centuries, Venice enjoyed a greater political and economic autonomy from the Western Empire than other imperial mints such as Verona, Milan and Pavia. Whilst under imperial rule, Venetian coinage from the ninth to the second half of the twelfth century had lost around 90% of its silver content. In 1156–1172, the Venetian mint obtained a complete and formal autonomy from the Western Empire, when the name of the ruling doge was struck on the coins instead of that of the emperor. New denarii (pennies), bowl-shaped and less debased than the previous coins of Vitale II Michiel, which weighed c. 0.41 g and had a fineness of 270/1000 in silver, were struck by the successive doges Sebastiano Ziani, Orio Malipiero and Enrico Dandolo (1172–1205). These pennies competed successfully against their Veronese rivals and dominated the coin circulation in north-eastern Italy [1–3].The nineteenth-century scholar Nicolò Papadopoli suggested [4] that the new pennies were again debased during the reign of Enrico Dandolo (1192–1205) because they contained slightly less silver than those issued under his predecessor Orio Malipiero (1178–1192). This suggestion, however, only relied on a single analysis. In these centuries, the coins were produced by casting, mechanical working and annealing. The silver-copper alloy of the required chemical composition was solidified in ingots and the final thickness was obtained by cold working and annealing. The single flans were produced by cutting plates with scissors or hollow punches, and their weight finely adjusted by cutting the edges. The high copper content of the Venetian pennies produced during the tenth and twelfth centuries resulted in red/pink-coloured flans [5]. An additional step needed to be performed, therefore, in order to obtain a silver-coloured surface, as required by the issuing authority. In the present study, 52 Venetian pennies (denarii) minted in the twelfth century were examined to determine their bulk and surface chemical compositions and investigate their surface enrichment process.

Experimental section

The following private collection coins were examined: 30 examples issued by Malipiero and 22 examples issued by Dandolo. The specimens have diameters in the range 0.5–1.1 cm, thicknesses of 0.4–0.6 mm and weights of 0.3–0.4 g. All the coins were analysed by X-ray fluorescence (XRF) (Kevex Delta 770, anode rhodium, secondary target gadolinium, 55 kV, 2 mA) without any preparation; the chemical composition was determined by a standardless fundamental parameter (FP) method (silver, copper, lead and antimony) on both sides. A selected group of coins was cut and their bulk and surface microstructures investigated on metallographic cross-sections by optical microscopy (OM) and scanning electron microscopy with energy dispersive X-ray analysis (SEM-EDX) (Stereoscan 440 equipped with a Falcon Fei UTW microanalyser). The instrument operated at 25 kV, with a working distance of 23 mm and the ZAF correction was applied for quantitative analysis (copper, silver and lead).

Results

The chemical composition of all samples was determined by XRF and is reported in Table 1. The data in Table 1 confirm that the coins are composed of a copper-silver alloy with a low lead content. Traces of mercury have been detected in one sample only, belonging to the Malipiero group.

The presence of mercury can be related to the amalgam plating method [6], which can produce very thin plating. After the application of a silver-mercury amalgam on the surface of

Table 1 Composition of the alloy (average values of wt% content, measured on both sides).

	Cu	Ag	Sb	Pb
Malipiero	71.8 ± 3	26.0 ± 3	0.15 ±0.05	2.04 ± 0.5
Dandolo	75.7 ± 3	21.8 ± 3	0.34 ± 0.05	2.14 ± 0.5

Table 2 Bulk and surface chemical composition (wt%), determined with EDS.

	Malipiero bulk	Malipiero surface	Dandolo bulk	Dandolo surface
Cu	70.8 ± 1.5	63.8 ± 1.8	80.2 ± 2.0	61.0 ± 1.6
Ag	26.9 ± 1	31.5 ± 1.5	18.6 ± 1.1	36.8 ± 2.0
Pb	2.3 ± 0.2	4.7 ± 0.2	1.4 ± 0.1	2.2 ± 0.1

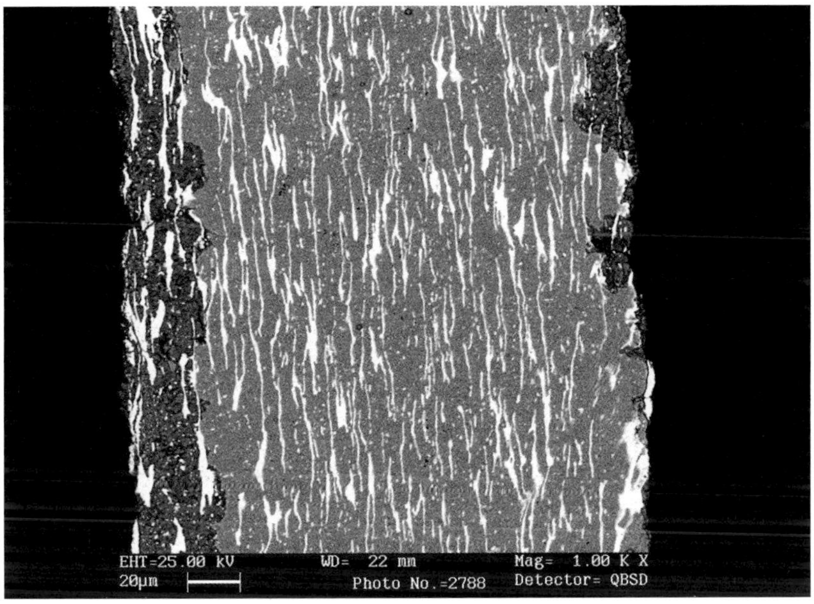

Figure 1 SEM-BSE image of a Malipiero coin showing the bulk microstructure typical of a copper-silver alloy.

the object to be plated, mercury is evaporated by heating, and silver remains to form a thin layer (mercury boiling point: 357°C). As mercury is never completely removed during heating and its presence, in most cases, is easily identified by XRF, we may conclude that only one of the coins attributed to Malipiero has been submitted to that process. Due to the high value of the coin, sectioning for the purposes of characterising its bulk composition and the microstructure of the surface layer was not allowed.

Taking the XRF thickness (estimated at 50–100 μm) of analysis into account, which can overestimate the surface composition in comparison to the bulk compositions, the emission from the Dandolo coins seems to indicate that they are more debased than the Malipiero coins, as expected by numismatics. Papadopoli's result is thus confirmed by a more meaningful number of samples. The silver content of such an alloy suggests a red/pink colour [7], which is very different from the shiny colour of silver-rich alloys, confirming that the coins would have been 'visually improved' before being circulated.

The SEM-BSE micrograph of a Malipiero coin in Figure 1 shows that the bulk microstructure is typical of a copper-silver alloy with the pro-eutectic alpha phase (7.9% silver) and the alpha + beta eutectic (92% silver and 5–7% lead).The elongated grains confirm that mechanical work was performed after casting in order to obtain the required thickness. The micrograph also reveals the presence of an exfoliated and porous surface layer, with different morphology and thickness in the range 10–70 μm. The bulk and surface chemical compositions, determined by EDX on two samples, are reported in Table 2. The silver contents of surface layers are higher than the bulk layers. These data and morphology can be explained by a modification process quite similar to that described by Cope regarding late Roman coinage [8]. They are a consequence of the heating of the coins during fabrication and particularly during the preparation of the coins from buttons and blanks, in the intermediate stages of conventional Roman minting practice [8]. Cope described the formation of a silver surface on copper-debased silver alloy in five successive steps: (a) casting the silver-copper alloy button, (b) heating the button in air to form a layer of copper oxides, (c) acid-pickling the flan to remove the copper oxides and reveal the silver-enriched phase, (d) hammering the blank for spreading the silver-rich phase laterally and (e) striking the blank. The morphology of the layers detected in the present coins suggests that a relatively similar process has been applied.

Conclusions

The results, from a numismatic and historical point of view, seem to demonstrate that Enrico Dandolo debased his coins, compared to those of his predecessor Orio Malipiero, and that the silver content of the Venetian coinage at the end of the twelfth century was around 25%, a figure very close to that (27%) obtained by some previous analyses. Such silver content made the Venetian coins probably the cheapest available on the monetary market of Italy, which could explain their great success. The coins were produced by casting and mechanical work, with chemical etching of the flans that resulted in a modified surface 10–80 μm thick.

XRF was useful to separate each emission through non-destructive examination of many samples, while the SEM-EDX was more appropriate in the investigation of the bulk and surface properties and the study of the metallographic microstructure; it also confirmed the XRF data [9,10].

References

1. Saccocci, A., 'La moneta in Italia e nell'occidente medioevali', in *Alle radici dell'euro: Quando la moneta fa la storia*, Canova, Treviso (2001) 112.
2. De Ruitz, M., *Monete a Venezia nel tardo Medioevo: Un Ritorno alle Fonti*, Canova, Treviso (2001) 163.
3. Rolnick, A.J., Velde, F.R. and Weber, W.E., 'The debasement puzzle: an essay on medieval monetary history', *Federal Reserve Bank of Minneapolis Quarterly Review* **21**(4) (1997) 8–20.
4. Stahl, A.M., *Zecca: The Mint of Venice in the Middle Ages*, Johns Hopkins University Press, Baltimore (2000).
5. Papadopoli, N., *Le monete di Venezia*, Vol. I, Venezia (1893).
6. La Niece, S., 'Silvering', in *Metal Plating and Patination: Cultural, Technical and Historical Developments*, ed. S. La Niece and P. Craddock, Butterworth-Heinemann, Oxford (1993) 201–210.
7. *ASM Handbook Volume 2. Properties and Selection: Nonferrous Alloys and Special-Purpose Materials*, ASM International (1990-11-01).
8. Cope, L.H., 'Surface-silvered ancient coins', in *Methods of Chemical and Metallurgical Investigation of Ancient Coinage*, ed. E.T. Hall and D.M. Metcalf, Royal Numismatic Society, London (1972) 261–278.
9. Beck, L., Bosonnet, S., Reveillon, S., Eliot, D. and Pilon, F., 'Silver surface enrichment of silver-copper alloys: a limitation for the analysis of ancient silver coins by surface techniques', *Nuclear Instruments and Methods in Physics Research B* **226** (2004) 153–162.
10. Calliari, I., Magrini, M., Zambon, A., Guerriero, P. and Martini, R., 'Microstructural and compositional characterization of Roman coins', X-*Ray Spectrometry* **28** *(*1999) 86–90.

Authors' addresses

- Irene Calliari, Department of Industrial Engineering, University of Padua, Padua, Italy (irene.calliari@unipd.it)
- Michele Asolati, Department of Archaeological Science, University of Padua, Padua, Italy
- Andrea Saccocci, Department of Cultural Heritage History and Conservation, University of Udine, Udine, Italy
- Francesco Grazzi, Consiglio Nazionale delle Ricerche, Istituto Sistemi Complessi, Sesto Fiorentino (FI), Italy
- Antonella Scherillo, Italy and Science and Technology Facility Council, ISIS Facility, Chilton, UK

Using stereo imaging in a scanning electron microscope study of the anatomical changes to *Mimosa* wood from traditional kilns in Pernambuco, north-east Brazil

Caroline Cartwright, Chris Jones, Peter Gasson and Claudia Luizon Dias Leme

KEYWORDS stereo imaging, charcoal, wood anatomy, calcium oxalate crystals, kilns, Brazil

Introduction

Wood anatomists routinely use the optical microscope when observing the detailed cellular structure revealed in transverse (TS), radial longitudinal (RLS) and tangential longitudinal (TLS) thin-sections to identify wood species. Variable pressure scanning electron microscopy (VP-SEM) and field emission scanning electron microscopy (FE-SEM) have provided the optimal means whereby all the qualitative characters one would expect to find in wood can be recognised in the charcoal. Wood retains most of such features when charred, but the dimensions and appearance of the cells change in various ways. Wood density, anatomical structure, moisture content, duration and temperature all influence wood behaviour during charring. Here we describe one specific aspect – stereo imaging in the SEM – which augments the published study of the qualitative changes that take place in the wood of *Mimosa ophthalmocentra* and *Mimosa tenuiflora* (Fabaceae family) when charred artificially at temperatures of 400°, 600° and 800°C, and then compared with charcoal produced in a traditional temporary kiln in Pernambuco in north-east Brazil [1]. The combined results can be applied to interpreting the conditions in which charcoal has been produced, and can also be extrapolated to characterise and reflect the range of temperatures occurring archaeologically in open fires and in kilns.

Methods, results and discussion

SEM examination of charcoal samples

Charcoal samples were fractured along the TS, RLS and TLS planes, mounted onto aluminium SEM stubs with Leit-C cement, coated with platinum for four minutes using an Emitech 550 sputter coater and examined using the secondary electron (SE) detector in the Hitachi S-4800 FE-SEM and in the Hitachi S-3700N VP-SEM, using the backscattered electron (BSE) detector. The detailed methodology and results will not be repeated here as they have been fully published [1]. This contribution focuses on new aspects of the study where we have used stereo imaging in the SEM to characterise changes in calcium oxalate crystals (in the charcoal) induced by different charring temperatures both in the experimental and in the traditional kiln.

Stereo imaging in the SEM

In this section we describe the principles and procedures of stereo imaging, briefly highlight the role played by the presence of crystals in the fissuring of wood structure according to temperature increases, and outline the significance of observing the changing features on prismatic crystals of calcium oxalate as an index of temperature change during the charring processes.

Stereo-paired images, often described as anaglyph images, are generated to provide a stereoscopic 3D view of an object. The effect is best observed using stereo glasses, where the two lenses are chromatically opposite. Colours commonly utilised include red (left eye) and cyan or green (right eye). The two images generated to form the stereo pair usually capture the same apparent field of view, but using a change of tilt between the two images to give the effect of a small lateral shift or parallax effect. When the two images are overlain and viewed through the stereo glasses an integrated image can be observed, giving the effect of three-dimensionality. For the imaging technique to work successfully it is important to retain the same field of view, magnification and focus for both images. Changing the SEM focus between capturing the two images may have the effect of introducing some rotation in one of the images, thereby compromising the finished result. Using an SEM with eucentric stage movement, allowing the stage to be tilted around the eucentric centre or focal point will help considerably. The tilt axis of the microscope should preferably give a left to right (lateral) movement to the field of view. If the stage tilt makes a north–south movement, the eye will not see the lateral angular change between the two captured images and the technique will not work. However, this can be

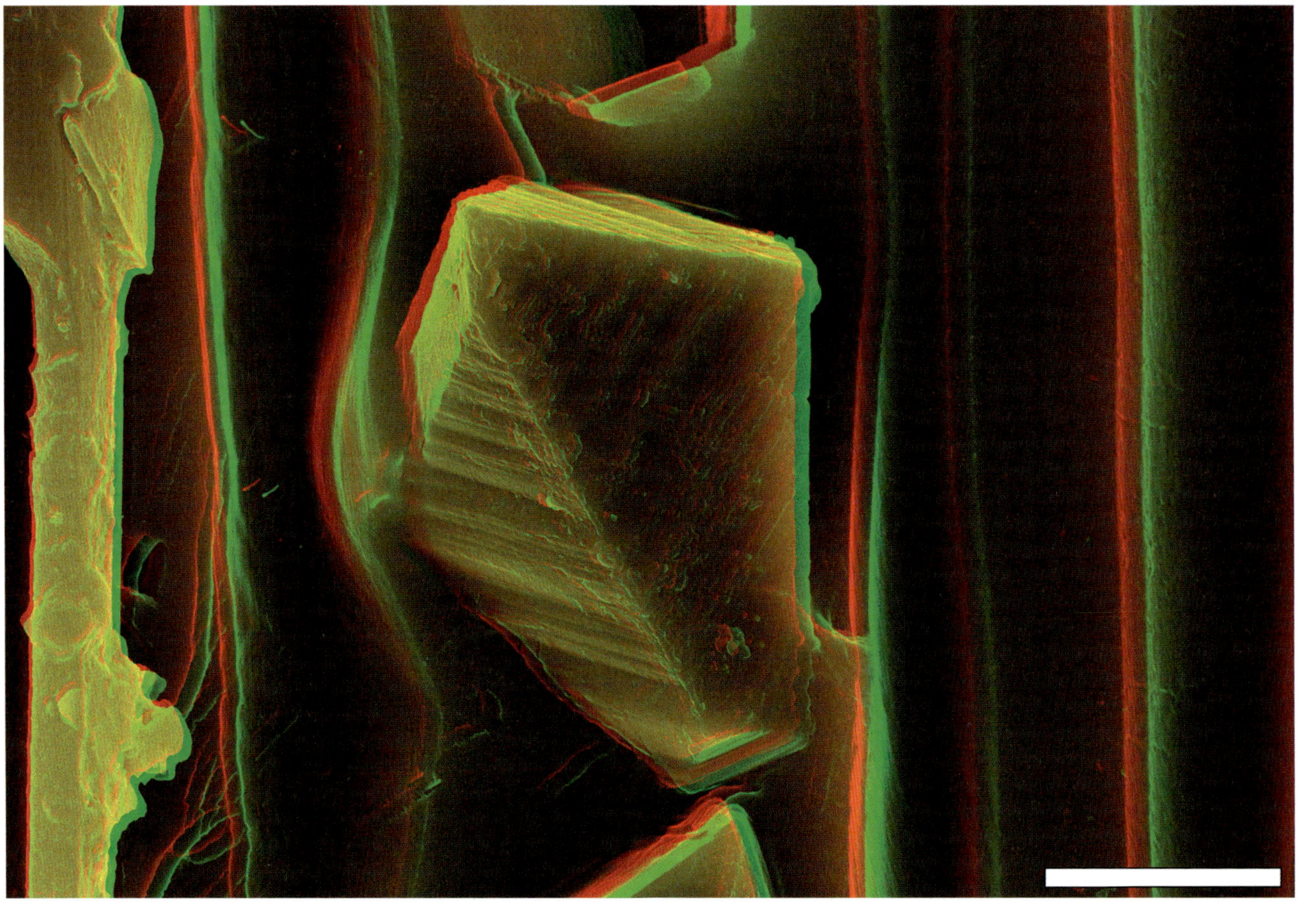

Figure 1 TLS view of a prismatic calcium oxalate crystal located in an axial parenchyma cell in *Mimosa tenuiflora* charcoal. Viewing this crystal in stereo mode in the FE-SEM (Hitachi S4800) emphasises the details of the characteristic temperature-induced striations that were only seen on Pernambuco kiln charcoal. Scale bar = 5 µm.

addressed by applying 90° raster (scan) rotation to the image, making the image on the SEM viewing monitor appear to shift left–right when tilt is applied. If the sample must be in a specific orientation, perhaps to illustrate a morphological feature, stage rotation may be used to set an appropriate view before scan rotation is applied. The degree to which the stage is tilted between capturing the two near-identical images is ultimately a process of trial and error. For comparatively modest magnifications up to a few hundred times, the first image will be captured using zero tilt, followed by a second image captured with the stage tilted usually somewhere between +6° and +10°. For higher magnifications in the order of a few thousand times less tilt is required. For example, a stereo image of the charcoal sample captured at an original magnification of 10× required the two images to be taken using zero tilt and +3° tilt. Software is commonly available to assist in the overlay of the two images to form a single stereo image. The software, which may already be available within the microscope's operating system or available as a freeware package, will often apply a colour hue to one or both images prior to creating the final stereo view.

Observing temperature-induced crystal structure deterioration

Applying the above procedures, stereo imaging was used in the SEM to enhance the features present in the prismatic crystals of calcium oxalate (Figure 1) that are located in chambered axial parenchyma strands and are particularly abundant in marginal parenchyma where they are tangentially aligned. These crystals have proved useful in evaluating changes in structure according to the different temperatures of the charring process. For example, at 400°C, radial to oblique fissures begin to occur amongst the fibres of *M. ophthalmocentra*, and tangential fissures occur along the marginal axial parenchyma where there are abundant crystals. We speculate that the presence of crystals may influence this fissuring because it rarely occurs in the paratracheal and vasicentric parenchyma, where crystals are much less abundant. The radial and tangential fissuring that began at 400°C becomes very apparent at 600°C; tangential fissures are plainly visible along the marginal axial parenchyma, especially where there are very abundant crystals. With regard to the changing morphology of the crystals, some new and important information has emerged from this study. At 400°C, their shape is usually retained although some have begun to crack. At 600°C small pits start to appear in some crystals and at 800°C crystals can appear quite spongy. In *M. tenuiflora* kiln charcoal, some crystals have begun to disintegrate and take on a granular texture when broken, whilst others appear striated or foliated (Figure 1). This suggests that the extended charring time in the kiln may have had an effect on crystal structure, although it is important to note that calcium oxalate crystals do not physically disintegrate until they reach a temperature of 1600°C, which far exceeds our charring temperatures.

Conclusions

The SEM examination of *M. ophthalmocentra* and *M. tenuiflora* has confirmed that the qualitative characters one would expect to find in wood are preserved in charcoal, and that it is possible to infer the conditions under which the charcoal was formed. Despite the fact that in highly fired charcoal, anatomical changes make the cellular structure dissimilar to that of reference wood thin-sections, by experimentally replicating [1], documenting and characterising anatomical changes induced in the charcoal through rising kiln temperatures, it has been possible to establish an augmented set of key features for identification. In this regard, high resolution FE-SEM examination and stereo FE-SEM imaging of the progressive deterioration of the structure of calcium oxalate crystals have provided useful new data. With increasing use of native and plantation-grown wood for charcoal in parts of the world with subsistence economies, there may be an enhanced demand for the identification of charcoals in similar case studies to help determine whether or not the wood has been legitimately harvested. Such results are also of use to those interested in renewable biomass resources, management of forests for fuel wood and charcoal, and the ultimate sustainability of fuel wood resources.

Reference

1. Leme, C.L., Cartwright, C.R. and Gasson, P.G., 'Anatomical changes to the wood of *Mimosa ophthalmocentra* and *Mimosa tenuiflora* when charred at different temperatures', *IAWA Journal* **31** (2010) 333–351.

Authors' addresses

- Caroline Cartwright, Department of Conservation and Scientific Research, The British Museum, London, UK (ccartwright@thebritishmuseum.ac.uk)
- Chris Jones, Electron Microscopy and Mineral Analysis (EMMA), Department of Mineralogy, Natural History Museum, London, UK (formerly at Hitachi High-Technologies Europe GmbH, Maidenhead, UK)
- Peter Gasson, Jodrell Laboratory, Royal Botanic Gardens, Kew, UK
- Claudia Luizon Dias Leme, Universidade Federal de Bahia, Depto de Botanica, IBIO – UFBA, Salvador, Brazil

Microstructure and impurities of bronze mirrors fabricated in the Koryo period (tenth–fourteenth century)

Jung Eun Choi and Masahiro Kitada

KEYWORDS Koryo, bronze mirror, copper sulphide, microstructure, alloy

Introduction

Different types of metal and metal craft technologies were extensively developed in Korea. Koryo (高麗) was the name of the Korean dynasty that lasted from the tenth to the fourteenth century. Koryo bronze mirrors were widely used in religious ceremonies and by the noble class as accessories. The backs of the mirrors were enhanced with various decorative patterns. Until the middle of the Koryo dynasty, such decorative patterns were influenced by the Chinese [1].

A study of these mirrors was considered useful in order to understand the metal craft technologies of the Koryo period, particularly as no detailed studies of the microstructures of Koryo bronze mirrors have yet been published. This article investigates the microstructures and impurities of cast bronze mirrors that were fabricated during the Koryo period.

Experimental

Optical microscopy (OM) and scanning electron microscopy (SEM) were used to observe the microstructures of the Koryo bronze mirrors. Polished mirror samples were made in order to identify each phase of the microstructure in the SEM. Energy dispersive X-ray (EDX) analysis was performed to determine the sample composition. Transmission electron microscopy (TEM) (Hitachi, H-9500) was used in combination with SEM to identify the impurities and observe the microstructures. A focused ion beam (FIB) (Hitachi, Fb-2100) was used to prepare samples of the mirrors for TEM analysis.

Results

Koryo Somoon mirror

The Koryo (tenth–fourteenth century) Somoon mirror (高麗素紋鏡 in Chinese) is an undecorated old Korean Somoon mirror [2]. The diameter of this mirror is 174 mm, and its weight is 0.656 kg. The Koryo Somoon mirror shown in Figure 1a is composed of copper (Cu) with 15.2 wt% lead (Pb), 9.3 wt% tin (Sn), and 0.5 wt% sulphur (S). The bronze mirror is plated with tin because the amount of tin contained in the mirror is too low to give it its silver colour. A study of the bronze mirrors used during the Edo period (seventeenth–nineteenth century) revealed that a tin layer existed on their surfaces, and that they were manufactured using a mercury amalgam [3]. In addition, 24 bronze mirror samples, which were recovered from the Korean peninsula, were subjected to a special treatment; their surfaces were typically coated with tin by three different processes referred to as wipe-tinning, amalgam-tinning and mercury polishing [4] in order to give a clearer image. In this study it was found that surface treatment of about 10% of tin bronze mirrors (e.g. in this sample mirror) was carried out, but high-tin bronze mirror surface coatings were not observed. The microstructure of the Koryo Somoon mirror contains the following four phases: dendritic copper-rich α phase, δ phase, Pb, and Cu-S grains. The shape of the dendritic α phase shows that this sample was made by the cast structure method. The δ phase is responsible for making the mirror white. As the tin content increases, the mirror becomes whiter because the δ phase is proportional to the concentration of tin.

The composition of the δ phase is copper with 26.7 wt% Sn and 1.3 wt% Pb. X-ray diffraction (XRD) analysis revealed that the chemical formula of the δ phase is $Cu_{41}Sn_{11}$. Lead enhances the fluidity of the molten bronze [5] and such high quantities of tin make the bronze alloy white. The fact that polygonal Cu-S grains exist around lead phases has not yet been reported. An SEM image of the Cu-S grains shows they are 3~30 μm in length and 3~5 μm in width. According to the results of the EDX analysis, the molar ratio of Cu:S is 2:1, and iron, lead, and zinc are detected as trace elements. This grain's electron diffraction pattern agrees with that of monoclinic Cu_2S (JCPDS 33-0490).

Next, the grain boundary between lead and the Cu_2S phase was studied using a TEM. It was observed that very small crystals exist between the lead and Cu_2S grains as shown in Figure 1b. These very small crystals cannot be observed using

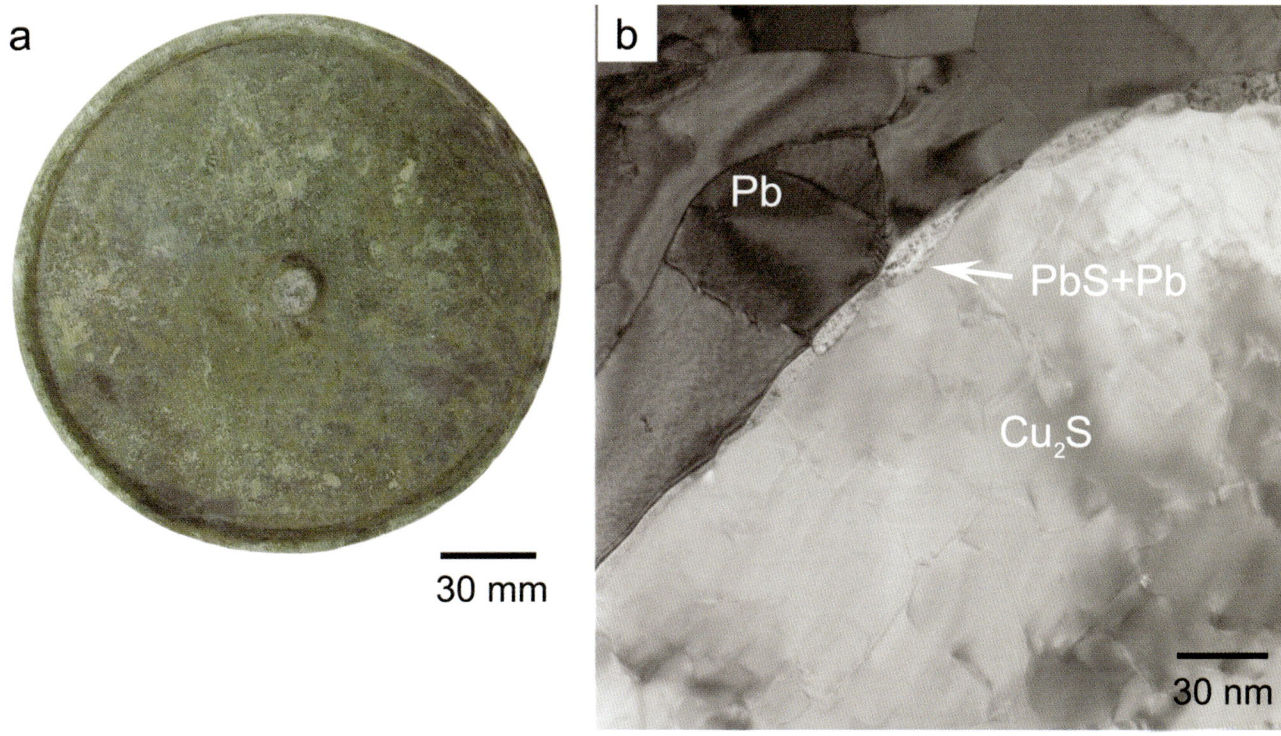

Figure 1 (a) Photograph of the Koryo Somoon mirror used as a specimen; (b) a magnified TEM image of an area containing Pb, Cu_2S and small crystals of PbS + Pb. (Photo: M. Kitada.)

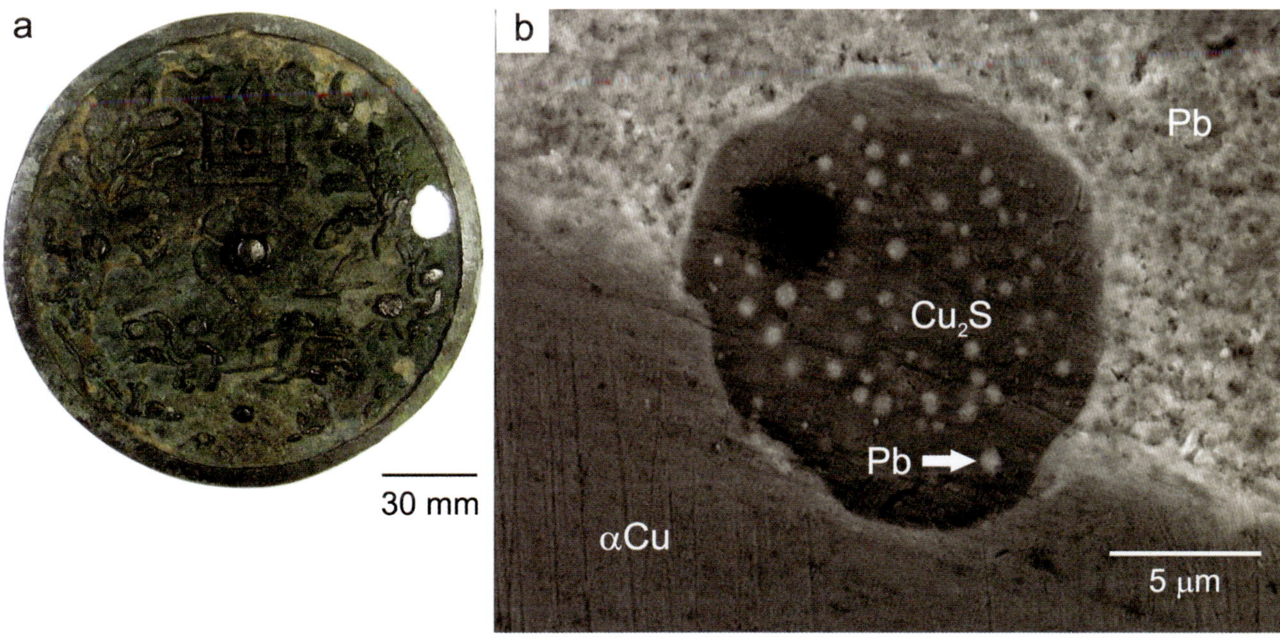

Figure 2 (a) Photograph of the Koryo Yuasosangmoon mirror used as a specimen; (b) an SEM image of a Cu_2S grain. The bright grains (one of which is indicated by the arrow) are Pb. (Photo: M. Kitada.)

SEM. The magnified TEM image of a crystal shows that it is approximately 200 nm wide and 100 nm long. The composition of the very small crystal is mainly lead with 10.0 wt% (41 mol%) S and 0.9 wt% (2 mol%) Cu. According to the EDX analysis, the molar ratio of Pb:S is approximately 1:1, and the electron diffraction pattern agrees with that of PbS (JCPDS 05-0592). It is thought that PbS is formed by a reaction between Cu_2S and Pb.

Generally, chalcopyrite ($CuFeS_2$) is used for making copper. Chalcopyrite needs to be roasted to oxidise the iron, and then smelted to form fayalite/olivine slag and copper- and iron-rich matte. All the iron is removed into the slag by oxidation. Subsequently, the copper sulphide starts to oxidise and then becomes Cu. In this case, copper sulphide is an intermediate product in the copper refining process. It is thought that the reasons behind the remaining copper sulphide are low refining temperature, the short reaction time between copper sulphide and O_2 and insufficient O_2 gas [6].

Koryo Yuasosangmoon mirror

The Koryo Yuasosangmoon mirror (高麗遊兒書像紋鏡 in Chinese) is a decorated old Korean mirror with a design depicting boys (Yuasosangmoon) playing with a musical instrument [2]. The diameter of this mirror is 102 mm, and its weight is 0.093 kg. The composition of the Yuasosangmoon mirror shown in Figure 2a is copper with 9.7 wt% lead, 10 wt% tin and 0.6 wt% sulphur. The microstructure is similar to that of the Koryo Somoon mirror and shows the copper-rich α phase, δ phase ($Cu_{41}Sn_{11}$), Pb and Cu-S grains. The Cu-S grain shown in Figure 2b is circular with a diameter of 2~20 μm and contains very small bright grains, as indicated by the arrow. The bright grains are smaller than 1 μm in diameter. The molar ratio of Cu:S was determined by EDX to be 2:1, and lead and silver were detected as trace elements.

The XRD pattern of the Cu-S grain agrees with that of monoclinic Cu_2S (JCPDS33-0490). This diffraction pattern is the same as that of the Cu_2S grains observed in the Koryo Somoon mirror. However, the Cu_2S grains of the Koryo Somoon mirror do not contain any inclusions, whereas the Cu_2S grains of the Koryo Yuasosangmoon mirror contain bright phases as shown in Figure 2b. The composition of these bright grains is lead with 13.6 wt% Cu and the XRD pattern agrees with that of Pb. Moreover, PbS crystals are observed to be present between the Cu_2S and the Pb grains.

The difference in the shapes of the Cu_2S grains of the two abovementioned mirrors is considered to be attributed to different solidification processes during their fabrication. Cu_2S has the highest melting point of 1280°C and is considered to be a very important compound for understanding the copper refining technology used during the Koryo period. If more data on Cu_2S are accumulated, regional and historical differences might be found, and classification may become possible.

Preliminary conclusions

The microstructures of the two Koryo bronze mirrors have four phases: dendritic copper-rich α phase, δ phase ($Cu_{41}Sn_{11}$), lead, and copper sulphide (Cu_2S), and exhibited cast microstructures. Cu_2S is present in both Koryo bronze mirrors examined. The copper sulphide (Cu_2S) compound is considered to be an intermediate product obtained during the copper refining process. The Cu_2S grains of the Koryo Somoon mirror have a single phase, whereas those in the Koryo Yuasosangmoon mirror contain Pb. It is thought that the difference in shapes of the Cu_2S grains of the two Koryo bronze mirrors is due to differences in their solidification processes. At the boundary between the Cu_2S and Pb grains, PbS crystals are observed. PbS is formed by a reaction between Cu_2S and Pb.

Acknowledgements

We gratefully acknowledge the assistance in transmission electron microscopy observation provided by Hitachi High-Technologies Corporation.

References

1. Lee, N.Y., *Koryokyong Youngu*, Shinyou, Seoul (2003)103–109.
2. Whang, H.G., *Kankokumonyousi*, Isshi, Kyoto (1978) 41.
3. Kitada, M., Kirino, F. and Yamamoto, K., 'Surface structures of metallic mirror fabricated in the late Edo era', *Journal of the Japan Institute of Metals* **71** (2007) 85–89.
4. Jeon, I.H., Lee, J.S., Beak, J.H. and Park, J.S., 'Bronze mirrors; surface-treatments; wipe-tinning; amalgam-tinning; mercury-tinning', *Journal of Conservation Science* **22** (2008) 87–98.
5. Valério, P., Silva, R.J.C., Monge Soares, A.M., Araújo, M.F., Braz Fernandes, F.M., Silva, A.C. and Berrocal-Rangel, L., 'Technological continuity in Early Iron Age bronze metallurgy at the south-western Iberian peninsula: a sight from Castro dos Ratinhos', *Journal of Archaeological Science* **37**(8) (2010) 1811–1819.
6. Azakami, T. and Awakura, Y., *Kinzoku Kagaku Nyumon Series 3, Kinzoku Seiren Kogaku (Extractive Metallurgy)*, The Japan Institute of Metals, Sendai (1996) 25–32.

Authors' addresses

- Jung Eun Choi, Tokyo University of the Arts, Tokyo, Japan (jechoi1203@gmail.com)
- Masahiro Kitada, Tokyo University of the Arts, Tokyo, Japan

Characterisation of an eighteenth-century Meissen plate from the Götzendorf-Grabowski service using VP-SEM and HV-SEM

Kelly Domoney, Andrew Shortland and Sebastian Kuhn

KEYWORDS SEM, XRF, Meissen, porcelain, eighteenth century

Introduction

Variable pressure and high vacuum scanning electron microscopy (VP-SEM and HV-SEM) and handheld X-ray fluorescence spectroscopy (HH-XRF) analyses are being carried out on two major collections of eighteenth-century hard-paste porcelain produced by the Meissen factory, Saxony. These form part of a collaborative doctoral research project between Cranfield University and Bonhams, London. The project aims to characterise the composition of overglaze enamel decoration, glazes and porcelain bodies through time in order to identify objects with later decoration and fakes, copies and reproductions of eighteenth-century objects.

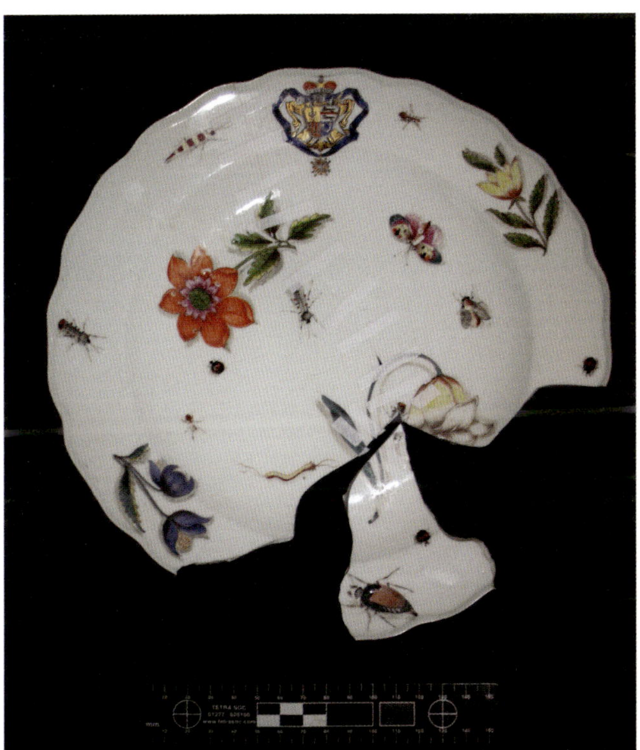

Figure 1 Meissen armorial plate from the Götzendorf-Grabowski service, 1740.

Previous studies [1–5] have traditionally used chromium-based green enamels and the presence of zinc in blue enamels to date Meissen porcelain. Body formulations also have the potential to be used as a dating tool; contemporary records suggest that the formula for porcelain changed some time after 1719 from using a lime flux to a potassium-rich feldspar flux [6].

This study presents compositional results of the porcelain body, glaze and overglaze enamel colours from a fragmented armorial plate from the Götzendorf-Grabowski service produced by the Meissen factory in 1740 (Figure 1). The plate is representative of the early high status services produced for the electors of Saxony given as diplomatic gifts to royal courts throughout Europe. The plate later formed part of the von Klemperer collection, Dresden, sometime between 1890 and 1926 [7]. The collection suffered irreparable war damage in 1945; breakage occurred to the plate and five other objects investigated in this project. The broken nature of the plate and group as a whole provided a rare opportunity to sample and analyse securely dated early eighteenth-century Meissen sherds in the SEM.

Methodology

The investigations comprise three stages based on the availability of collections with specific date ranges. Stage 1 aims to establish the composition of early porcelains securely dated between 1710 and 1763. Two sample groups are being used: the von Klemperer collection, which consists of six fragmented objects including the plate from the Götzendorf-Grabowski service, and the Hoffmeister collection, which comprises over 300 complete objects.

Due to its fragmentary nature, the von Klemperer collection was selected for analysis by SEM-EDX. Flat surface chips of enamel, polished cross-sections and large fragments were analysed under HV and VP in a Leo 435VP SEM fitted with an EDAX energy dispersive spectrometer (EDX) at 20 kV, 200 μA and between 2000 and 3000 counts per second

Table 1 Results of HV and VP SEM-EDX analyses of glass standard Corning C prepared three ways and body, glaze and enamels of a Meissen armorial plate from the Götzendorf-Grabowski service. (Wt% oxides normalised to 100%; –: below detection limit; n: number of individual analyses in the average results.)

Sample	Colour	Munsell colour	*n*	SiO_2	Al_2O_3	CaO	MgO	Na_2O	K_2O	Fe_2O_3	TiO_2	CoO	CuO	MnO	NiO	ZnO	As_2O_3	SnO_2	Sb_2O_3	BaO	PbO	Au_2O_3
Corning C																						
Target composition*	n/a	n/a	n/a	34.8**	0.9	5.1	2.8	1.1	2.8	0.3	0.8	0.2	1.1	0.8**	0.0	0.1	–	0.2	0.0	11.4	36.7	–
Polished section; coated; HV	n/a	n/a	3	35.6	1.0	4.8	2.3	1.0	2.8	0.3	1.5	0.2	1.1	0.0	0.0	0.0	–	0.3	0.0	11.0	37.5	–
Fragment; uncoated; VP	n/a	n/a	3	36.6	1.1	4.8	2.4	1.1	2.6	0.3	1.8	0.2	1.2	0.1	0.1	0.1	–	0.0	0.0	10.9	36.2	–
Surface chips; coated; HV	n/a	n/a	4	35.9	1.0	4.8	2.4	1.0	2.7	0.3	1.6	0.2	1.2	0.1	0.0	0.1	–	0.0	0.0	11.1	37.3	–
Meissen Plate from Götzendorf-Grabowski service																						
Body																						
GG1, GG3[a]	n/a	n/a	4	60.0	35.5	0.5	0.1	0.6	2.9	0.4	–	–	–	–	–	–	–	–	–	–	–	–
Glaze																						
GGc1, GGc2[b]	n/a	n/a	4	72.0	16.2	2.8	0.1	0.8	4.5	1.0	0.1	–	–	–	–	–	–	–	–	–	2.2	–
Enamels																						
GG_G1[c]	Yellow	10 Y 8/8	2	38.1	5.8	4.1	0.2	0.2	0.5	1.0	–	–	–	–	–	–	–	–	7.2	–	42.7	–
GG_G1[c]	Green (yellow)	5 GY 5/6	2	25.9	4.0	2.6	0.2	0.2	0.4	1.2	–	0.1	4.9	–	–	–	–	0.1	10.9	–	49.3	–
GG_2[b]	Green (yellow)	5 GY 5/6	3	27.3	3.9	3.0	0.1	0.2	3.0	1.0	–	–	6.1	–	–	–	–	–	6.5	–	51.3	–
GG_B6[c]	Turquoise (dark)	10 GY 4/6	2	39.6	6.3	3.4	0.2	–	0.5	0.5	–	5.0	7.3	–	1.2	–	0.6	0.4	–	–	35.0	–
GG_B6[c]	Blue	7.5 PB 3/8	2	38.1	2.5	3.3	0.1	1.0	1.6	0.7	–	2.7	–	–	0.3	–	0.4	8.3	–	–	40.9	–
GG_L[c]	Purple (mid)	10 P 3/10	2	50.0	8.5	4.2	0.4	0.4	0.7	0.9	–	0.1	–	0.1	0.1	0.2	–	5.2	–	–	25.9	3.0
GG_C[c]	Red (brick)	10 R 3/10	2	52.6	10.0	4.1	0.2	0.4	0.9	13.3	0.1	–	–	–	–	–	–	–	–	0.1	18.1	–
GG_4[b]	Brown (dark)	5 YR 2/4	2	39.1	5.2	2.4	0.3	0.4	1.0	12.5	–	–	–	0.6	–	–	0.2	–	2.0	–	35.9	–
GG_C[c]	Black	N1.75	2	40.0	5.6	2.9	0.4	0.3	0.3	6.0	–	1.4	0.2	5.7	0.3	–	–	–	–	–	36.7	–

* Target composition of Corning C [12]

** Unpublished data from Brill cited in Vincenzi [13]; Mn not found in replicate analyses of Corning C by Umpire and Control Services Inc. using $LiBO_2$/ICP [12].

[a] Samples prepared as polished cross-sections, carbon coated and analysed using HV SEM-EDX at 20 kV and 200 μA.

[b] Samples consist of flat surface chips, stub mounted, carbon coated and analysed using HV SEM-EDX at 20 kV and 200 μA.

[c] Samples consist of large fragments analysed using VP SEM-EDX at 20 kV and 200 μA.

using EDAX Genesis software. Glass standard 'Corning B' was used to monitor analytical drift. At least two analyses were conducted on each sample, and the mean composition calculated. The instrument allowed analysis of elements with an atomic number of 6 (carbon) and above, and a detection limit of 0.1% was applied to all elements. In order to validate the results obtained from polished sections, fragments and surface chips using VP or HV mode, samples of glass standard 'Corning C' were prepared and analysed in exactly the same manner as the porcelain samples. Three to four analyses were carried out on each sample and the results averaged. As shown in Table 1, the results of the three preparation techniques and two conditions were virtually identical and consistent with the published values.

HH-XRF is being conducted on the larger von Klemperer fragments and complete objects from the Hoffmeister collection using an Oxford Instruments X-MET 5100 analyser at 45 kV and 15 kV with an 8 mm diameter X-ray beam spot size. The relationship between concentration and XRF peak area data will be evaluated using a range of lead glass standards close in composition to the enamel paints.

Stages 2 and 3 of this project aim to complete the chronology of Meissen porcelain technology up to c.1950. Objects securely dated to 1760–1800 and 1800–1950 in conjunction with known copies, fakes and reproductions will be analysed using HV-SEM, VP-SEM and HH-XRF.

Preliminary results: plate from the Götzendorf-Grabowski service

The mean oxide composition of the porcelain body indicates the paste consists of high potash (2.9%) and low lime (0.5%). Results are consistent with formulations dating to post-1731 produced using a potassium-rich feldspar flux [8–10]. The glaze consists of the same components as the body but with higher concentrations of silica, potash and lime to create the clear hard, glassy layer on which enamel colours were applied. Low levels of lead are also present in the glaze. Levels of silica (72.0%), alumina (16.2%) and potash (4.5%) are also consistent with glazes previously analysed dating to post-1731 [10].

Enamel colours consist of high levels of lead and silica with low levels of various metallic oxides. Results suggest that these metallic oxides are acting as pigments suspended in lead silicate fluxes that enable the decoration to be fused to the glaze surface. Yellow enamels are coloured with high levels of antimony and lead indicative of the pigment Naples yellow. Antimony is also present in green enamels with a yellow hue along with high levels of copper as the colour-forming minerals. Blue enamels are coloured with cobalt (2.7%) with high levels of tin (8.3%) most likely used as an opacifier. Cobalt is also present in the dark turquoise enamel in addition to copper. Nickel and arsenic are present in enamels with cobalt and are indicative of impurities within the cobalt ore. High levels of gold (3.0%) and tin (5.2%) occur in the purple enamel and are indicative of purple of Cassius, a colour preparation derived from the precipitation of gold particles with tin chlorides [11]. The red enamels are coloured with iron, as are the brown enamels in addition to antimony. Iron and antimony are recorded as the constituents of 'contour brown', a preparation described in a formula book dating to 1731 attributed to Meissen master painter Höroldt [6]. Analyses of the black enamels suggest that a combination of pigments was employed including iron, manganese, cobalt and copper.

Conclusions

This SEM-EDX study has accurately characterised the composition of the body, glaze and overglaze enamel colours of a securely dated high status Meissen object. Results are consistent with known eighteenth-century technologies and form the basis of a baseline compositional dataset. Continuing research aims to complete the characterisation study in order to elucidate changes in eighteenth-century porcelain technology and determine the potential for HH-XRF to be used as a non-destructive tool to attribute complete objects of unknown date.

Acknowledgements

We would like to thank the Hoffmeister family and the von Klemperer family for their kind permission in allowing us to use their collections for this project.

References

1. Bezur, A. and Casadio, F., 'Du Paquier porcelain – artistic expression and technological mastery: a scientific evaluation of the materials', in *Fired By Passion: Vienna Baroque Porcelain of Claudius Innocentius Du Pacquier*, ed. M. Chilton, Arnoldsche Art, Hartford and Stutgaart (2009) 1165–1214.
2. Page, J. and Chilton, M., 'Scientific analysis of Meissen Commedia dell'Arte sculpture in the Gardiner Museum Collection', in *Harlequin Unmasked: The Commedia dell'Arte and Porcelain Sculpture*, ed. M. Chilton, Yale University Press, New Haven (2001) 323–326.
3. Lee, L., Emerin K. and Hess H., *Semi-quantitative XRF Analysis of Meissen Porcelain*, technical report, Busch-Reisinger Museum Collection (2008).
4. Halfpenny, P. and Mass J., 'Two converging perspectives on one Meissen tureen', *American Ceramic Circle Journal* **XV** (2009) 71–77.
5. Stiegelschmitt, A. and Tomandl, G., 'Untersuchung der Materialzusammensetzung alter Porzellane im Hinblick auf ihre Kunsthistorische Zuordnung', *Sprechsal* **118**(10) (1985) 975–978.
6. Heintze, Dr., 'The development of the chemico-technical management from the beginning to the present time', in *Meissen China an Illustrated History*, ed. K. Berling, Dover Publications, New York (1972) 117–143.
7. Kuhn, S., 'Collecting culture: the taste for eighteenth-century German porcelain', in *The Arnhold Collection of Meissen Porcelain 1710–50*, ed. M. Cassidy-Geiger, The Frick Collection, New York (2008) 22–118.
8. Schulle, W. and Ullrich, B., 'Ergebisse gefügeanalytischer Untersuchungen an Böttgerporzellan', *Silikattechnik* **33** (1982) 44–47.
9. Hornig-Sutter, M., 'Meißener Porzellane des 18. Jahrhunderts im Visier von Kunstgeschichte und Naturwissenschaft. Teil III', *Welkunst* **55**(1) (1985) 18–23.
10. Schulle, W. and Ullrich, B., 'Orientierende Untersuchungen an der Gasuren von historischen Meißener Porzellanproben', *Silikattechnik* **36** (1985) 170–173.
11. Shaw, K., *Ceramic Colours and Pottery Decoration*, Maclaren and Sons Ltd., London (1968).
12. Brill, R.H., *Chemical Analyses of Early Glasses*, vol. 2, The Corning Museum of Glass, Corning, NY (1999).

13. Vincenzi, E.P., Eggins, S., Logan, A. and Wysoczanski, R., 'Microbeam characterization of Corning archaeological reference glasses: new additions to the Smithsonian Microbeam Standard Collection', *Journal of Research of the National Institute of Standards and Technology* **107** (2002) 719–727.

Authors' addresses

- Kelly Domoney, Centre for Archaeological and Forensic Analysis, Cranfield University, Shrivenham, UK (k.domoney@cranfield.ac.uk)
- Andrew Shortland, Centre for Archaeological and Forensic Analysis, Cranfield University, Shrivenham, UK
- Sebastian Kuhn, British and Continental Ceramics Department, Bonhams, London, UK

An example of the application of SEM-EDX-XRF to the study of ancient gold artefacts

Daniela Ferro and Vania Virgili

KEYWORDS SEM, EDX, XRF, ancient jewellery, micro diagnostics

Introduction

The scientific diagnostics field is constantly striving towards new analytical methods for the acquisition of accurate results that can provide complementary data to recognise the manufacturing technology of historical and artistic objects. In the study of jewellery, elemental composition in specific areas (e.g. soldering areas and decoration items) gives useful information that contributes towards a better understanding of the goldsmith's techniques. Scanning electron microscopy with energy dispersive X-ray analysis (SEM-EDX) is commonly used now in the cultural heritage sector. An interesting improvement of this technique is the coupling of SEM-EDX with X-ray fluorescence (XRF).

Methodology

The SEM-EDX-XRF system was produced by inserting an X-ray tube (RX38) with a tungsten (W) anode into the SEM LEO450VP chamber. The X-ray tube was attached at a specific angle of 45° with respect to the sample and 90° with respect to the solid state Si(Li) EDX detector (FWHM of 70 eV for F) that detects the characteristic X-rays generated by the sample. The beam potential of both systems was 20 kV. In order to interface the SEM with the X-ray tube a flange was constructed consisting of a window of Kapton ($C_{22}H_{10}N_2O_5$) and a brass collimator that allows a 100 μm spot to be created. This fixed system, obtained by means of geometrical considerations and experimental tests, allows the focus of the electron and the X-ray beam on the same micro-area of the sample. EDX data quantification was performed using the ZAF4 corrective program of the Oxford Instruments INCA300 system. With regard to XRF qualitative analysis, XRF spectra were acquired with the same Si(Li) detector used for EDX, whilst for quantitative determination they were processed separately through peak deconvolution, and calibrated using a set of gold alloy standards.

This combined configuration allows us to exploit jointly the capabilities of the two kinds of excitation sources, electrons and X-rays, and of the two techniques, SEM-EDX and XRF [1–3]. On the one hand, XRF tends to be more analytically precise, provides quantitative analysis of major elements and qualitative analysis of minor and sometimes trace elements present at deeper layers than those detected by EDX, but it is limited to a larger analysed sample area (100 μm). On the other hand, SEM-EDX takes advantage of smaller analysed areas (1 μm) at very high magnification and enables the examination of the surface of the object at micro- and nano-scale, the determination of the concentration of the major elements of the alloy, and the acquisition of maps of the distribution of elements on the surface; yet it lacks the precision of XRF analysis.

Results

The complementary capabilities of SEM-EDX-XRF were applied to the study of samples produced by experimental archaeology and of ancient pieces of jewellery. In order to evaluate the different penetration depths of the two beams, preliminary measurements with both techniques were carried out on a multilayered specimen consisting of several layers of different materials and thicknesses. In this experimental test, EDX detected the elements from the first two layers (nickel-lanthanum of 1 mm and silicon oxide of 1 mm), whilst XRF identified those from the deeper layers, respectively the layer of gold of 10^{-5} mm and copper of 1 mm. The analytical results demonstrate that SEM-EDX-XRF is able to provide depth profiles of elemental composition from the outer surface (EDX) to the sub-surface regions (XRF).

SEM-EDX-XRF was applied to a modern gold griffin head, manufactured in the context of experimental archaeology. It was filled with clay in order to reproduce the ancient goldsmithing procedure for manufacturing a granulated three-dimensional figure. Some authors consider that clay was used to provide a malleable structural support that would also absorb heat during the soldering of gold granules [4]. In order to produce the reference sample, a gold alloy with 91.5 wt% Au, 7.0 wt% Ag and 1.5 wt% Cu was used; these values have been considered as a calibration test for the analytical instrument. The comparison between the EDX and XRF spectra (Figure 1) highlighted the significant advantages of using SEM-EDX-XRF. In particular, the XRF spectrum reveals traces of iridium, which was not detected by EDX during an acquisition time of five measurements of 100 s each for both the systems. The increased sensitivity of XRF for

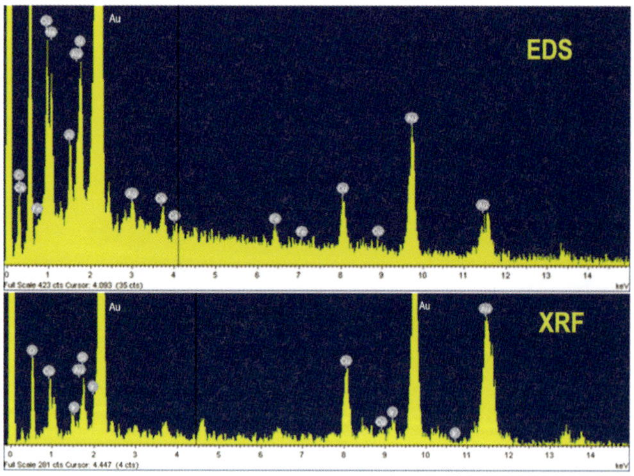

Figure 1 Comparison between EDX and XRF spectra from the same area of analysis of a gold alloy sample.

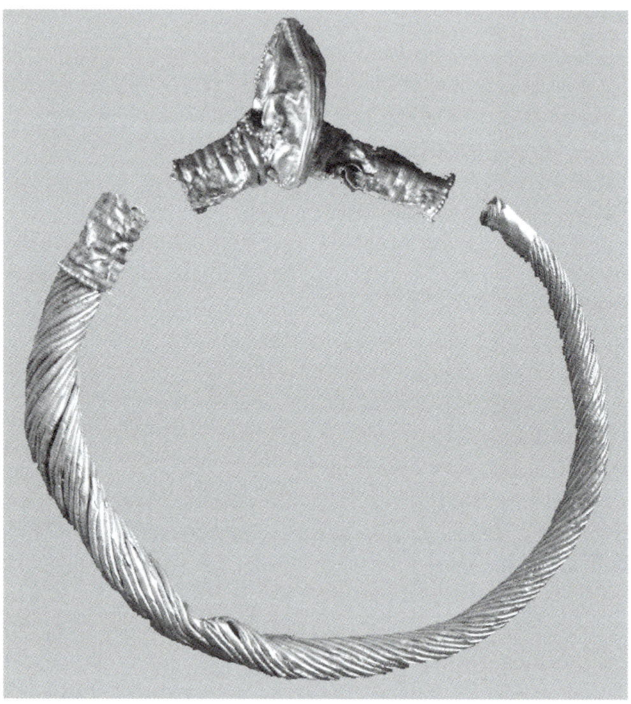

Figure 2 Hoop earring, fourth century BC, Archaeological Museum of Adria, Italy.

minor and trace elements is due to the improved signal-to-noise (peak-to-background) ratio of the XRF spectrum over the EDX spectrum where the higher Bremsstrahlung background makes it harder to detect smaller peaks, i.e. elements in minor and trace concentrations. Figure 1 illustrates the capability of XRF to detect elements in the spectrum low energy region. The vacuum environmental chamber of SEM provides increased efficiency with respect to conventional air-path XRF analysis by avoiding the absorption of low energy X-rays and permitting the enlargement of the range of detected elements. In addition, it is possible with SEM-EDX-XRF to take advantage of the fact that EDX microanalysis systems have a high quality Si(Li) detector, with an ultra-thin window that allows the analysis of critical light elements. In this way, SEM-EDX-XRF can use the electron beam excitation for lighter elements below 2 kV, and the more powerful XRF excitation for X-ray lines above 2 kV.

The capabilities of SEM-EDX-XRF described above allowed the understanding of the manufacturing process of a gold hoop earring from Adria (Figure 2), dated to the fourth century BC, from the Archaeological Museum of Adria. The earring is composed of a hoop and a clasp, and shows an elaborate manufacturing process with different parts made using several techniques.

A filling material was visible in several breaks in the artefact, especially in the inner part of the hollow hoop and inside the two semi-spheres of the bulla. X-ray radiography confirmed the presence of a continuous compact matrix in both components of the earring. Both EDX and XRF detected the presence of elements characteristic of clay (silicon, aluminium, calcium, magnesium and iron). XRF and EDX were used to evaluate a possible different composition of the filling material from the semi-quantitative estimations. The collected information could suggest the presence of an intentional filling material inside the bulla that the goldsmith used for technological reasons.

Reference literature states that the material usually used by ancient goldsmiths to fill hollow gold objects was an organic material such as bitumen or resin [5]. In the case of the Adria earring, the material found inside the bulla is an inorganic filling material such as clay, the use of which in the ancient world is yet to be confirmed.

Conclusions

The above case studies show the information that the SEM-EDX-XRF system was able to provide in the study of manufacturing technologies of ancient jewellery. From the preliminary results, it is possible to foresee its future application in several other types of studies such as that of bronze patinas or surface alterations of inorganic materials.

Acknowledgements

We are grateful to David Loepp for having manufactured the precious piece through experimental archaeology.

References

1. Ferro, D., Formigli, E., Pacini, A. and Tossini, D., *La saldatura nell'oreficeria antica: indagini archeometriche e archeologia sperimentale*, Edizioni Kappa, Rome (2008).
2. Cross, B.J. and Witherspoon, K.C., 'Integrated XRF in the SEM', *Microscopy and Microanalysis* **10**(2) (2004) 104–105.
3. Procop, M., Hodoroaba, V.-D. and Rackwitz, V., 'A microfocus X-ray source for improved EDS and XRF analysis in the SEM', *Microscopy and Analysis* **25**(4) (2011) 11–13.
4. Grimwade, M., *Introduction to Precious Metals: Metallurgy for Jewelers and Silversmiths*, Brynmorgen, Brunswick (2009).
5. Connan, J., 'Use and trade of bitumen in antiquity and prehistory: molecular archaeology reveals secrets of past civilizations', *Philosophical Transactions of the Royal Society of London B* **354** (1999), 33–50.

Authors' addresses

- Daniela Ferro, National Research Council of Italy - Institute of Nanostructured Materials, Rome, Italy (daniela.ferro@cnr.it)
- Vania Virgili, Department of Chemistry, 'Sapienza' University of Rome, Rome, Italy (vania.virgili@uniroma1.it)

Preliminary archaeometric study of the metallic grave goods from a rich late Roman burial at Torrejón de Velasco (Madrid, Spain)

Oscar García-Vuelta, Alicia Perea, Fabián Cuesta, Marc Gener, Ignacio Montero-Ruiz, Mercedes Murillo-Barroso and Martina Renzi

KEYWORDS archaeometallurgy, Late Roman jewellery, grave goods, Iberian Peninsula, SEM-EDX

Introduction

The site of Camino de Seseña (Torrejón de Velasco, 26 km south of Madrid) was excavated between September 2006 and February 2008.[1] An Imperial Roman villa was documented as well as other structures, among which stands out an isolated Late Roman tomb (fourth-fifth century AD), which is unique in the Iberian Peninsula in terms of its characteristics and funerary remains. The tomb consists of a rectangular pit (2.4 × 0.8 m, 1.5 m deep) without human remains. It contained a rich set of grave goods that included weapons, ritual and sumptuous objects of different materials (pottery, glass, bone and metal). This paper presents the preliminary study of the non-ferrous metallic materials using scanning electron microscopy with energy dispersive X-ray analysis (SEM-EDX) carried out at the Laboratorio de Microscopía Electrónica y Microanálisis (Microlab) at the Centro de Ciencias Humanas y Sociales, Consejo Superior de Investigaciones Científicas (CCHS-CSIC) in Madrid. This study is part of an ongoing wider research project.

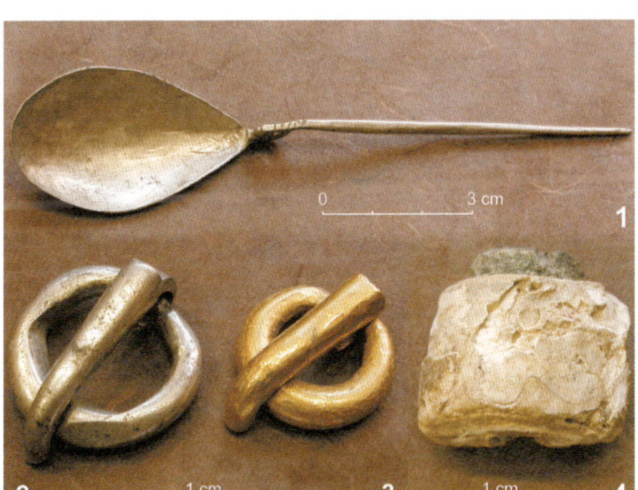

Figure 1 1: Silver spoon (TV5). 2: Silver belt buckle (TV9). 3: Gold belt buckle (TV11). 4: Sword amulet/pommel (TV8). (Images: O.G-Vuelta.)

Grave goods studied

The grave goods were found in two locations: at the bottom of the pit and in a lateral niche in the east wall of the tomb. The following objects were recovered from the bottom of the pit: a situla; two belt fittings with rivets; a brazier of copper-based alloy; three silver buckles and one in gold; and a possible ivory ring amulet (Figure 1). In the niche there were two silver objects (a spoon and a bowl). Some types of objects, such as the buckles and the silver spoon are not commonly found in Roman Iberia and it is the paucity of archaeometric studies for this kind of material in the Iberian Peninsula that prompted this study.

Methods

The equipment used in this study was a variable pressure scanning electron microscope (VP-SEM) (Hitachi 3400N-II), provided with a secondary electron (SE) and a backscattered

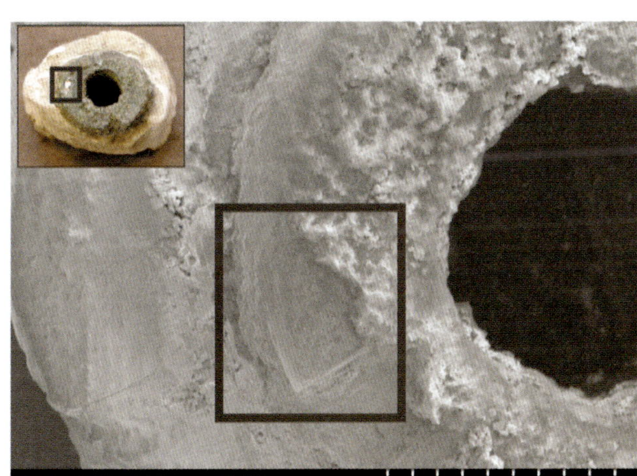

Figure 2 Sword amulet/pommel TV8. Micrograph of the bone/ivory body and the analysed central silver ring (polished). SEM-SE image, 15 kV, ×10. Left: object detail. (Image: O.G-Vuelta.)

Table 1 Analytical results for the tomb objects. (Det: element identified in the spectrum but not quantified (< 1%); – : not detected.)

Object	Inventory number	Ag	Cu	Au	Zn	Pb	Sn
Bowl	50393/4	93.3	6.6	Det	–	–	–
Spoon	50393/5	92.4	7.5	–	–	–	–
Sword pommel	50395/8	79.3	18.6	–	2.1	–	–
Large buckle (prong)	50395/9	95.6	4.4	–	–	Det	–
Large buckle (body)	50395/9	95.3	4.7	–	–	Det	–
Buckle (prong)	50395/10	96.3	3.7	–	–	–	–
Buckle (body)	50395/10	99	Det	Det	–	–	–
Buckle (prong)	50395/11	1.5	–	98.5	–	–	–
Buckle (body)	50395/11	1.7	–	98.3	–	–	–
Buckle (body)	50395/15	95.7	4.3	–	–	–	–
Buckle (prong)	50395/17	90.3	8.2	–	1.6	–	–
Situla	50395/4	–	87	–	–	7.2	6.1
Brazier	50395/3	–	73	–	21.1	3	2.8
Belt fitting	50395/12	–	78.4	–	3.6	9.9	7.2
Belt fitting (rivet)	50395/12	–	90.9	–	–	Det	9.1
Belt fitting	50395/13	–	87.3	–	4.3	–	8.3

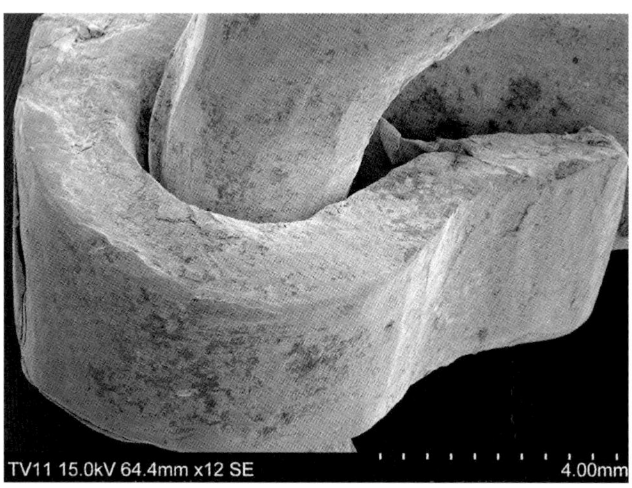

Figure 3 Plastic deformation working traces. Gold belt buckle TV11 (prong and body). SEM-SE image, 15 kV, ×12.

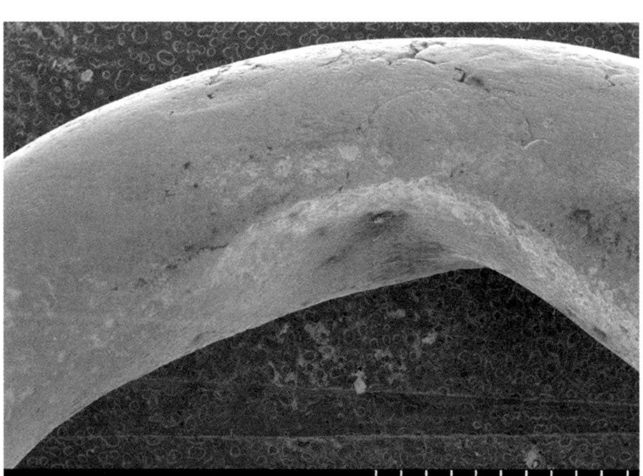

Figure 4 Wear traces at the inner face of the silver buckle TV9 body. SEM-SE image, 15 kV, ×10.

electron (BSE) detector. The EDX analysis was carried out with a Bruker Quantax 200 Xflash 4010 detector (SDD) and a Bruker SVE III Xflash analytical signal processing unit. The analytical results were processed using Bruker Quantax Sprit v.2.1 software.

Once the corrosion layer was mechanically removed from a small area (Figure 2), the objects were placed directly in the SEM chamber. Sampled metallic flakes were used to analyse objects that were too large for the chamber. Examination and analyses of the samples were done at high vacuum, using the SE detector.

Results are given in weight % (Table 1), as the average of three analyses for each sample (×300 window). Only elements over 1% were quantified. Those elements identified in the spectrum but not quantified are marked as 'Det' (< 1%).

Results

Copper-based alloys

It is worth noting that each object was made using different alloys: a tin-bronze rivet, a leaded bronze situla, a brass brazier with high levels of lead and tin, and belt fittings of mixed alloys (gunmetal and leaded gunmetal). The rivet and the plates of each of the two belt fittings were also made from different alloys. The predominance of leaded bronze and mixed alloys seems to be characteristic of Late Roman metallurgy [1, 2].

Gold and silver alloys

High purity silver (>95%) is used predominantly in all silver objects, as is usual for Late Roman silver objects in the Iberian

Peninsula studied thus far, although silver of less purity is also detected [3]. In addition, two items are noteworthy for being made of a silver-brass alloy, confirmed by the presence of zinc. The copper/zinc ratio varies between 84/16 and 90/10. Although silver-brass alloys were known since the second century AD in other areas of the Roman Empire [4], this is the earliest silver-brass alloy item currently identified in the Iberian Peninsula. Low levels of lead and gold have also been detected in some objects.

The gold buckle (50395/11) is also notable because of the purity of its metal (>98% Au). Topographic examination of the surface of artefacts with SEM shows tool marks (Figure 3) and use wear (Figure 4) in both gold and silver buckles.

Although the research regarding this grave is still in progress, and will include lead isotope analysis, the results obtained to date confirm the exceptional nature of this find, which provides new and important analytical data that augments what is currently available from the research of Late Roman metallurgy in the Iberian Peninsula.

Acknowledgements

This research was supported by the CONSOLIDER-INGENIO 2010 Program (CSD2007-00058) and the Comunidad Autónoma de Madrid (PADCAM Project S2007/HUM-543).

Note

1. The archaeological reports of this excavation are still in progress by the directors, R. Flores and P. Sanabria. All the objects have been deposited at the Museo Arqueológico Regional de la Comunidad de Madrid (Alcalá de Henares, Madrid).

References

1. Dungworth, D., 'Roman copper alloys: analysis of artefacts from northern Britain', *Journal of Archaeological Science* **24** (1997) 901–910.
2. Montero Ruíz, I., 'Estudios sobre metalurgia antigua en la provincia de Toledo: el proyecto Arqueometalurgia de la Península Ibérica', in *II Congreso de Arqueología de la provincia de Toledo. La Mancha occidental y la Mesa de Ocaña*, vol. I, Diputación provincial de Toledo, Toledo (2001) 275–301.
3. Rovira, S., Blázquez, Mª L., Ballester, A. and González, F., 'Estudio metalógráfico', in *El disco de Teodosio*, ed. M. Almagro Gorbea, Real Academia de la Historia, Madrid (2000) 139–150.
4. Mortimer, C., 'Early use of brass in silver alloys', *Oxford Journal of Archaeology* **5**(2) (1986) 233–242.

Authors' addresses

- Oscar García-Vuelta; Arqueometal Research Group, Instituto de Historia, Centro de Ciencias Humanas y Sociales (CCHS-CSIC), Madrid, Spain
- Alicia Perea, Arqueometal Research Group, Instituto de Historia, Centro de Ciencias Humanas y Sociales (CCHS-CSIC), Madrid, Spain (alicia.perea@cchs.csic.es)
- Fabián Cuesta; Arqueometal Research Group, Instituto de Historia, Centro de Ciencias Humanas y Sociales (CCHS-CSIC), Madrid, Spain
- Marc Gener, Arqueometal Research Group, Instituto de Historia, Centro de Ciencias Humanas y Sociales (CCHS-CSIC), Madrid, Spain
- Ignacio Montero-Ruiz, Arqueometal Research Group, Instituto de Historia, Centro de Ciencias Humanas y Sociales (CCHS-CSIC), Madrid, Spain
- Mercedes Murillo-Barroso, Arqueometal Research Group, Instituto de Historia, Centro de Ciencias Humanas y Sociales (CCHS-CSIC), Madrid, Spain
- Martina Renzi, Arqueometal Research Group, Instituto de Historia, Centro de Ciencias Humanas y Sociales (CCHS-CSIC), Madrid, Spain

Characterisation of metal threads from the decorations of the audience room in Dresden castle using SEM-EDX analysis

Sylvia Hoblyn and Christoph Herm

KEYWORDS metal threads, corrosion products, silver plated with gold

Introduction

Dresden castle was destroyed by fire in 1701 and King Augustus II the Strong had it rebuilt in the Baroque style. Some interiors survived the destruction of the castle at the end of World War II, as they had been taken to several safe sites. For the reconstruction of the state rooms these interiors are going to be restored. Part of the project is the throne ensemble of the audience room. Parts of the throne ensemble were manufactured in Leipzig by G & E Andreas Dietrich Apel in 1718–1719. A major cleaning procedure was carried out in 1891, but it is still unclear which parts of the decoration of the state rooms were cleaned and which cleaning materials were used. It is assumed that the throne ensemble was treated as one unit and therefore exposed to the same conditions [1]. The decorations consist of various textiles, many of which contain metal threads made of flattened wire strips wound around a silk core. The aim of the investigation using scanning electron microscopy with energy dispersive X-ray analysis (SEM-EDX) was to determine the metal used for these strips and to characterise their corrosion compounds. The investigations were complemented by cleansing tests carried out by Schramm [2] to compare the cleaned and uncleaned surfaces.

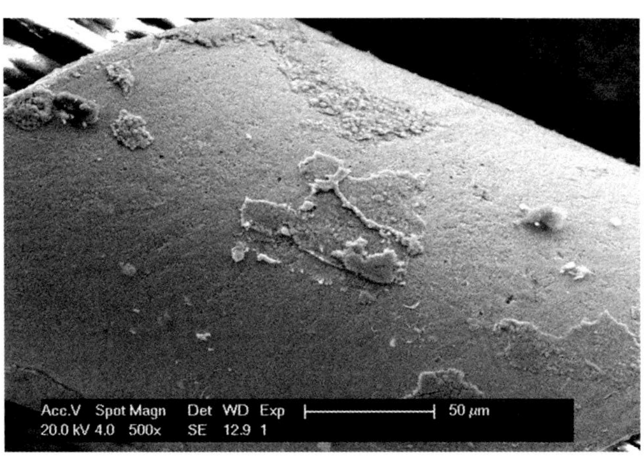

Figure 1 SEM-SE image of uncorroded sample with few deposits.

Experimental

The samples were mounted on aluminium plates with double-sided carbon tabs and tested without any further coating by using a Philips XL 30 SEM equipped with an Ametec EDX spectrometer. The following instrumental settings were used: 20 kV accelerating voltage, a spot size of 6 for the EDX analysis and various spot sizes (3, 4, 6) for imaging, and a working distance of 13 mm. The secondary electron detector was used for the SEM micrographs. The EDX spectrometer software EDAX Genesis (version 4.6) includes peak deconvolution to allow the identification of chlorine in the presence of silver.

For the cleansing tests Schramm used a commercial acidic thiourea solution (Einfach Brillant Silberbesteck Tauchbad) [3].

Samples

Samples were taken from ornaments of the baldachin (canopy) of the throne ensemble. Three typical examples are discussed in this paper:

1. With few or no obvious signs of corrosion, uncorroded (Figure 1).
2. With a dark layer of corrosion (Figure 2).
3. After the cleansing procedure with thiourea solution (Figure 3).

Results

The results show that the metal strips are made of silver, plated with a gold layer. The surface exhibited different types of corrosion. In uncorroded areas without any deposits (Figure 1), gold and silver were detected. The corrosion can be seen as fine deposits on the uncorroded sample and was visible as bulky crusts on the corroded sample (Figure 2).

Corrosion products were found on the outer and inner surfaces of the metal strips. The type of corrosion was the

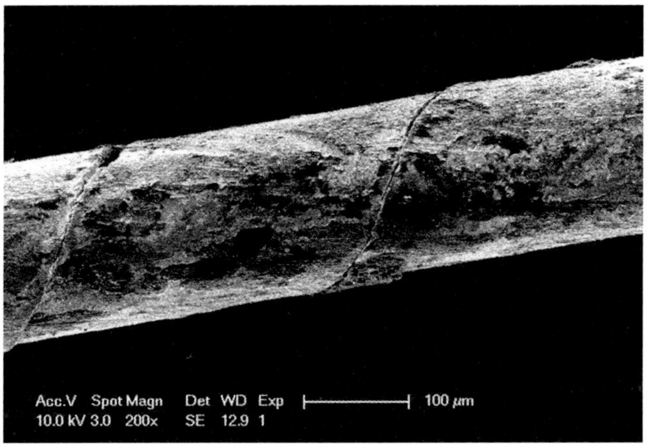

Figure 2 SEM-SE image of sample with layer of corrosion.

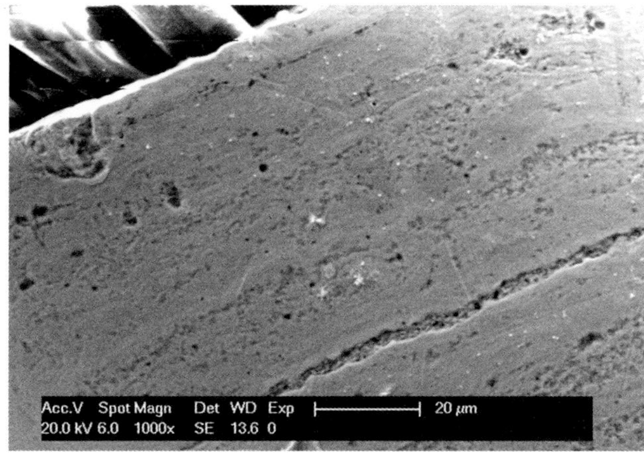

Figure 3 SEM-SE image of sample after treatment with the thiourea solution.

same on both sides of each sample. These corrosion products identified by EDX analysis consist mainly of silver sulphide and silver chloride, and have also been reported elsewhere on textiles decorated with silver threads [4]. The formation of silver sulphide by hydrogen sulphide in the air and the creation of a black surface are very common. The mechanism and the origins of the formation of silver chloride on surfaces of metal threads are still not clarified and need further investigation.

After treatment with the thiourea solution, a shiny metallic surface was obtained. However, substantial amounts of silver sulphide on the surface have still been detected by SEM-EDX, especially where the cleaning solution had no direct access. Parts of the cleaned surface show silver and gold without further elements. Other parts are porous and contain mainly silver. These pores became visible due to the cleaning process and show a streaky pattern (Figure 3).

Conclusions

The different corrosion processes resulting in fine deposits (uncorroded surface) or bulky deposits (corroded surface) indicate that the original metal thread surfaces were slightly different from one another. The uncorroded sample not treated with the cleaning solution shows scattered pores without a pore pattern (Figure 1). The porous surface of the cleaned sample shows a streaky pattern according to the direction of the roll during manufacturing (Figure 3). This indicates that these pores are a result of the corrosion process rather than the cleaning treatment. Obviously these metal threads were damaged more during the rolling process than the uncorroded threads. This damage evidently initiated the corrosion process with the subsequent formation of the bulky crust (Figure 2).

References

1. Schneider, S., Dr. Schneider & Kuester, Leipzig, personal communication (2010).
2. Schramm H.-P., Laboratory Prof. Schramm, Dresden, unpublished report nr. 11c/07 and 77/07 (2007).
3. www.brillant-schmuckpflege.de/produkte/bestecktauch.htm (accessed 10 July 2011).
4. Anheier, B. and Stauffer, A., 'Erkenntnisse und Fragen bei der rasterelektronenmikroskopischen Untersuchung von Metallfäden aus der Zeit zwischen 1600 und 1800', *Zeitschrift für Kunsttechnologie und Konservierung* **12**(2) (2003) 378–386.

Authors' addresses

- Sylvia Hoblyn, Hochschule für Bildende Künste Dresden, Labor für Archäometrie, Dresden, Germany (hoblyn@servl.hfbk-dresden.de)
- Christoph Herm, Hochschule für Bildende Künste Dresden, Labor für Archäometrie, Dresden, Germany

Metal threads: evaluation of a cleaning method

Ingrid Karina Jiménez Cosme, Carolusa González Tirado, Jannen Contreras Vargas and José Luis Ruvalcaba Sil

KEYWORDS gilding, gilded thread, thiourea, corrosion, silver

Introduction

One of the most popular international methods for cleaning metal threads was the use of a commercial silver dip containing an organosulphur chelating compound called thiourea ($SC(NH_2)_2$) and hydrochloric acid, to reach pH 1. Due to the potential harmful nature of this material, a research project for alternative cleaning methods was undertaken in 2008 (Figure 1).

Although some studies have been carried out on the damage caused by thiourea residues on silver [1–3], an experiment using different thiourea solutions and a gel was designed in order to provide reliable data to assist the decision-making on the cleaning of silver corrosion on gilded silver threads. A solution of thiourea-formic acid at pH 5.5, and a purpose-made thiourea gel with the same characteristics, were tested and compared against a commercial silver dip available in Mexico under the name of Limpiasteg 130, which has a very similar composition to that of Goddard's silver dip used in the UK.

Methods

Thirty-eight samples of metal strip wound around silk were taken from an eighteenth-century cloak from the Franz Mayer Museum collection – half from two threads twisted together and half from single threads. This garment was selected after non-destructive analysis; first using an optical microscope (OM) to verify the wires' appearance and to verify that core fibres were silk, and then using a portable X-ray fluorescence (XRF) device that allowed us to identify the wires as gilded silver because of the amount of each element on the different layers, core and plating.

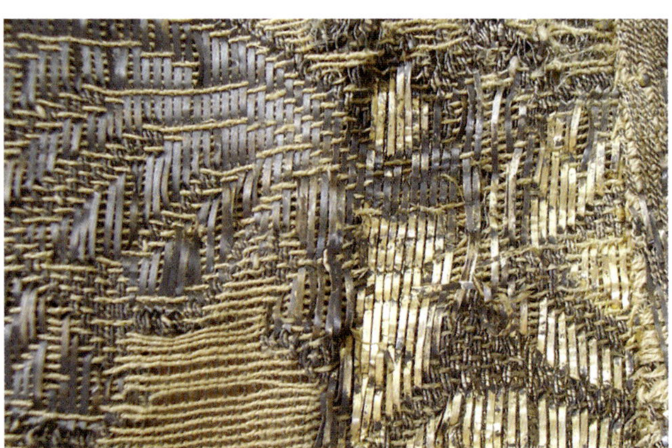

Figure 1 Example of a cleaning process affecting silk threads. (Image: Ingrid Jimenez.)

First, the manufacturing technique of each sample was identified and its conservation status was established and digitally recorded using OM photography. The second step was to clean the threads with three different thiourea mixtures using the commercial silver dip Limpiasteg 130, which contains hydrochloric acid and pH 0, a thiourea solution with formic acid with pH 5.5, and a thiourea gel with formic acid, pH5.5, hydroxyl propyl methyl cellulose (HPMC) as thickener and Triton X100 as surfactant. The third step was to evaluate the appearance of the metal surface by OM and scanning electron microscopy (SEM).

Application was carried out using cotton buds at environmental temperature; time of exposition was different for each mixture: for liquid solutions less than five seconds rubbing with a cotton bud, for gel 30 seconds of exposure; in all cases distilled water was used for clearing.

Results

The SEM images show that both double and single yarns have parallel line markings from the drawing or pulling manufacturing technique. All 19 samples of single yarn threads show a very compact corrosion layer of small crystals, whilst the remaining 19 samples of double thread have a porous layer of needle-shaped sulphide crystals. The main difference between both threads was the type of corrosion; the first group showed pitting corrosion, and the second, corrosion over the gilt layer but poor adhesion between metallic layers (Figures 2 and 3). Single yarn threads were more resistant to the silver sulphide dissolution method, but cleaning was heterogeneous, and required more time for reaction.

The two types of corrosion exhibited by both groups of samples (single and double yarn threads) suggest that the

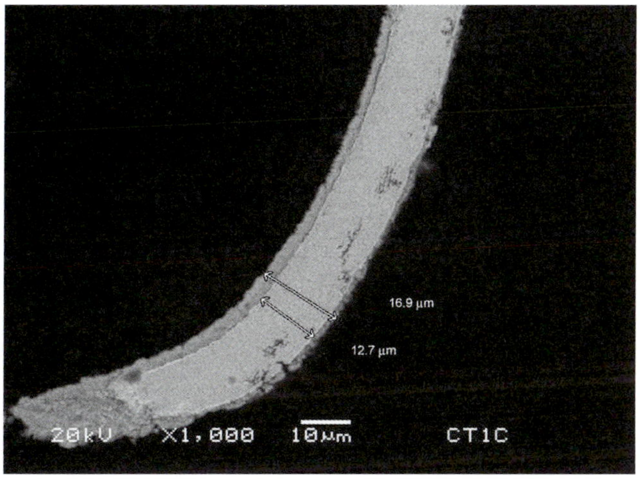

Figure 2 Pitting corrosion in a single yarn. (SEM image: Gerardo Villa.)

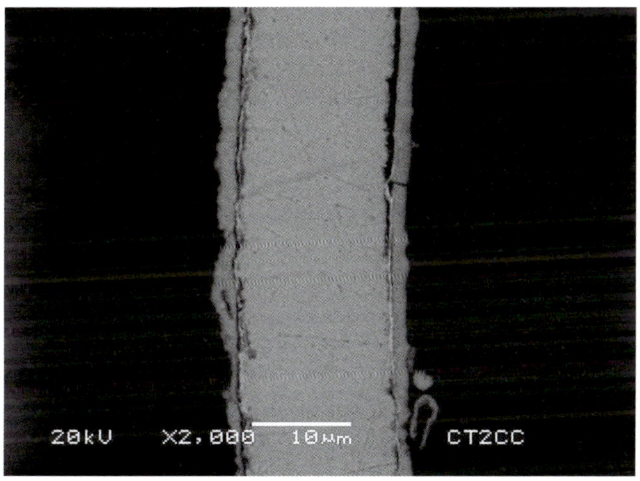

Figure 3 Corrosion over the gilt layer. (SEM image: Gerardo Villa.)

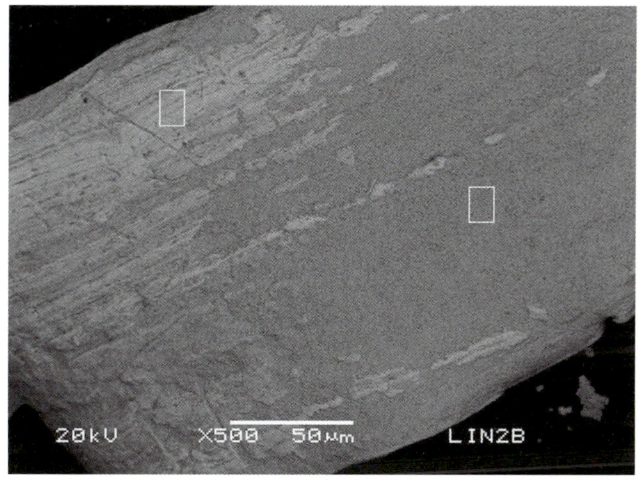

Figure 4 The exposed silver nucleus. (SEM image: Gerardo Villa.)

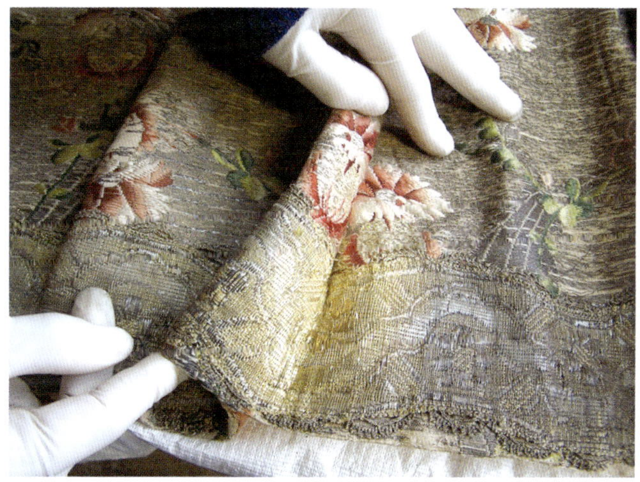

Figure 5 Example of metal thread deterioration. (Image: Ingrid Jimenez.)

deterioration is directly related to the subtle differences between the manufacturing techniques of both metal threads. It is important to emphasise that both groups were exposed to exactly the same conditions as they originate from the same cloak.

Samples cleaned with the commercial silver dip (pH 0) showed areas where silver sulphides were completely stripped. However in other areas, metallic silver had also dissolved, causing the collapse of the thin gilding. Gold particles scattered on the surrounding fabric were also detected. The silk fibres display fractures and loss of mechanical strength.

Threads cleaned with the thiourea gel (pH 5.5) exhibited an intermediate cleaning level; a very thin layer of silver sulphide remained and it was very difficult to remove gel residues without affecting the structure of the threads. However, the use of a gel allowed a single-spot application of the reagent, and the contact time with the corrosion products increased. This improved the reaction under less acidic and more controlled conditions, which was safer for silk fibres. In both cases it was necessary to rinse well in order to remove residues and to apply a slight mechanical action on the thread, which unfortunately may have caused distortion and even loss of fibres. The thiourea-formic acid solution (pH 5.5), however, being a liquid, was absorbed by the fibres. During cleaning, the effect on the corrosion products was similar to that obtained by using a gel.

Conclusions

None of the applied methods produced a complete elimination of silver sulphides; cleaning was heterogeneous in all cases, with the silver nucleus exposed next to the corrosion products (Figure 4). This may trigger new corrosion processes due to the presence of newly formed anodic and cathodic areas. The 5.5 pH procedures were not as aggressive to fibres as the pH 1 commercial silver dip, which easily promotes silk hydrolysis.

None of the cleaning methods used in this experiment can be considered safe for both the organic and the inorganic components of metal threads. In most cases the manufacturing technique and the conservation status of this kind of object is a question of compromise between the most efficient method of cleaning the metals and the most satisfactory treatment for silk. Thus, if the main objective of cleaning silver sulphide from metal threads is purely aesthetic, it is necessary to consider the potential harm to the object due to the possible irreversible negative effects of the treatment in the short, medium or long term (Figure 5).

Acknowledgements

We would like to thank conservator Ricardo Pérez Alvarez of the Museo Franz Mayer.

References

1. Berger, S. and White, W.B., *The Daguerreotype: Nineteenth Century Technology and Modern Science*, Smithsonian Institution, Washington, DC (1991).
2. Contreras Vargas, J., 'El camino de la fórmula', *Intervención, Revista Internacional de Conservación, Restauración y Museología, INAH* **1** (2010) 25–33.
3. Kurth, D.G., 'The reaction of thiourea to dicyandiamidine sulfate on silver surfaces investigated by reflection-absorption infrared spectroscopy', *Fresenius Journal of Analytical Chemistry* **365**(5) (1999) 465–466.

Authors' addresses

- Ingrid Karina Jiménez Cosme, Escuela Nacional de Conservación, Restauración y Museografía, San Diego Churubusco, Mexico (ingridkjc@gmail.com)
- Carolusa González Tirado, Escuela Nacional de Conservación, Restauración y Museografía, San Diego Churubusco, Mexico (rcarolusagt@gmail.com)
- Jannen Contreras Vargas, Escuela Nacional de Conservación, Restauración y Museografía, San Diego Churubusco, Mexico (jannencontreras@gmail.com)
- José Luis Ruvalcaba Sil, Instituto de Física-UNAM, Mexico (sil@fisica.unam.mx)

Tetitla, Porch 25: scientific analysis applied to the preservation of a mural painting from Teotihuacan

Ingrid Karina Jiménez Cosme

KEYWORDS Teotihuacan, mural painting, archaeometry, manufacturing technique

Introduction

Tetitla was discovered by French archaeologist Laurette Sejourne in 1963. It is a settlement complex, occupied from AD 200 to 700, and its mural paintings show the technical and stylistic evolution of Teotihuacan painting. The depiction of the Red Eagles belongs to the Early Tlalmimilolpan phase, about AD 200–300 (Figure 1).

A conservation project, coordinated by the National School of Conservation, Restoration and Museology (ENCRyM-INAH, Mexico), began in 2007. During the treatment of the floor, some fragments of wall paintings were found in the pavement. These fragments, which belong to Porch 25, revealed the dramatic image change and showed that a black pigment was still present in the fragments, although it could no longer be seen in the painting on the wall (Figure 2).

Methodology

The aim of this research is to understand all the causes of deterioration of the black pigment, and establish the materials and manufacturing techniques used by the craftsmen of Tetitla. Based on the information derived from literature research about the materials and manufacturing techniques of the paintings of Teotihuacan, small models were made to reproduce some of the techniques, using the original materials, in order to compare their characteristics with those of the original fragments. The fragments selected were analysed by optical microscopy (OM), petrographic analysis and scanning electron microscopy with energy dispersive X-ray analysis (SEM-EDX) (OM-Leica, SEM JEOL JSM-6460 low vacuum).

Discussion and results

Cross-sections from the unburied fragments show a distinct separation between the plaster and black paint layer, ruling out the hypothesis of a *fresco* technique mentioned by Villagra, Torres and Littmann [1–3]. Examination by SEM-EDX revealed a difference in texture between the layers; the paint layer is more porous than the plaster. There were no traces of burnishing, no clays were found and the strata showed no traces of compaction (Figures 3 and 4). The elemental analysis was carried out without confirmation of the presence of muscovite mica found by Magaloni in 2001 [4]; however lampblack

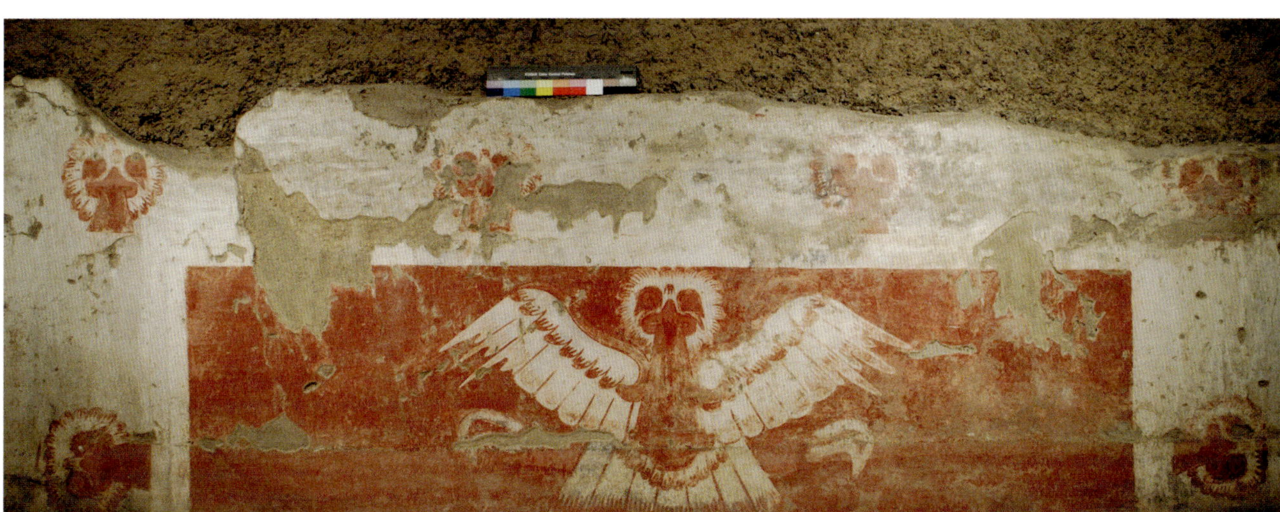

Figure 1 South wall of Porch 25, September 2007: before treatment. (Image: Gerardo Ruiz Hellion.)

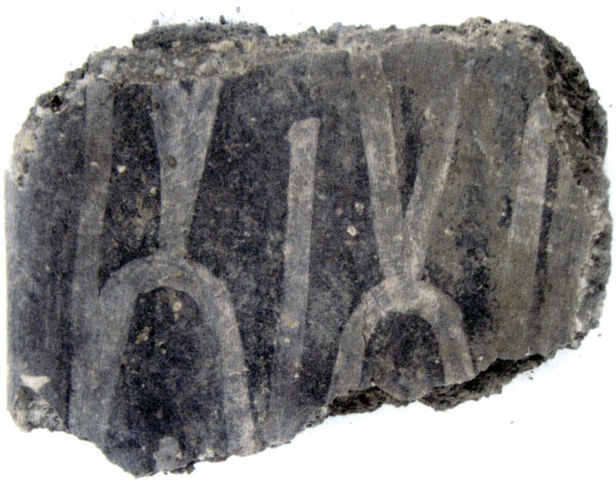

Figure 2 Unburied fragment, July 2008. (Image: Ingrid Jimenez.)

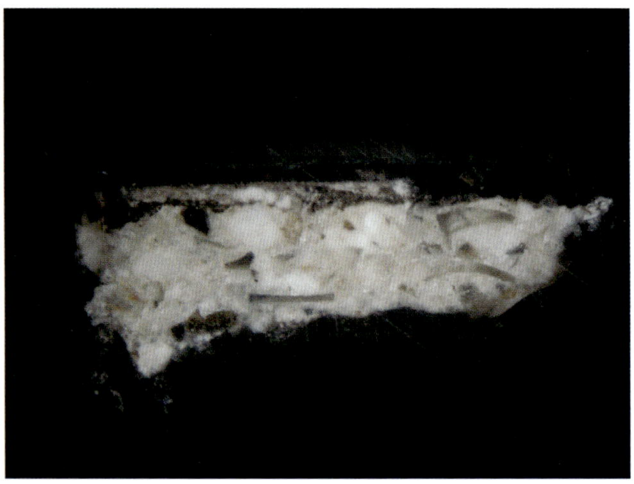

Figure 3 Cross-section from one of the fragments discovered in 2008, OM image, ×20. (Image: Ingrid Jimenez.)

and a high percentage of calcium were identified in the paint layer. Petrographic analysis revealed the presence of volcanic materials such as quartz, tuff, basalt, pyroxene, plagioclase, and traces of obsidian. Calcium carbonate and silica-rich fillers were found in the plaster.

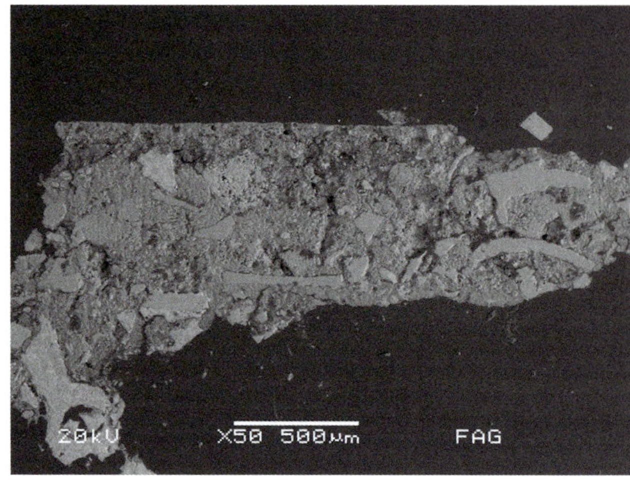

Figure 4 Cross-section detail in SEM: the difference between porosity and thickness, February 2008. (Image: Jose Antonio Alba.)

After identification of the original materials, model replicas of the paintings were made using limestone, silica sand, commercial lampblack (carbon black), lampblack from ocote wood, mica and polishers made of obsidian, in order to understand the manufacture of the paintings (Figure 5). Ocote is a conifer (*Pinus montezumae*) which, according to paleoenvironmental studies by McClung [5], was common in the Teotihuacan area. Based on this information we made replicas using a black pigment from ocote. The results of experiments on ocote lampblack carbon showed that it has superior hiding power compared to that of a commercial carbon black, resulting in a highly saturated colour even if mixed with lime. To prepare a paint with carbon black the use of a surfactant or emulsifier was necessary; this may indicate that Teotihuacan painters had knowledge of the use of such materials in wall painting.

Porosity and poor adhesion of the black layer can be associated with the use of a degraded binder, but the high calcium content suggests that the *secco* technique was used for the black paint, although it had only been previously reported in the green and blue colours (malachite and azurite due to sensitivity to the alkalinity of lime) (Figure 6). In Mexico that type of technique is commonly prepared with an organic binder.

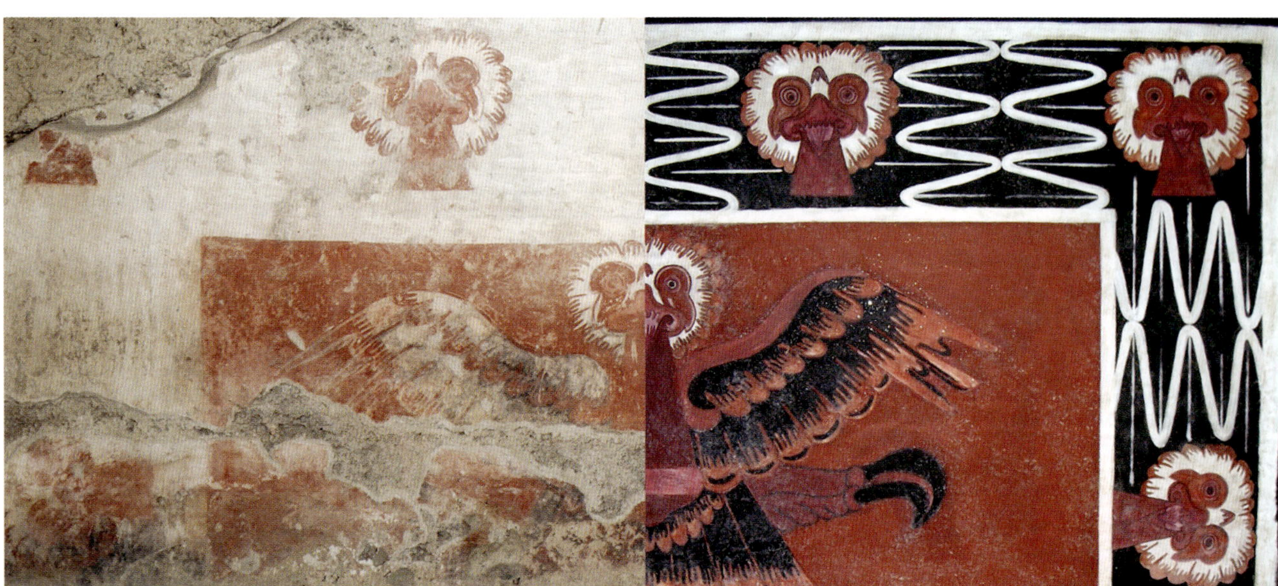

Figure 5 Comparison between the mural painting in 2007 and the museographic reproduction at the Museum Beatriz de la Fuente in Teotihuacan.

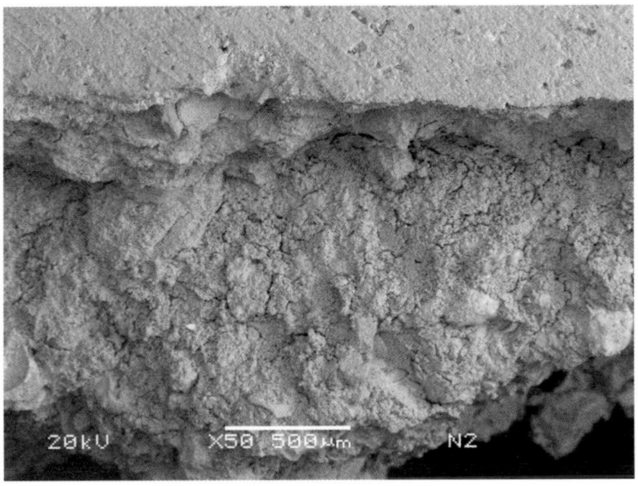

Figure 6 Unburied fragment stratigraphy, February 2009. (Image: Jose Antonio Alba.)

Burnishers can be used on *fresco* or *secco* and there is no difference in the use of clays for obtaining a polished surface.

Conclusions

From these results we may conclude that not only were green and blue colours applied using the *secco* technique, but also more colours than previously thought. The investigation on the specific decay of the black pigment remains to be carried out as well as the identification of the organic additives; the research is still in progress.

Acknowledgements

I would like to thank Jose Antonio Alba, Jaime Torres, Lilia Felix, Arturo León, Jaime Cama Villafranca and students from the BA in Conservation at the ENCRyM.

References

1. Villagra, A., 'Murales prehispánicos, copia, restauración y conservación', in *Homenaje al Dr. Alfonso Caso*, ed. J. Comas, Imprenta Nuevo Mundo, Mexico (1951) 425–426.
2. Torres, L., 'Materiales y técnicas de la pintura mural de Teotihuacan', in *IX Mesa redonda: Teotihuacan*, Sociedad Mexicana de Antropología, Mexico (1976) 17–42.
3. Littmann, E., 'Appendix Two: The physical aspects of some Teotihuacán murals', in *The Mural Paintings of Teotihucan*, ed. A. Miller, Dumbarton Oaks, Washington, DC (1973) 176–186.
4. Magaloni, D., 'El espacio pictórico teotihuacano, tradición y técnica', in *La pintura mural prehispánica en México, I Teotihuacan*, ed. B. De la Fuente, vol. 1, part 2, IIE-UNAM, Mexico (2001) 187–225.
5. McClung, E., 'Paleoenvironmental studies in Teotihuacan Valley', in *Teotihuacan: Identity and Heritage of Mexico, Conference held at the National Museum of Anthropology, Mexico City, 30 July 2009.*

Author's address

Ingrid Karina Jiménez Cosme, ENCRyM-INAH, Mexico (ingridkjc@gmail.com)

Decorations of Pīr-i Hamza Sabzpūš's tomb in Iran

Amir-Hossein Karimy and Parviz Holakooei

KEYWORDS stucco, SEM, gypsum, mural paintings, Seljuq

Introduction

Pīr-i Hamza Sabzpūš's tomb, the Seljuq monument situated in the town of Abarkūh (central Iran) built in the twelfth century AD, is a domed square edifice constructed of mud bricks. Its façade is covered with simple straw clay rendering, and painted stucco ornaments and mural paintings decorate the interior. There is a fragmentary historical inscription in the mihrab that includes the words '... and five hundred'. Stucco relief decorates the dado of 1.10 m. Godard and Afšār have closely related this stuccowork to that of twelfth-century monuments such as Gonbad-i Alaviān (in Hamedān), Heydari-e School (in Qazvin) and Masjid-i Jāmē (in Varāmin) [1, 2].

There are eight blind arcades in four sides of the monument (each side has two), each of which is decorated with seven medallions painted in red, green, white, blue and black on a white ground layer on a fine straw clay rendering.

This poster paper attempts to discriminate between different types of stuccowork in this monument by discussing technical differences and similarities. By means of chemical spot tests and scanning electron microscopy with energy dispersive X-ray analysis (SEM-EDX), the chemical composition of the pigments applied to the mural paintings and the mihrab are identified. SEM-EDX was carried out using a Philips XL30 and the XRD with a Bruker AXS D8.

Pigment identification

SEM-EDX was used for the identification of pigments used in the mihrab and mural paintings. From the analytical results shown in Table 1 it can be seen that chalk (calcium carbonate) was used as a white pigment on both the mural paintings and stuccos (the presence of carbonate anions was determined by wet chemical spot tests [3]). However, there is a slight difference between the chemical composition of white pigment in the mural paintings and the stuccos. Red ochre [4] and green malachite [5] were used in both the mihrab and relief stuccos.

However, there are some differences amongst the pigments that were used in the mihrab and the stuccos. Yellow ochre [6] and lampblack (identified using the ignition test [7]) were used in the mihrab, but these pigments were not found on the mural paintings. Azurite [8] was used as a blue pigment on the mural paintings, but ultramarine [9] was used in the mihrab.

Technological studies of the stuccos

Preliminary observations distinguished two strata of stucco, one on top of the other, both on the mihrab and the dado. The lower stratum has a greyish appearance whilst the upper one is

Table 1 SEM-EDX results of chemical composition of pigments.

Semi-quantitative elemental analysis (wt %)	Al	Si	Ca	Zn	Pb	Na	Mg	K	S	Fe	Cu
White pigment of mihrab	7.03	17.35	70.2	1.54	3.88	0	0	0	0	0	0
White pigment of mural paintings	0.71	14.06	85.24	0	0	0	0	0	0	0	0
Blue pigment of mihrab	10.9	24.3	35.31	0	0	2.47	5.4	3.92	15.15	1.82	0.74
Blue pigment of mural paintings	12.49	33.95	39.7	0	0	2.16	0	5.24	6.44	0	0
Red pigment of mihrab	21.64	64.46	0	0	0	0	0	0	0	13.9	0
Red pigment of mural paintings	8.42	23.95	36.14	0	0	5.27	0	0	10.93	15.48	0
Green pigment of mihrab	4.54	11.67	47.44	0	0	0	0	0	19.79	0	16.57
Green pigment of mural paintings	12.09	0	45.16	0	0	0	0	0	10.92	0	31.83
Black pigment of mihrab	4.13	12.25	40.36	0	0	0	0	0	43.25	0	0
Yellow pigment of mihrab	12.42	36.06	0	0	0	0	0	0	0	51.52	0

Table 2 XRD results from the phase composition of stuccos.

Crystalline phases (wt %)	Upper stucco layer of the dado	Upper stucco layer of the mihrab	Stucco layer underneath the dado	Stucco layer underneath the mihrab
Gypsum ($CaSO_4 2H_2O$)	32.8	37.4	25.9	2.5
Anhydrite ($CaSO_4$)	15.6	19.9	5.3	64.8
Bassanite ($CaSO_4 0{\cdot}5H_2O$)	0	0	0	0
Quartz (SiO_2)	5.4	9.2	3.9	2.4
Biotite ($KMg_3(Si_3Al)O_{10}(OH)_2$	3.4	2.6	2	1.8
Tridymite (SiO_2)	8.9	0	17.5	0
Albite ($NaAlSi_3O_8$)	2.8	2.5	0	1.4
Illite ($K_{0.7}Al_2(Si,Al)_4O_{10}(OH)_2$	0	0	0	3.4
Calcite ($CaCO_3$)	31.1	23.8	21.1	21.6
Natrite (Na_2CO_3)	0	0	4.2	0

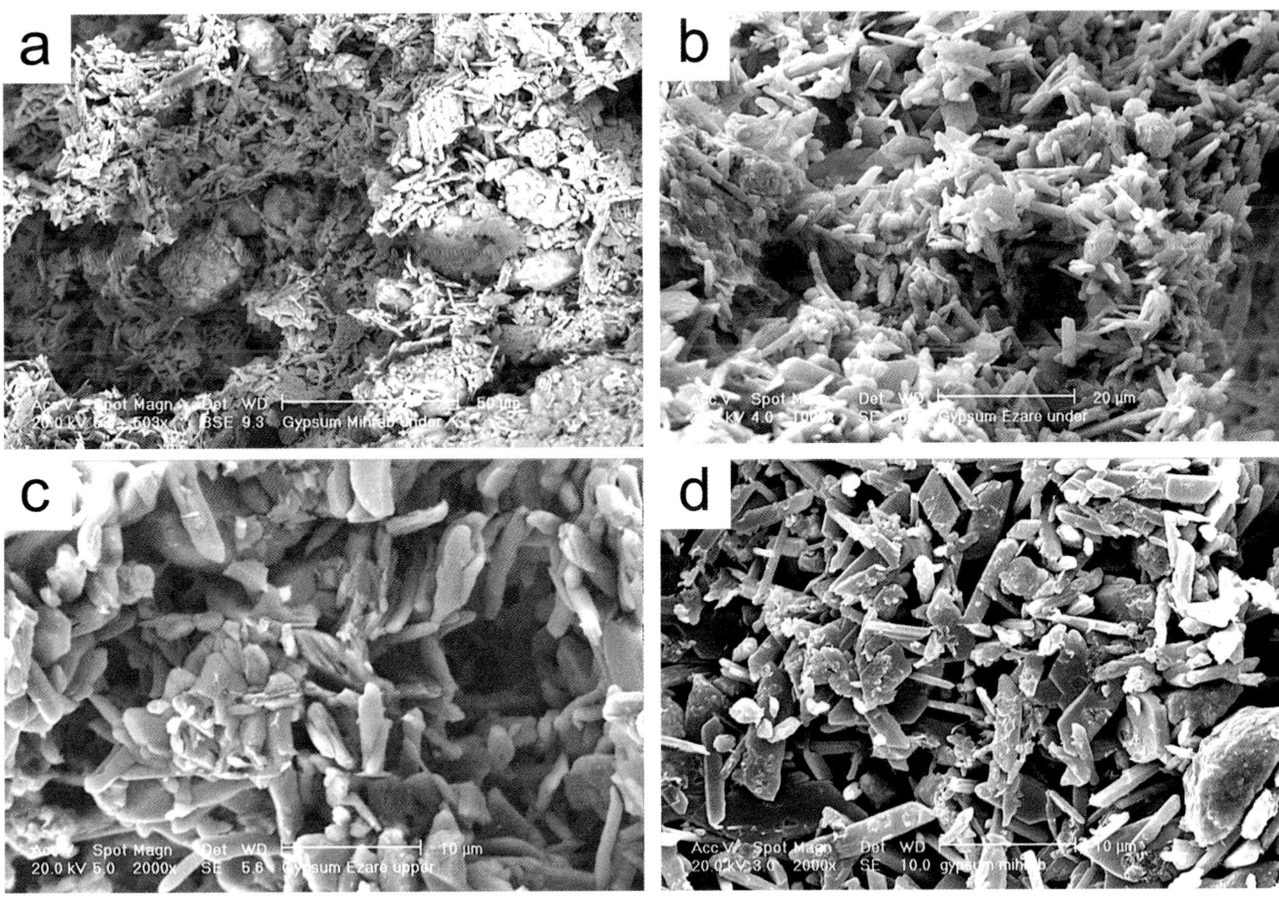

Figure 1 (a) Crystalline structure underneath the stucco of the mihrab together with charcoal particles. (b) Gypsum crystals underneath the layer of the dado and voids in between. (c) Crystalline structure of the upper stucco layer of the dado. (d) Gypsum and calcite lamellar crystals in the upper stucco layer of the mihrab.

yellowish. Although the upper layer has high relative mechanical strength, the lower stratum is quite fragile and powdery. These two layers were analysed by X-ray diffraction (XRD) in order to determine crystalline phases (Table 2). SEM was used as an effective tool for observing the crystalline morphology of the samples (Figure 1).

The SEM micrographs show that environmental deterioration factors (such as humidity and temperature) have altered and separated the gypsum crystals in the lower layers of the stucco, leading to a reduction in mechanical strength (so these layers seem relatively fragile). Under higher magnification, well-developed gypsum crystals can be seen, which are perfectly interlocked, providing greater strength to the upper layer.

Gypsum is heated to around 170°C to produce plaster of Paris (calcium sulphate hemihydrate or $CaSO_4{\cdot}0{\cdot}5H_2O$) [10]. Small charcoal fragments were observed in the SEM micrographs and natrite (Na_2CO_3) was detected in the lower stratum of the dado stucco (Table 2). Charcoal and natrite can be interpreted as the byproducts of the firing process of gypsum. In fact, overheated gypsum, in the procedure of firing, collapses onto the fire because the volume of gypsum decreases [10], and consequently the final product is mixed with ash (a combination of charcoal and soda ash (natrite)). Since the lower stratum is not visible in the final appearance of the stuccos, the contaminated plaster of Paris was used underneath, probably for reasons of economy (gypsum mines in the Abarkūh district are quite restricted). In addition, the diminution of gypsum content and increasing anhydrite in the lower stratum of the stucco shows that gypsum was exposed to excessively high temperatures during the making of the plaster of Paris. It also confirms that this overheated gypsum was used.

Conclusions

The *in-situ* observations and experimental investigation have shown that the stuccowork and wall paintings were executed in different phases of the construction and decoration of the building. The results of SEM-EDX microanalysis show that some of the pigments used on the stucco are different from the pigments used on the wall paintings and, consequently, wall painting and stuccowork were executed in different periods. The data obtained from SEM observations (Figure 1) and XRD results (Table 2) suggest that the chemical composition and crystalline morphology of two strata of stuccowork in the dado and the mihrab are not the same. Longer crystal size and the greater distance gap between the crystals in the stuccoworks of the dado (see Figure 1) would imply that the execution technique in the mihrab and the dado is probably different. Consequently, the dado and the mihrab were probably executed in different periods. Overheated gypsum was used to create the underlying layer in the stuccowork; gypsum-based waste was used to reduce consumption of plaster of Paris.

Acknowledgements

We would like to express our sincere appreciation to A.H. Rezaei, Tarbiat Modarres University, Tehran, Iran for SEM micrographs and SEM-EDX microanalysis and G. Jozanikohan, University of Tehran, Iran for XRD analyses.

References

1. Godard, A., *Athar-é Iran III, Abarkūh*, Paris (1937) 54–56.
2. Afšār, I., *Yādgārhā-i Yazd I*, Tehran (1969–1975) 342–344.
3. Odegaard, N, Carroll, S. and Zimmt, W.S., *Material Characterization Tests for Objects of Art and Archaeology*, 2nd edn, Archetype Publications, London (2005) 102.
4. Elias, M., Chartier, C., Prévot, G., Garay, H. and Vignaud, C., 'The colour of ochres explained by their composition', *Materials Science and Engineering B* **127** (2006) 70–80.
5. Gettens, R.J. and FitzHugh, E.W., 'Malachite and green verditer', *Studies in Conservation* **19**(1) (1974) 2–23.
6. Hradil, D., Gryga, T., Hradilová J. and Bezdicka P., 'Clay and iron oxide pigments in the history of painting', *Applied Clay Science* **22** (2003) 223–236.
7. Winter, J. and West FitzHugh, E., 'Pigments based on carbon', in *Artists' Pigments: A Handbook of their History and Characteristics*, vol. 4, ed. B.H. Berrie, Archetype Publications, London (2007) 1–38.
8. Gettens, R.J. and FitzHugh, E.W., 'Azurite and blue verditer', *Studies in Conservation* **11**(2) (1966) 54–61.
9. Plesters, J., 'Ultramarine blue, natural and artificial', *Studies in Conservation* **11**(2) (1966) 62–91.
10. Singh, N.B. and Middendorf, B., 'Calcium sulphate hemihydrate hydration leading to gypsum crystallization', *Progress in Crystal Growth and Characterization of Materials* **53** (2007) 57–77.

Authors' addresses

- Amir-Hossein Karimy, Art University of Isfahan, Isfahan, Iran (ahkarimy@gmail.com)
- Parviz Holakooei, University of Ferrara, Ferrara, Italy (parviz.holakooei@unife.it)

A study of the microstructure and condition of thin sheets of painting support ivory using SEM and ESEM

Satomi Kitano, Alan Derbyshire, Nigel Meeks, Caroline Cartwright, Chris Stain and Geoffrey Mitchell

KEYWORDS ivory, portrait miniatures, warping, relative humidity, dentinal tubules

Introduction

This paper outlines a study of the microstructure of thin sheets of ivory used as a painting support for portrait miniatures. Warping of the ivory support is one of the main problems commonly found in such miniatures from the late eighteenth and early nineteenth century. Portrait miniatures from this period are painted on very thin sheets of ivory that are often only 0.2 mm thick. Warping can lead to cracking of the ivory and can also accentuate flaking of the paint layer. The problem of warping in ivory has therefore been of long-term interest to conservators who deal with portrait miniatures, including those at the Victoria and Albert Museum in London. The causes of warping are complex, not least because artists normally adhered the thin ivory sheets onto paper or card before painting. The possible causes of warping are thought to relate to the differential reactions of the ivory/adhesive/paper or card layers to changes in relative humidity (RH). It is well known that ivory is hygroscopic and anisotropic [1]. However, only a few scientific studies have been undertaken related to this subject and systematic analyses of the morphological and microstructural changes due to changes in RH or moisture in such thin sheets of ivory have yet to be fully carried out.

In order to understand the behaviour of the ivory substrates of portrait miniatures, scanning electron microscopy (SEM) and environmental scanning electron microscopy (ESEM) were used to study the microstructure and condition of ivory samples.

Samples and methods

Reference ivory samples were sourced from old, small elephant tusks belonging to the Furniture Conservation section at the Victoria and Albert Museum. Thin tangentially oriented sheets of ivory were cut down from the tusks as described in a nineteenth-century book on turning and cutting ivory [2]. The average thickness of the samples was approximately 0.4 mm. The surfaces of ivory sheets were polished using cotton swabs and pumice powder mixed with ethanol. This was applied in order to reduce the saw marks that could interfere with SEM observations.

Field emission (FE) SEM (Hitachi S-4800) and variable pressure (VP) SEM (Hitachi S-3700N with low vacuum mode) were used at the Department of Conservation and Scientific Research at the British Museum, London. The polished ivory sheet was cut into four pieces (up to 20 mm long) using a scalpel, mounted in resin, which was recommended in order to keep the thin samples mechanically stable during SEM examination, and further polished down to 1 μm.

Each of the three resin mounts comprised different orientations: transverse and longitudinal sections of the standard sheet (mount 1), transverse and longitudinal sections of the warped sheet (mount 2), and the tangentially cut flat surface of the standard sheet (mount 3). Additionally, the recto and verso of the warped sheet were stuck directly to an aluminium stub.

ESEM (FEI Quanta 600F) was used at the Centre for Advanced Microscopy of the University of Reading to test the effects of changing RH on the ivory with *in-situ*, real-time experiments using water vapour in the SEM chamber. A tangentially cut flat surface of ivory was used for the experiments. A polished ivory sheet was cut into small pieces (6 × 6 mm) in order to fit on small aluminium stubs.

Results

In the SEM micrographs showing the dentinal tubules (part of the natural structure of ivory), there appear to be cracks on both the transverse and the longitudinal sections (i.e. all the cross-sections on both mounts 1 and 2). It seems that the sharp ends of the tubules have the potential to become longer by extending and merging with adjacent tubules.

On the tangentially cut flat surface (which is also a longitudinal view) that represents the painted surface of an ivory miniature, the tubules appear as short and round holes. This sample exhibits the form of the dentinal tubules. Some of the concentrically oriented tubules that are more open provide natural 'weak spots', which might create cracking.

On the longitudinal sections of both the standard and the warped sheets, collagen filling the dentinal tubules is visible (Figure 1). This image shows degradation of collagen (arrows in Figure 1) and enlargement of the dentinal tubules. Fine cracks, which are not the dentinal tubules, can also be seen at right angles to the directions of the tubules. These are believed to

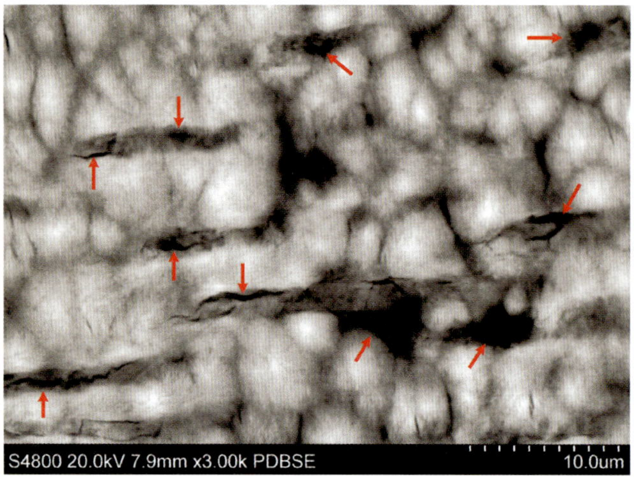

Figure 1 FE-SEM micrograph, ×3000. Longitudinal section of the artificially warped sheet from resin mount 2 showing fine cracks in the ivory as well as degradation of collagen in the dentinal tubules. (Image: © The Trustees of The British Museum.)

be due to dehydration and are considered to be a major cause of physical deterioration of ivory [3].

Experiments using the ESEM suggest that changes in RH affect the microstructure of ivory. SEM micrographs were captured on the same area of a sample while the RH was increased and decreased in the microscope in 'live' experiments. Increasing the RH from 1% to 100% did not produce significant changes in the sample, i.e. while the ivory would be absorbing water vapour. However, when the RH was then lowered from 100% to 50%, narrow, open tubules opened widely in seconds as the ivory released the absorbed moisture and began to dry and shrink in the vacuum chamber of the SEM. This indicates that changes in RH do affect the microstructure of ivory. However, the possible damaging effect or accelerated changes caused by the electron beam and vacuum cannot be disregarded.

Conclusions

The use of SEM techniques has enabled the observation of the microstructure of thin sheets of ivory and its potential deterioration process. Enlargement and cracking of the tubules and degradation of the collagen were seen and identified as the most likely deterioration processes, and ESEM experiments showed that changes in RH accentuate the problem. The potential for damage caused by electron beam/sample interactions, low pressure and temperature in the SEM, however, must be taken into account. Hence the best solution for stabilising thin ivory miniatures is to control the storage/display RH and temperature to prevent even small physical changes in shape that occur with repeated atmospheric cycling. This research promotes further study towards a better understanding of the material at the microscopic level at which incipient degradation can be observed. The results are useful for conservators, curators and scientists who work with ivory and for future research in this area.

This study provided a number of informative results that illustrate the natural structure of ivory and the potential effect of changes in RH to the microstructure of ivory. The details revealed in the longitudinal sections compared with those visible in the transverse have yielded important structural information. Collagen fibrils filling the dentinal tubules on the longitudinal sections were also observed. The structure of the dentinal tubules was clearly visible and a possible deterioration process was identified through their elongation and merging. Although the original aim of this project was to fully characterise the causes of warping and cracking of ivory substrates in relation to RH changes, the complexity of the issues encountered did not allow clear confirmation of the detailed mechanisms involved within the timeframe permitted.

Acknowledgements

We would like to thank Bob Wood (R.J.W. Products, Kent, UK) for cutting ivory sheets, Sonia O'Connor (University of Bradford, UK), Rees Rawlings (Imperial College London, UK), Alex Ball (Natural History Museum, London, UK) and William Lindsay (Royal College of Art/Victoria and Albert Museum Conservation, London, UK).

References

1. Lafontaine, R.H. and Wood, P.A., 'The stabilization of ivory against relative humidity fluctuations', *Studies in Conservation* **27**(3) (1982) 109–117.
2. Holtzapffel, C., *Turning and mechanical manipulation. Intended as a work of general reference and practical instruction, on the lathe, and the various mechanical pursuits followed by amateurs*, vol. 1, Holtzapffel & Co., London (1843) 148.
3. O'Connor, S., University of Bradford, UK, personal communication (March and April 2009).

Authors' addresses

- Satomi Kitano, Royal College of Art/Victoria and Albert Museum Conservation, London, UK (satomi.kitano@network.rca.ac.uk)
- Alan Derbyshire, Paper, Books and Paintings Conservation section, Victoria and Albert Museum, London, UK (aland@vam.ac.uk)
- Nigel Meeks, Department of Conservation and Scientific Research, The British Museum, London, UK (nmeeks@thebritishmuseum.ac.uk)
- Caroline Cartwright, Department of Conservation and Scientific Research, The British Museum, London, UK (ccartwright@thebritishmuseum.ac.uk)
- Chris Stain, Centre for Advanced Microscopy, University of Reading, UK (c.j.stain@reading.ac.uk)
- Geoffrey Mitchell, Centre for Advanced Microscopy, University of Reading, UK (g.r.mitchell@reading.ac.uk)

White grounds applied to the main altarpiece of the Coimbra Old Cathedral: historical technology and material characterisation

Agnès Le Gac, Ana Isabel Seruya, Maria José Oliveira and Isabel Ribeiro

KEYWORDS *gesso grosso, gesso sottile*, priming, gilding, polychrome altarpiece

Introduction

As inert material, gesso was of great importance in gilded and painted artworks from the southern European countries. White calcium sulphate-based priming was the basis for the application of metal leaf and pigments [1, 2]. The types and numbers of preparatory layers were dictated by sculpture size and shape [3, 4]. Gesso's specific physical and chemical properties, determined by its manufacturing process, were crucial for the final appearance and durability of these artworks [5]. Although gesso grounds have been well studied for easel painting [6, 7], little is known regarding three-dimensional artworks.

In the Iberian Peninsula during the Baroque period, old masters' treatises recommended a double gesso ground for indoor wooden altarpieces and sculptures for the application of gold leaf (mostly water gilding) and colour paints [1, 2]. This was confirmed in recent studies and compilations of deeds written by notaries for polychrome artworks [3, 4, 6].

The main altarpiece of the Coimbra Old Cathedral provided the opportunity to characterise the inert material used in preparatory layers in Portugal, and to study the practices of painter-gilders working there. The sixteenth-century colour scheme of this altarpiece, commissioned by Bishop D. Jorge de Almeida and completed in 1502, was renovated several times. According to historical records, four campaigns were undertaken by different painters:

1502 – Gothic and original polychromy: Jean d'Ypres, Flemish painter [8]
1583 – Cleaning and partial refurbishment: Fernão da Costa [9]
1685 – Baroque polychromy: Manoel da Costa Pereira [10]
1900 – Complete restoration: Mestre Augusto Gonçalves and others [9]

Objectives

As each intervention was dated, a complete study of the ground layers found on the architectonic structure and statues was carried out to verify:

(1) which inert material was used during each intervention;
(2) in which phase the calcium sulphate was used;
(3) the presence of other compounds mixed with the main inert material;
(4) the number of coats constituting each white ground and its thickness, according to the physical and aesthetical parameters involved.

By cross-checking the whole dataset, the present work aimed at characterising the priming technological aspects implemented in Portugal over time, and to verify if they match the written sources.

Methodology

One hundred micro-samples were prepared as cross-sections to perform a preliminary study on all layered structures and then to select the most representative of the ground layers applied in 1502, 1685 and 1900. Twenty relevant samples were selected and embedded in Epofix resin, hand dry-polished on Micromesh cloths (P1500-12000), and examined and analysed using the following methods:

- Optical microscopy (OM), instrument Zeiss Axioplan 2 Imaging.
- Scanning electron microscopy with energy dispersive X-ray analysis (SEM-EDX), instrument JEOL, JSM 7001F/ detector Oxford, INCA 250.
- X-ray diffraction (XRD), instrument BRUKER AXS, D8 Discover, for data on the crystalline phases of calcium sulphate [11].

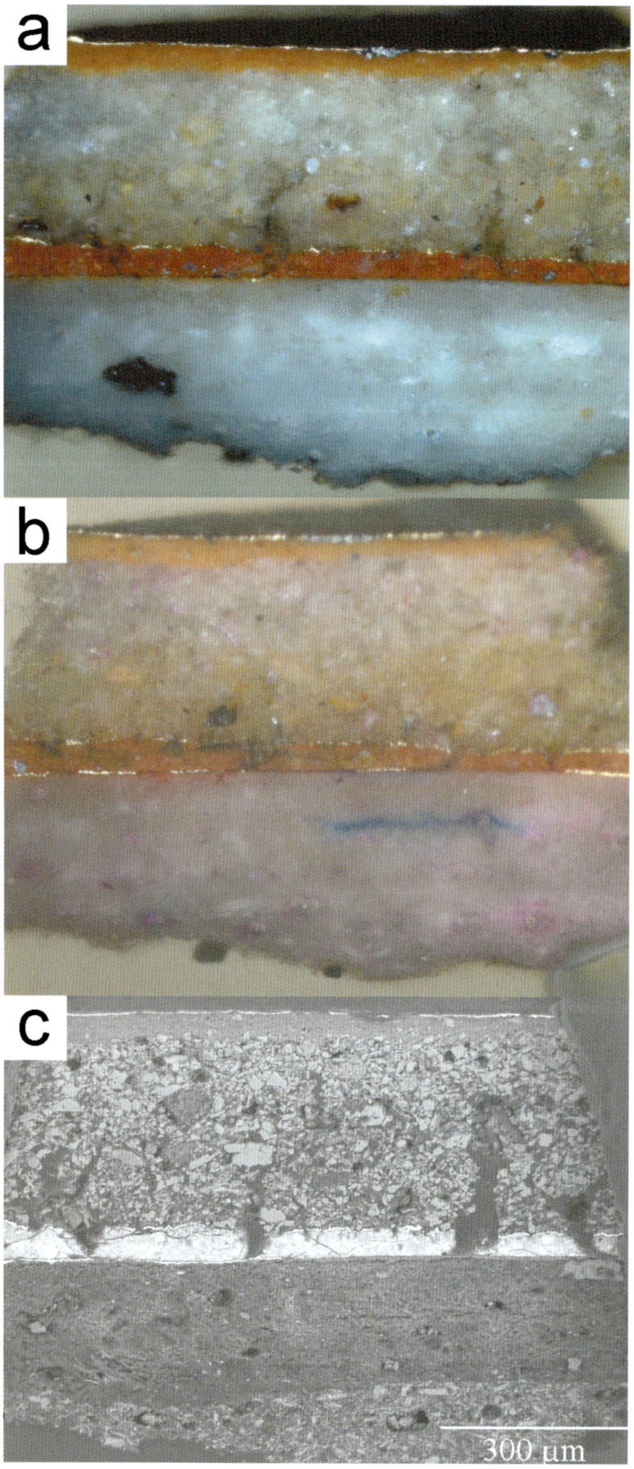

Figure 1 Sample (R-01) from a gilded carved element of the main altarpiece showing the 1502 triple ground and the 1685 double ground: (a) MO (×100), reflected light; (b) chemical test: acid Fuchsina = proteins (pink colour); (c) SEM-BSE image (×100), 25 keV.

Results and discussion

1502 – Gothic polychromy

A triple white ground was observed on the water-gilded carved elements of the architectonic structure (Figure 1) and on the figures' shiny flesh tone. Both techniques (gilding and flesh imitation) required thick preparatory layers (c.400 µm thick) offering better compressive strength during the burnishing process [12].

This priming was executed using burnt anhydrous calcium sulphate (anhydrite, $CaSO_4$) coarsely ground (*gesso grosso*) for the lower coat, anhydrite mixed with a little gypsum and black particles for the coat in between, and mainly dihydrate calcium sulphate (gypsum, $CaSO_4{\cdot}2H_2O$), finely ground (so-called *gesso sottile* or *gesso mate*) for the upper coat. The aciculate-shaped particles found at that level are consistent with fine gesso, purified by the slaking and screening processes.

In the middle of the altarpiece, only two coats were found on the silvered panel against which the Virgin of the Assumption stands out. This part of the structure is flatter, thus easier to cover and polish prior to the silver leaf attachment. Both coats were executed mainly with fine gesso creating a smooth ground layer, 250 µm thick.

The surfaces covered with appliqué relief brocade, typical of a flamboyant Gothic polychromy influenced by the techniques from Brabant, had only one layer of fine gesso, c.20–40 µm thick. This ensured attachment of the raised ornaments.

No ground layers were applied over the architectonic structure painted with azurite. In fact, a carbon black underlayer was laid directly over the wooden support for the coarse ground blue pigment to achieve an intense colour. A white reflective ground would have conflicted with the intended optical effect.

1685 – Baroque polychromy

The water-gilded elements of the architecture and main statues were coated with a double ground (Figure 1). However, on the statuettes, which were also gilded, and on the parts of the structure painted blue, the ground never exceeded one layer. This helped to preserve the tiny figures' plastic expression on a smaller scale and to improve the smalt application required for the blue surfaces at that time.

Calcium sulphate was used everywhere. The XRD results showed the crystalline phases are always anhydrite, irrespective of the coat analysed. The coarse particles of calcium sulphate are of a heterogeneous size and morphology in both the base and top coats. The application of a double ground is clearly discernible by OM (Figure 1a), since the lower coat, quite orange in colour, was mixed with clay minerals and quartz. It is well differentiated from the whiter upper coat that does not contain these additives. Such compositions are perfectly highlighted by SE and BSE imaging (Figure 1c) because of the very fine lamellate crystals of aluminosilicate aggregates in the lower coat. In the cases where a single ground was applied, inert materials and clay minerals were mixed together in variable proportions.

1900 – Polychromy on new elements

Only a thick single ground was applied to the new statuettes. Taking advantage of the industrial era, the white preparatory layer applied was mostly mechanically and finely ground gypsum. Using SEM, very uniform *gesso sottile*, made up of particles with a prismatic, tabular and acicular morphology was observed. XRD analysis indicates that it was mixed with a little calcium carbonate.

Conclusions

The inert material found in each priming of the altarpiece, as expected in Portugal, was gesso; even the Flemish painter employed it in 1502 to achieve the Gothic polychromy. Calcium sulphate was bound with a proteinaceous medium [1–4, 12–13], put in evidence with acid fuchsin (acidic dye for staining protocol), giving the priming a pink coloration (Figure 1b). This is consistent with traditional old masters' recipes and available historical records.

XRD analysis revealed an unexpected result: in 1685, Manoel da Costa Pereira prepared both coats of the double ground with anhydrite, mixing just a little amount of gypsum with it. He followed only the principle of building up a double ground with no regard for the consecutive application of *gesso grosso* and *gesso sottile*.

Further research is needed regarding the presence of chalk in the ground dated 1900 to understand whether this other inert material was a deliberate choice by the restorer or corresponded to an admixture in a marketed gesso for gilding.

Acknowledgements

We thank Isabel Nogueira of the ICEMS-IST for providing the SEM-EDX data, and Teresa Madeira Amorim for content editing.

References

1. Nunes, P., *Arte da pintura, symmetria, e perspectiva* [1615], Editorial Paisagem, Porto (1982).
2. Pacheco, F., *Arte de la pintura* [1649], Las Ediciones de Arte, Barcelona (1982).
3. Serck-Dewaide, M. et al. [Belgium], Campoy Naranjo, M. et al. [Spain] and Le Gac, A. et al. [Portugal], 'Les techniques utilisées dans l'art baroque religieux des XVIIe et XVIIIe siècles au Portugal, en Espagne et en Belgique', in *Policromia*, ed. A.I. Seruya, IPCR, Lisbon (2004) 119–155.
4. Le Gac, A., *Le retable majeur de la Sé Velha de Coimbra et la polychromie dans le diocèse de Coimbra à l'époque baroque. Aspects techniques et esthétiques*, PhD dissertation, Universidade Nova de Lisboa (2009).
5. Mecklenburg, M.F., 'Some mechanical and physical properties of gilding gesso', in *Gilded Wood: Conservation and History*, ed. D. Bigelow, Sound View Press, Madison, WI (1991) 163–170.
6. Véliz, Z., 'Wooden panels and their preparation for painting from the Middle Ages to the seventeenth century in Spain', in *The Structural Conservation of Panel Paintings*, ed. K. Dardes, The Getty Conservation Institute, Los Angeles (1998) 136–148.
7. Townsend, J.H., Doherty, T., Heydenreich, G. and Ridge, J. (eds), *Preparation for Painting: The Artist's Choice and its Consequences*, Archetype Publications, London (2008).
8. Garcia, P.Q., *Artistas de Coimbra. Documentos para as suas biografias*, [s.n.], Coimbra (1923).
9. Vasconcelos, A., *A Sé Velha de Coimbra*, Arquivo da Universidade de Coimbra, Coimbra (1930).
10. Louro, J.M.S., 'O restauro do retábulo da capela-mor da Sé Velha no século XVII', *Mundo da Arte* **13** (1983) 60–63.
11. Bezou, C., Nonat, A., Mutin, J.C., Nørlund-Christensen, A. and Lehman, K.S., 'Investigation of the crystal structure of γ-$CaSO_4$, $CaSO_4 \cdot 0{\cdot}5H_2O$ and $CaSO_4 \cdot 0{\cdot}6H_2O$, by powder diffraction methods', *Journal of Solid State Chemistry* **117** (1995) 165–176.
12. Le Gac, A., Seruya A.I., Lefftz, M. and Alarcão, A., 'The main altarpiece of the Old Cathedral of Coimbra (Portugal): characterization of gold alloys used for gilding from 1500 to 1900', *Revue d'Archéométrie* **33** (2010) 423–432.
13. Le Gac, A., 'Preliminary research on glues made of tawed leather, parchment and rabbit-skin', in *Parchment and Leather: Research, Conservation-restoration, Craft*, ed. ICOM-CC, Copernicus University Toruń, Toruń (2012) [in press].

Authors' addresses

- Agnès Le Gac, Departamento de Conservação e Restauro, Faculdade de Ciências e Tecnologia, Universidade Nova de Lisboa; Centro de Física Atómica, Universidade de Lisboa, Lisbon, Portugal (alg@fct.unl.pt)
- Ana Isabel Seruya, Centro de Física Atómica, Universidade de Lisboa, Lisbon, Portugal
- Maria José Oliveira, Laboratório de Conservação e Restauro de José de Figueiredo, Instituto dos Museus e da Conservação-IP, Lisbon, Portugal
- Isabel Ribeiro, Laboratório de Conservação e Restauro de José de Figueiredo, Instituto dos Museus e da Conservação - Instituto Português, Lisbon, Portugal

Characterisation of lapis lazuli for a provenance study by means of SEM-EDX and SEM-cathodoluminescence

Alessandro Lo Giudice, Alessandro Re, Debora Angelici and Giovanni Pratesi

KEYWORDS lapis lazuli, SEM-EDX, cathodoluminescence, provenance, archaeometry

Introduction

Lapis lazuli has been used for more than 7000 years for the manufacture of precious objects and jewellery. The main quarry for this rock in Afghanistan (Badakhshan) is still active (Figure 1), but there are other quarries that could have been exploited in antiquity [1–4]. Sources that have been suggested include: the Pamir mountains (Lyadzhuar Dara, Tajikistan) [1, 5], Pakistan (Chagai Hills) [4, 5], Siberia (Irkutsk, near Lake Baikal) [1], Iran [1, 3] and Sinai [6]. The last two are not geologically confirmed and are still debated [2, 5]. For this reason, a provenance study of lapis lazuli could provide answers to some important issues, in particular for in-depth study of its use and distribution through historic commercial routes.

In the present work a systematic study has been performed on lapis lazuli from four quarries in Afghanistan, Chile, Siberia and Tajikistan using energy dispersive X-ray analysis (EDX) and cathodoluminescence (CL) coupled with scanning electron microscopy (SEM). The main goal of this characterisation was to find at least one marker for each provenance. These markers could be the presence or absence of a mineralogical phase, or one or more elements present in a particular mineral, or the different luminescence of the same mineral.

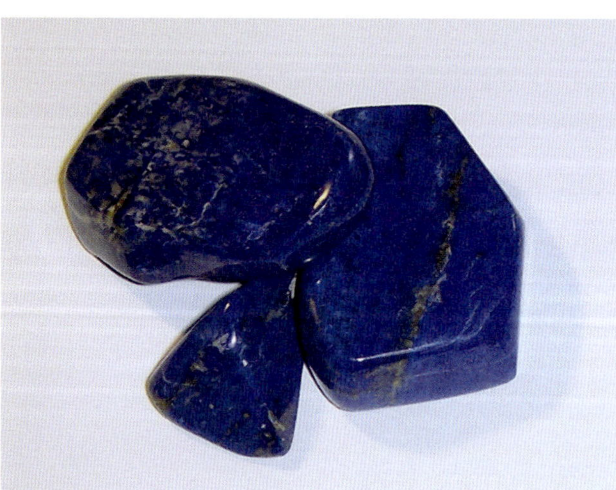

Figure 1 Lapis lazuli from Afghanistan.

Experimental

The analysed lapis lazuli samples are from the Museum of Natural History (Università di Firenze) and have been collected from four sources: Sar-e-Sang Badakhshan (Afghanistan), Ovalle (Chile), Irkutsk near Lake Baikal (Siberia) and the Pamir mountains (Tajikistan). A more detailed description of the samples under investigation has been reported elsewhere [7]. Fifteen semi-thin polished sections were prepared and a preliminary characterisation of the samples, identifying the distribution of the main mineral phases present in the rocks, was carried out by means of optical microscopy (OM) and cold-CL (CL8200 Mk3). After carbon coating (less than 0.5 μm in thickness) selected grains were studied by means of SEM-EDX (Leica Stereoscan 420 with an Oxford PentaFET EDX) and SEM-CL (Oxford MonoCL). All the SEM-CL measurements were performed in the range 350–750 nm at 15 kV energy, with a probe current of 1 nA, an investigated area of 20×25 μm^2 and 16 mm working distance. All the SEM-EDX measurements were performed at 20 kV energy with a probe current of 200 pA, an investigated area of 10×15 μm^2 and 19 mm working distance. To obtain the chemical concentration of each phase we also measured a set of 10 reference standards (calcite, apatite, jadeite, diopside, plagioclase, marcasite, tugtupite, sanidine, wollastonite and titanium oxide).

CL is a powerful technique used to rapidly identify the mineral phase distribution [8]. Luminescence is activated by defects in the lattice structure of the mineral. In some cases substitutional trace elements in the crystal can activate the process so that this technique is able to reach a sensitivity of some parts per million (ppm). Unfortunately it is not usually possible to quantify the elemental contents by analysing the intensity of the CL signal due to competition between luminescence centres and chemical elements that work as quenchers. However, by means of SEM-EDX and SEM-CL it is possible to obtain both the elemental composition of a small area (a few μm^2) and its luminescence properties in the ultraviolet (UV) visible spectral range.

Due to the heterogeneous nature of lapis lazuli (which may include lazurite, diopside, pyrite, calcite, wollastonite and K-feldspar, amongst other minerals), it is very difficult to

Table 1 Markers for provenance recognition of lapis lazuli with SEM-EDX and SEM-CL.

Provenance	Marker
Chile	WOLLASTONITE. Main mineralogical phase observed only in Chilean samples; it is characterised by a double broad band of luminescence at 560 nm and 620 nm.
	Many samples have shown a contemporaneous low potassium and high sulphur concentration in lazurite (K/S<0.03).
Asian	DIOPSIDE. Main mineralogical phase observed only in Asian samples under investigation; it is characterised by a broad band of luminescence at about 585 nm.
Afghanistan	Many samples have shown a contemporaneous low sulphur and high potassium concentration in lazurite (K/S>0.2).
	Small quantity of titanium in diopside (0.1%w<Ti<1%w).
Pamir	Diopside shows a weak additional luminescence band at 690 nm.
	CANCRINITE. Main mineralogical phase observed only in Tajikistan samples; it is characterised by a strong UV band and a regular multi-peak emission from 480 nm to 600 nm.

identify provenance markers by analysing the elemental composition of the whole rock; hence investigation has focused on the mineral phases. SEM-CL spectra were collected from all the main luminescent phases, whereas SEM-EDX investigation focused on two mineral phases: lazurite (responsible for the blue colour and present in all the samples) and diopside (present only in the lapis lazuli samples from the three Asian provenances).

Results and conclusions

Despite the limited number of samples analysed, results so far are promising and the experimental differences among rocks from different sources are significant [7, 9]. Table 1 summarises markers for provenance recognition of lapis lazuli.

Chilean lapis lazuli can be distinguished from others by the presence of wollastonite ($CaSiO_3$) that can be rapidly recognised by means of the intense manganese-related luminescence bands at 560 nm and 620 nm. Moreover, many samples have shown lazurite with a coinciding lower quantity of potassium and higher quantity of sulphur than other lapis lazuli (K/S<0.03) [9].

Lapis lazuli from Pamir is characterised by a cancrinite phase (not yet identified) with a strong UV emission and a regular multi-peak emission from 480 to 600 nm (six peaks separated by about 24 nm). Moreover, the luminescence spectra of Pamir diopside show an additional weak luminescence band at 690 nm in comparison with the others that has not as yet been successfully attributed to a luminescence activator.

In lapis lazuli from Afghanistan a particular mineral phase has not been recognised but many samples have shown a coinciding higher quantity of potassium and lower quantity of sulphur in lazurite than other lapis lazuli (K/S>0.2) [9]. All the rocks show a titanium content of about 0.1–1 wt% in diopside.

Siberian lapis lazuli is characterised by a relatively high content of barium that allows the discrimination of this provenance [10]. The amount of barium is too low to be detected by means of SEM-EDX, but it is easily detectable using other techniques such as micro-particle-induced X-ray emission (PIXE) analysis [9]. It is interesting to note that all the markers obtained in this work by means of SEM techniques are also observable with in air ion beam analysis (micro-PIXE, micro-particle-induced gamma ray emission (PIGE) and micro-ionoluminescence) that are applicable not only to lapis lazuli rocks but also to works of art [11, 12]

It is worth stressing that all the results in this work are restricted to a limited quantity of samples, and that the reliability of the present conclusions is to be confirmed by future additional statistics and work in progress.

Acknowledgements

This work has been carried out and supported financially within the framework of the INFN experiment 'FARE'. We gratefully acknowledge the Compagnia di San Paolo for the funding of a PhD scholarship on physics applied to cultural heritage. We warmly thank Paolo Olivero, Elisa Conz, Maria Albonico and Chiara Avataneo for their contributions to this work.

References

1. Herrmann, G., 'Lapis lazuli: the early phases of its trade', *Iraq* **30**(1) (Spring 1968) 21–57.
2. Herrmann, G. and Moorey, P.R.S, 'Lapis lazuli', in *Reallexikon der Assyriologie und Vorderasiatischen Archäologie*, vol. 6, ed. W. de Gruyter, Berlin (1983) 489–492.
3. Moorey, P.R.S., *Ancient Mesopotamian Materials and Industries: The Archaeological Evidence*, Eisenbrauns, Winona Lake, IN (1999).
4. Bulgarelli, G.M. and Tosi, M., 'La lavorazione ed il commercio delle pietre semipreziose nelle città dell'Iran protostorico (3200–1800 a.C.)', *Geo-Archeologia* **1–2** (1977) 37–50.
5. Delmas, A.B. and Casanova, M., 'The lapis lazuli sources in the ancient East', in *Association of South Asian Archaeologists in Western Europe, 9th International Conference, Venice*, ed. M. Taddei, Istituto Italiano per il Medio ed Estremo Oriente, Rome (1990) 493–505.
6. Nibbi, A., *Ancient Egypt and some Eastern neighbours*, Noyes Press, Park Ridge, NJ (1981).
7. Lo Giudice, A., Re, A., Calusi, S., Giuntini, L., Massi, M., Olivero, P., Pratesi, G., Albonico, M. and Conz, E., 'Multitechnique characterization of lapis lazuli for provenance study', *Analytical and Bioanalytical Chemistry* **395** (2009) 2211–2217.
8. Pagel, M., Barbin, V., Blanc, P. and Ohnenstetter, D., *Cathodoluminescence in Geosciences*, Springer-Verlag, Berlin and Heidelberg (2000).
9. Re, A., Lo Giudice, A., Angelici, D., Calusi, S., Giuntini, L., Massi, M. and Pratesi, G., 'Lapis lazuli provenance study by means of micro-PIXE', *Nuclear Instruments and Methods in Physics Research B* **269**(20) (2011) 2373–2377.
10. Ivanov, V.G., Samoylov, V.S., Petrov, L.L. and Yaroshenko, S.K., 'Geochemical features of the genesis of lazurite-bearing metasomatites in the South Baikal Region', *Doklady Earth Science Section* **227** (1976) 200–202.

11. Ballirano, P. and Maras, A., 'Mineralogical characterization of the blue pigment of Michelangelo's fresco *The Last Judgment*', *American Mineralogist* **91** (2006) 997–1005.
12. Calusi, S., Colombo, E., Giuntini, L., Lo Giudice, A., Manfredotti, C., Massi, M., Pratesi, G. and Vittone, E., 'The ionoluminescence apparatus at the LABEC external microbeam facility', *Nuclear Instruments and Methods in Physics Research B* **266** (2008) 2306–2310.

Authors' addresses

- Alessandro Lo Giudice, Dipartimento di Fisica Università di Torino and INFN Sezione di Torino, Torino, Italy (alessandro.logiudice@unito.it)
- Alessandro Re, Dipartimento di Fisica Università di Torino and INFN Sezione di Torino, Torino, Italy
- Debora Angelici, Dipartimento di Fisica Università di Torino and INFN Sezione di Torino, Torino, Italy
- Giovanni Pratesi, Dipartimento di Scienze della Terra and Museo di Storia Naturale, Università di Firenze, Firenze, Italy

Secondary phases in archaeological and historical materials: a microstructural approach for interpreting the correct sequences of crystallisation

Lara Maritan, Michele Secco, Claudio Mazzoli and Gilberto Artioli

KEYWORDS secondary phases, microstructural analysis, crystallisation sequence, cultural heritage materials, environmental conditions

Introduction and methods

The occurrence, type and distribution of secondary phases in archaeological objects and construction materials of historical buildings has been analysed to evaluate the conservation conditions of the objects and to model the alteration processes in different environmental conditions [1]. X-ray powder diffraction (XRD) and Fourier transform infrared spectroscopy (FTIR), commonly used for recognising the alteration of mineral phases, mineral profiling and bulk geochemical determination of element leaching, only provide information on the bulk changes, without recognising the evolution of the alteration process, as indicated by microstructural relations among alteration products. The precipitation of small quantities of secondary phases, hardly detectable by bulk techniques, can be successfully studied by scanning electron microscopy (SEM). The combination of microstructural and microchemical analyses make this technique a powerful tool for the identification of secondary phase crystallisation sequences and stratigraphy in historical materials, providing information both on the composition of circulating fluids and pollutants, and changes in the chemical-physical conditions of the burial environments.

This paper describes case studies of alteration in ceramic materials from both terrestrial (Frattesina, north-eastern Italy) and lagoon-like environments (Olbia, Sardinia, Italy), and concretes from a 1930s fertiliser production plant located in a heavily polluted industrial area (Porto Marghera, Venice, north-eastern Italy), for which SEM analysis was crucial to define the crystallisation sequence of secondary phases.

Microstructural and microchemical analyses were carried out on carbon-coated polished thin-sections, as well as on fragments, using a CamScan MX 2500 SEM with LaB_6 source, coupled with an energy dispersive X-ray (EDX) spectrometer for semi-quantitative analysis, operating at 20 kV, 160 mA and a working distance of about 30 mm.

Results and conclusions

Ceramics

SEM microstructural analysis of the Iron Age pottery from Frattesina shows the presence of secondary aggregates of vivianite ($Fe_3(PO_4)_2{\cdot}8H_2O$) and mitridatite ($Ca_2Fe_3(PO_4)_3O_2{\cdot}3H_2O$) within large and micro-sized pores often displaying a geode-type structure (Figure 1a) and invading micron-sized pores in the surrounding groundmass. In some cases they form fan-shaped aggregates of prismatic crystals with pyramidal facets and perfect cleavage on [010] (Figure 1b), or encrustations on the pottery surface. They are always located in specific micro-structural sites and do not impregnate the whole potsherd.

The mobility of phosphorus is related to the presence, regime and chemistry of groundwater and to bacterial activity; these factors control local pH and Eh conditions [1]. The occurrence of a corona-like structure of mitridatite on relicts of vivianite indicates progressive chemical-physical changes of the terrestrial burial environment. In more detail, vivianite precipitated at neutral to low pH and low Eh [2], and reacted with Ca-rich groundwater at increasing pH and Eh, producing mitridatite, probably in relation to a progressive increase of burial depth and ascent of the water table, as observed in canal bed sediments and in other archaeological contexts [3].

An additional case study concerns fragments of Roman African Red Slip (ARS) ware from the ancient harbour of Olbia, buried in a lagoon-like environment for nearly 19 centuries. Secondary crystals of pyrite (FeS_2) and jarosite ($(K,Na)Fe_3(SO_4)_2(OH)_6$) were identified by SEM. Besides rare euhedral micro-sized crystals, pyrite forms abundant framboidal crystal aggregates in the meso-pores, in some cases with a thin rim of jarosite (Figure 1c), with a variable Na/K ratio, although K-rich terms prevail [4]. The concentration of this mineral phase increases towards the surface, and is particularly abundant in the first 50 μm, where pyrite drastically decreases. Jarosite

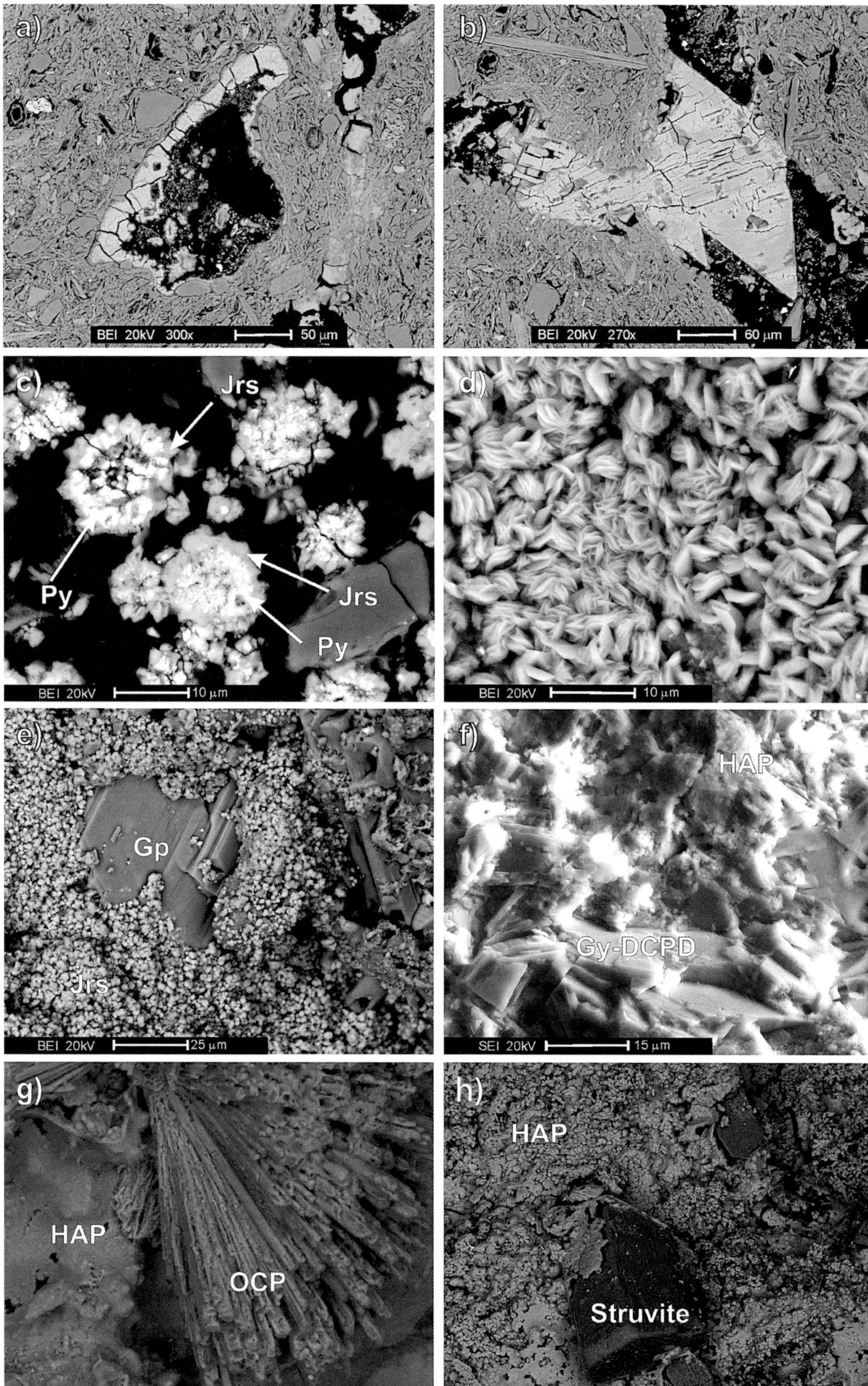

Figure 1 Backscattered (BSE) image and secondary electron (SE) image of: (a) the geode-type structure of phosphate crystals; (b) prismatic crystals of phosphates; (c) rim of jarosite (Jrs) on framboidal crystal of pyrite (Py); (d) jarosite crystals; (e) microcrystalline aggregates of jarosite (Jrs) and tabular crystals of gypsum (Gp); (f) tabular gypsum-brushite (Gy-DCPD (dicalcium phosphate dihydrate)) crystals with hydroxylapatite (HAP) grown epitaxially; (g) elongated octacalcium phosphate (OCP) crystals on a poorly crystalline hydroxylapatite (HAP) matrix; (h) euhedral struvite crystals on a poorly crystalline hydroxylapatite (HAP) matrix.

also occurs on the altered surface, where it forms idiomorphic bladed crystals a few micrometres in size, arranged in rosette-like aggregates, sometimes closely packed and covering the entire surface (Figure 1d), associated with rare gypsum crystals (Figure 1e). The altered portions of these ceramic materials do not contain any trace of hematite, which is normally very abundant in ARS ware. Therefore, pyrite formed after hematite due to the intense activity of sulphate-reducing bacteria, in general low Eh and basic conditions of burial [4]. This transformation caused pervasive changes in the original colours of the potsherds, from reddish-yellow to grey. The crystallisation of jarosite and gypsum after pyrite is here related to the oxidation of iron sulphide [4] during progressive consumption of organic matter and increase in sea level, resulting in a significant change in the pH and Eh conditions. Differences in the distribution of the secondary phases in the potsherds suggest that pervasiveness of the alteration process is controlled by the structure of the pore system.

In both cases, the abundance of the secondary phases in the pottery was not sufficiently high to be detected by XRD or to affect bulk chemical composition, so that these potsherds could not be discriminated as post-burial contaminated samples on a geochemical basis only. Moreover, the identification of a corona-like structure by SEM allowed the recognition of the crystallisation sequence of the secondary phases as a consequence of changes in the chemical-physical burial conditions.

Concrete

Over the past decades many studies have focused on the alteration of concrete structural elements due to the action of external chemical degrading agents, mainly focusing on the effects of external sulphate attack [5]. Yet, there are few references in the literature to phosphate attack in cementitious materials. A further case study concerns concretes from a building of industrial archaeological interest at Porto Marghera, which suffered from severe atmospheric pollution, with emissions of sulphur and nitrogen linked to industrial activities in the area during the last 50 years. The production of superphosphate- and ammonium phosphate-bearing fertilisers in the industrial plant also resulted in intense soil pollution. According to the SEM microstructural profiling, tabular crystals of a gypsum-brushite ($CaSO_4 \cdot 2(H_2O) - Ca(HPO_4) \cdot 2(H_2O)$) solid-solution phase formed in the external strata of the materials under the superficial crust (Figure 1f). Microanalyses indicated a high chemical variability of this phase, with the prevalence of gypsum-rich terms and stoichiometrical ardealite ($Ca_2(SO_4)(HPO_4)\times4(H_2O)$). Moreover, microcrystalline aggregates of hydroxylapatite ($Ca_5(PO_4)_3(OH)$) grew epitaxially on gypsum-brushite seed crystals (Figure 1f). Elongated octacalcium phosphate ($Ca_8(HPO_4)_2(PO_4)_4\times5(H_2O)$) crystals, surrounded by a matrix of poorly crystalline hydroxylapatite with limited OH^-–Cl^- and SO_4^{2-}–HPO_4^{2-} substitutions, were observed in the inner portions (Figure 1g). The core of the altered matrix contained poorly crystalline apatite, associated with euhedral struvite ($(NH_4)MgPO_4\times6(H_2O)$) and dittmarite ($(NH_4)MgPO_4\times(H_2O)$) crystals characterised by a scarce NH^{4+}–K^+ substitution (Figure 1h).

On the basis of analytical results and background literature, a four-stage alteration model was created:

1. The pore network was permeated by fluid pollutants rich in phosphates, sulphates, ammonia and chlorides, enriched in sulphates and characterised by greater acidity in the external portions due to the contribution of the atmospheric pollutants.
2. CSH (calcium silicate hydrates) and CH (calcium hydroxide, portlandite) phases underwent dissolution, and gypsum and hydroxylapatite precursors precipitated according to pH and sulphate concentration gradients; gypsum-brushite solid-solution (acid precursor [5]) precipitated in the external portions and around the steel bars used for reinforcement (locally acidified by chloride attack); octacalcium phosphate (neutral precursor [6]) precipitated in the intermediate layers; amorphous calcium phosphate (basic precursor [6]) precipitated in the cores.
3. Precursors converted into hydroxylapatite (brushite by epitaxial growth [6], octacalcium phosphate by dissolution-reprecipitation [6], amorphous Ca-phosphate by longer range-order development [6]); this process was strongly inhibited in the external strata by pH gradients and stabilising ions in solution, in particular by Mg [5].
4. The terms of the gypsum-brushite solid-solution dehydrated on the surface due to the heating effect of radiation from the sun, and formed concretions of partially dehydrated sulphates (gypsum, bassanite, anhydrite) and crusts of anhydrous phosphates (monetite).

The microstructural analysis of the concretes showed that the materials underwent alterations causing heterogeneous sequences of secondary phase crystallisation, which were strongly influenced by pH and concentration gradients at a millimetric-centimetric scale.

In the study of the alteration processes in both potsherds and concretes, the use of SEM is fundamental to detect the secondary phases and reconstruct the alteration dynamics and sequences of crystallisation, especially when the secondary products are very low in quantity or scarcely crystalline, making them hardly detectable by XRD or spectroscopic investigation methods.

References

1. Maritan, L. and Mazzoli, C., 'Phosphates in archaeological finds: implications for environmental conditions of burial', *Archaeometry* **46** (2004) 673–683.
2. Nriagu, J.O. and Dell, C.I., 'Diagenetic formation of iron phosphates in recent lake sediments', *American Mineralogist* **59** (1974) 934–946.
3. Dodd, J., Large, D.J., Fortey, N.J., Kemp, S., Styles, M., Wetton, P. and Milodowski, A., 'Geochemistry and petrography of phosphorus in urban canal bed sediment', *Applied Geochemistry* **18** (2003) 259–267.
4. Secco, M., Maritan, L., Mazzoli, C., Lampronti, G.I., Zorzi, F., Nodari, L., Russo, U. and Pesavento Mattioli, S., 'Alteration processes of pottery in lagoon-like environments', *Archaeometry* **43** (2011) 809–829.
5. Skalny, J.P., Odler, I. and Marchand, J., *Sulfate Attack on Concrete*, Spon, London (2001).

6. Johnsson, M.S.A. and Nancollas, G.H., 'The role of brushite and octacalcium phosphate in apatite formation', *Critical Reviews in Oral Biology& Medicine* **3** (1992) 61–82.

Authors' addresses

- Lara Maritan, Department of Geosciences, University of Padua, Padua, Italy (lara.maritan@unipd.it)
- Michele Secco, Department of Geosciences, University of Padua, Padua, Italy
- Claudio Mazzoli, Department of Geosciences, University of Padua, Padua, Italy
- Gilberto Artioli, Department of Geosciences, University of Padua, Padua, Italy

Scanning electron microscopy investigation of Late Bronze Age high-tin socketed axes: hoards from Langton Matravers, Dorset, southern England

Nigel Meeks, Aude Mongiatti, Duncan Hook, Ben Roberts, Andrew Fitzpatrick and Peter Woodward

KEYWORDS high-tin bronze, Late Bronze Age, socketed axe, hoard, southern England

Introduction

In 2007, four large hoards of Late Bronze Age socketed axes dated to the Llyn Fawr phase, c.800–600 BC, were discovered in Langton Matravers, Dorset, and duly reported under the Portable Antiquities Scheme by metal detectorists Brian Thomas, Les Keith and Tom Peirce with the assistance of Alfy O'Connell of Putlake Adventure Farm. During subsequent excavation, no evidence of a nearby settlement or metal workshop was found. The axes were acquired by the Dorset Natural History and Archaeological Society for the Dorset County Museum (DORCM 2009.88). Together, the four hoards comprise the largest axe deposit known from this metalworking phase in Britain [1]. They contained a total of c.400 axes, with almost half of them broken, cracked or incomplete. Nearly all are decorated with three ribs ending in a single pellet, although a few have four or five ribs. Most are linear faceted of Type Portland, which has a local distribution in Dorset especially near the coast [2]. Two are of the rarer Type Blandford, examples of which have been found nearby at Blandford Forum.

The axes are as-cast: the casting flashes from bivalve moulds remain and the sockets are filled with sand and clay cores from the moulding process. No attempt has been made to 'finish' the axes by hammering, grinding or polishing. Many exhibit areas of the original silver-coloured surface, with varying degrees of shiny black or encrusted green corrosion products present either as small patches or as a more complete covering. These unusual surface features prompted the scientific investigation to explain their appearance, which suggested high-tin content as previously reported on axe finds from Dorset [3].

Methods

Scanning electron microscopy with energy dispersive X-ray analysis (SEM-EDX) was ideally suited for the technological examination of surfaces and polished cross-sections of these bronze axes. Several axes with both silvery and black surface colours were examined directly to study their surface microstructure and composition. Samples were taken from 13 representative axes at their haft ends, to avoid changing their profile. These samples were mounted in resin and polished down to 1 μm, providing cross-sections from the outer surface to the inside edge of the socket. This allowed the full characterisation of the surface and body metal and investigation of any variations in microstructure and composition within each axe and between axes.

Object surfaces and samples were examined using backscattered electron (BSE) imaging at low magnification (c.30×) to record the overall microstructures, macro inclusions and corrosion, and at high magnification (c.1000×) to show the finer phase structures. Metallography and X-ray fluorescence (XRF) applied to the polished sections provided complementary information, including the levels of trace elements present.

Results

Metallurgical structure and chemical composition

The axes are all high-tin bronze alloys of relatively low melting points (c.850–950°C), with tin contents between 16 wt% and 23 wt%, and variable concentrations of trace elements Ni, As, Sb, Ag, Co, determined by XRF. Their microstructures are typical of high-tin bronzes with some copper-rich α phase solid solution in various grain orientations and a dominant continuous network of the tin-rich intermetallic compound δ (32.6 wt% Sn) forming the characteristic $\alpha + \delta$ eutectoid (Figure 1) [4, 5]. This dense network determines the mechanical properties of the alloys making them hard and brittle, as exemplified by the number of fractured, cracked and incomplete axes. The brittle nature of the alloys precludes any cold hammering to finish the axes and thus they may not have been intended to be working tools. The axes also contain variable amounts of lead (between c.3 wt% and 18 wt%) as finely distributed globules in the alloys with a low lead content, but as

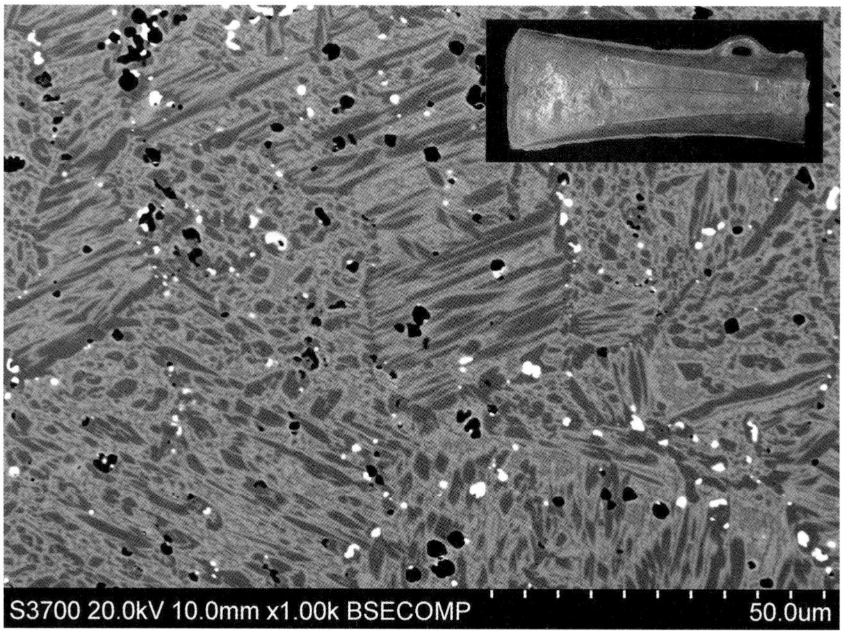

Figure 1 Polished cross-section from a typical axe with a silvery surface (inset), showing the eutectoid microstructure throughout the body metal (mid-grey) with a fine distribution of lead globules (white), α phase (darker grey), and porosity (black). Major element composition (by SEM-EDX): Cu 75 wt%, Sn 22 wt%, Pb 3 wt%.

large globules distributed heterogeneously in those with a high lead content. This wide concentration range does not suggest the addition of a precise quantity of lead to the alloy. The lead does not improve the mechanical properties of the alloys, but would improve their castability by increasing fluidity over the alloy freezing range.

Surface colour

The surfaces of the selected axes show no signs of grinding or polishing but display a variety of surface colours from silvery, green, grey to black (often on a single axe), as seen on other high-tin bronzes [6], but the silvery and black colours of some make them look very different from axes in other Bronze Age hoards [7]. The silvery colour is due to the dominant presence of the intermetallic compound δ in the eutectoid microstructure at the surface, which is reflected throughout the internal microstructure as explained above (Figure 1). The surfaces may have also been enriched in eutectoid during casting by inverse segregation, a phenomenon called 'tin-sweat' [5] that can occur in tin-rich alloys at fast cooling rates. As the alloy solidifies and cools, the remaining molten metal enriches in tin up to the eutectoid composition of 27 wt%, and at the same time the axe shrinks in the mould and can force this remaining tin-rich liquid metal to the surface via fine interdendritic pathways where it solidifies as a continuous eutectoid structure.

On the surface, the copper-rich α phase corrodes first leaving the usually corrosion-resistant intermetallic silvery compound network. However in certain cases, this intermetallic compound does corrode, turning to black tin oxides that remain as pseudomorphic ghost structures. Therefore the original physical surface and microstructure remain but appear black in colour. Some axes have both silvery and black areas due to varying local corrosion conditions, and others are green due to the presence of copper corrosion products.

Conclusions

The Langton Matravers axes were found to be unfinished, as-cast, high-tin bronzes, originally silvery and inherently brittle due to the presence of the δ intermetallic compound throughout the structure and at the surface. Their colour and mechanical properties (i.e. brittleness) would have made them distinctive at the time of manufacture, compared to the more usual low-tin bronze axes used as practical tools. They have not been 'tinned' by the deliberate application of metallic tin to their surface, as this would produce the same colour but a very different microstructure [8]. The axes vary slightly in size and cutting edge shapes, indicating the use of a variety of bivalve moulds. Although all axes are high-tin bronzes, the wide concentration range of major elements, especially lead, and trace elements suggest that no precise recipe was followed and that different batches of bronze were used.

This hoard deposition raises many questions. Why were the axes made of these compositions? Do they represent an attempt to produce a hard alloy, with the resulting brittleness being an unfortunate side-effect? Was the alloy chosen for its silvery colour alone? Why were the axes unfinished and apparently buried shortly after production? Were they a form of currency and linked to trade (they were found only 2 km from the sea)? Could they represent a precious, ritual deposit in a special place? Ongoing research will consider such questions in the wider context of other bronze axes, hoards and locations and the results and interpretations will be fully published elsewhere.

Acknowledgements

Mike Dinwiddy is deeply thanked for his extensive involvement in the excavation and the writing of the subsequent report. We are grateful to Dot Bruns and Nisha Doshi who investigated and recorded the axes when brought to the British Museum. We would like to thank Peter

Northover for his information and advice. Antony Simpson is thanked for help in producing the original poster and Catherine Higgitt for helpful comments on the text.

References

1. Huth, C., 'Metal circulation, communication and traditions of craftsmanship in Late Bronze Age and Early Iron Age Europe', in *Metals Make the World Go Round: The Supply and Circulation of Metals in Bronze Age Europe*, ed. C.F.E. Pare, Oxbow Books, London (2000) 176–193.
2. Pearce, S., *The Bronze Age Metalwork of Southwestern Britain*, British Archaeological Reports 120 (British Series), Oxford (1983).
3. Northover, J.P., 'Bronze in the British Bronze Age', in *Aspects of Early Metallurgy*, ed. W.A. Oddy, Occasional Paper 17, British Museum Publications, London (1980) 63–70.
4. Meeks, N.D., 'Surface characterization of tinned bronze, high-tin bronze, tinned iron and arsenical bronze', in *Metal Plating and Patination*, ed. S. La Niece and P.T. Craddock, Butterworth-Heinemann, London (1993) 247–275.
5. Hanson, D. and Pell-Walpole, W.T., *Chill-Cast Tin Bronzes*, Edward Arnold and Co., London (1951).
6. Meeks, N.D., 'Patination phenomena on Roman and Chinese high-tin bronze mirrors and other artefacts', in *Metal Plating and Patination*, ed. S. La Niece and P.T. Craddock, Butterworth-Heinemann, London (1993) 63–84.
7. O'Connor, B., 'Llyn Fawr metalwork in Britain: a review', in *The Earlier Iron Age in Britain and the Near Continent*, ed. C. Haselgrove and R. Pope, Oxbow Books, Oxford (2007) 64–79.
8. Meeks, N.D., 'Tin-rich surfaces on bronzes: some experimental and archaeological considerations', *Archaeometry* **28**(2) (1986) 133–162.

Authors' addresses

- Nigel Meeks, Department of Conservation and Scientific Research, The British Museum, London, UK (nmeeks@thebritishmuseum.ac.uk)
- Aude Mongiatti, Department of Conservation and Scientific Research, The British Museum, London, UK
- Duncan Hook, Department of Conservation and Scientific Research, The British Museum, London, UK
- Ben Roberts, Department of Prehistory and Europe, The British Museum, London, UK
- Andrew Fitzpatrick, Wessex Archaeology, Salisbury, UK
- Peter Woodward, Dorset County Museum, Dorchester, UK

Analytical characterisation of bole used in gilded plasterwork (Arab Room, Palácio da Bolsa, Oporto, Portugal)

Patricia Mestre, Fernando Rocha and João Coroado

KEYWORDS plasterwork, bole, gilding, SEM-EDX, XRD

Introduction

Gilded plasterwork has played an important role in the decorative arts of Portugal, in both religious and secular buildings, particularly in the north of the country, reaching paramount importance during the eighteenth and nineteenth centuries.

Several materials are involved in this artistic technique. Bole is used as the preparatory layer and gold leaf on its surface. Bole is an argillaceous material rich in iron oxides, used mainly in water gilding. Usually bound with animal glue, sometimes including other additives, it was applied as a gilding ground in order to provide a flat smooth 'laying-on' surface [1, 2] to the plasterwork prior to the application of the gold leaf.

The Arab Room (Salão Árabe) of Palácio da Bolsa, in Oporto (northern Portugal), built between 1862 and 1880, is an emblematic symbol of the city owing to the extensive decorative scheme of gilded and polychromatic plasterwork on its walls and ceilings. Throughout its history, this 135 m^2 room has undergone several restoration interventions, some of them without any adequate scientific supervision or technical support.

Making use of several analytical techniques, the aim of this study was to compare the chemical and mineralogical compositions of the preparatory layers used in burnished and unburnished gilded surfaces of the Arab Room in order to distinguish the techniques used in the original areas and the different restored areas.

Methods

Ten samples of gilded plasterwork were removed from selected areas of the Arab Room for this study (Table 1). Samples were chosen to represent different types of gilding, unburnished (samples 0-4-2, 0-10-2 and 0-19-7) and burnished gold leaf (the rest of the collection), in original and restored areas.

Samples were mounted in epoxy resin blocks (cross-sections). The identification of different layers was carried out using a Nikon Eclipse E400 POL microscope equipped with a Zeiss AxioCam MRc camera.

Chemical characterisation of the boles was performed on a Hitachi SU-70 high resolution-field emission (HR-FE) SEM equipped with the Peltier cooled XFlash 4010 B-U EDX detector of the Bruker QUANTAX 400 EDX system (SEM-EDX). A probe current of 32 μA was applied, with an acceleration voltage of 15 kV and with 60 seconds of acquisition time.

Extra samples were powdered together and analysed by X-ray diffraction (XRD). XRD was performed by an X'Pert-Pro MPD Philips/Panalytical apparatus, equipped with a PW3040 generator and a PW3050/60 goniometer. The radiation applied was copper K α (λ = 1,5405 Å) at 30 mA and 50 Kv, at a scan speed of 0,02°/s and made from 2° to 65°. Identification of crystalline phases was carried out using the International Center for Diffraction Data Powder Diffraction Files (ICDD PDF), but the gold was not detected due to the minute quantities present, compared to the rest of the constituents of the samples.

Results and discussion

The stratigraphic sequence and characteristics of the gilded layers were examined using reflected light polarised microscopy. Some samples showed a single gilded layer while others showed two layers (Figure 1). The thickness of the entire bole layer that preceded each gold leaf layer varied from 10 μm to 100 μm; their colour range varied from deep dark red to orange, apart from the top layer of sample 0-10-2 that had a light cream colour.

The scanning electron microscopy with energy dispersive X-ray analysis (SEM-EDX) of bole from the burnished gold samples removed from original and restored panels showed that their main elements are silicon, aluminium, iron, calcium and potassium (apart from the top layer in sample 0-13-8). XRD confirmed these results, showing the presence of calcite, goethite, gypsum and also muscovite (the latter on 0-T12-5 and 0-20-1) (Table 1). It is possible that the presence of chlorine

Table 1 Composition of the bole obtained through SEM-EDX and XRD and its relation with the gilding technique. (GF: ground floor; Burn: burnished; Unb.: unburnished.)

Sample ref.	Location area	Typology	Gold	Number of gilded layers		SEM-EDX results	XRD results
0-7-4	GF panel 7 wall	Original	Burn.	One		Silicon, aluminium, calcium, iron, sulphur, potassium and titanium	Calcite, goethite and gypsum
0-T12-5	GF panel 12 ceiling	Original	Burn.	One		Silicon, aluminium, iron, potassium, calcium, titanium and sulphur	Calcite, goethite, gypsum and muscovite
0-4-4	GF panel 2 wall	Original	Burn.	Two	Top	Silicon, aluminium, iron, calcium, potassium, and chlorine	Calcite, goethite and gypsum
					Inner	Silicon, aluminium, iron, potassium, calcium and sulphur	
0-13-8	GF panel 13 wall	Original	Burn.	Two	Top	Silicon, aluminium, lead, iron, potassium, calcium and titanium	Calcite, goethite and gypsum
					Inner	Silicon, aluminium, iron, potassium, calcium, sulphur and chlorine	
0-19-11	GF panel 19 wall	Restored	Burn.	One		Silicon, aluminium, iron, calcium, titanium and potassium	Calcite, goethite and gypsum
0-20-1	GF panel 20 wall	Restored	Burn.	One		Silicon, aluminium, iron, calcium, potassium, sulphur and titanium	Calcite, goethite, gypsum and muscovite
0-17-6	GF panel 17 wall	Restored	Burn.	Two	Top	Silicon, aluminium, iron, sodium, chlorine and calcium	Calcite, goethite and gypsum
					Inner	Silicon, aluminium, iron, calcium, chlorine, potassium and sodium	
0-4-2	GF panel 4 wall	Original	Unb.	Two	Top	Lead, sodium, calcium, aluminium, sulphur and silicon	Hydrocerussite, muscovite, calcite, goethite and gypsum
					Inner	Lead, calcium, silicon, sodium, iron, aluminium and sulphur	
0-10-2	GF panel 10 wall	Original	Unb.	Two	Top	Zinc, sodium, lead, aluminium, calcium and silicon	Cerussite, wurtzite, sodium alum and calcite
					Inner	Lead, zinc, sodium, calcium, iron and silicon	
0-19-7	GF panel 19 wall	Restored	Unb.	Two	Top	Lead, zinc, aluminium, calcium and sulphur	Unable to perform the XRD analysis
					Inner	Lead, zinc, calcium, silicon, aluminium and sulphur	

in the SEM-EDX results is due to the epoxy resin that has impregnated the samples.

The samples of bole from unburnished gold layers showed a different composition. SEM-EDX of both ground layers in sample 0-4-2 indicated the presence of lead, sodium, calcium, aluminium, silicon, sulphur (traces in the top layer) and iron in its inner layer (dark red colour). XRD identified hydrocerussite, muscovite, calcite, goethite and gypsum.

In sample 0-10-2, zinc, lead, sodium, calcium, aluminium, silicon (on the cream-coloured top layer) and iron on the inner layer (dark red colour) were identified. XRD detected cerussite, wurtzite, sodium alum and calcite. In sample 0-19-7, lead, zinc, aluminium, calcium, traces of sulphur and some silicon on the inner layer were detected with SEM-EDX. Due to insufficient material it was not possible to perform XRD analysis on this sample.

When comparing the results it becomes clear that there is a difference between the composition of the gilding ground used for the burnished gold leaf samples and the unburnished one.The bole on the burnished gold samples was composed mainly of a clay-like material, possibly a phyllosilicated material, illitic/muscovitic or kaolinitic type (due to the presence of silicon, aluminium and potassium) and iron oxi-hydroxides (iron identified by SEM-EDX and goethite by XRD) [1, 3].

The vestigial presence of titanium could be due to anatase, which is a minor phase mineral associated with clay minerals. The existence of calcium in all the samples could be due to the addition of calcium carbonate [2, 4], although its combination with sulphur on sample 0-7-4 could be related to the addition of gypsum [1, 5]. Due to the constituents of the powdered samples analysed by XRD it is almost certain that the presence of clay minerals was masked by the huge amount of gypsum (from the plaster substrate).

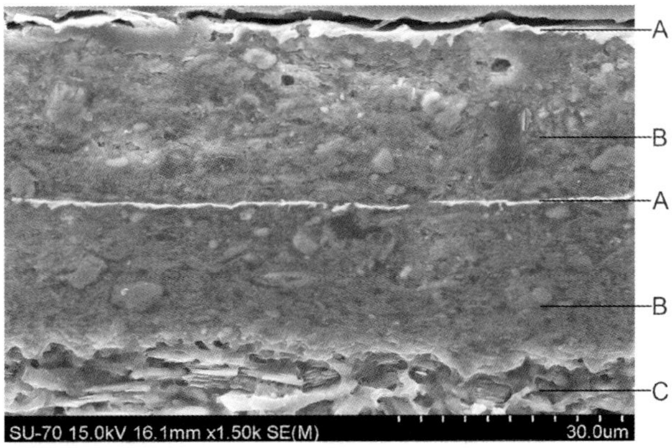

Figure 1 SEM secondary electron image of sample 0-4-4 showing two gold leaf layers (A), two bole layers (B) and gypsum plaster (C).

The gilding ground applied in areas of unburnished gold is composed largely of a mixture of lead- and/or zinc-based pigments (lead and zinc identified by SEM-EDX, hydrocerussite, cerussite and wurtzite by XRD), possibly some gypsum or calcite and also some clay-like material (silicon and aluminium identified by SEM-EDX and muscovite by XRD) but only in small amounts.

Conclusions

Gilding techniques applied to the decorative plasterwork in the Arab Room follow traditional methods as documented in several treatises and the literature [2, 4–8]. This comprises the technique of water gilding, where bole was habitually mixed with an organic binder and some other additives before being used as a laying-on base for a gold leaf intended to be burnished. Mordant gilding, where a preparatory layer of pigments is also mixed with an organic binder and other additives, is used for unburnished gold in smaller areas. Both these processes were also applied in the restoration interventions that have been carried out to date in the Arab Room.

Acknowledgements

This study is funded by the FCT – Fundação para a Ciência e Tecnologia (Portuguese Foundation for Science and Technology) within the scope of the Project PEst-C/CTE/UI4035/2011. Patricia Mestre wishes to acknowledge FCT – Fundação para a Ciência e Tecnologia for the PhD grant Ref.: SFRH/BD/47969/2008.

References

1. Hradil, D., Grygar, T., Hradilová, J. and Bezdička, P., 'Clay and iron oxide pigments in the history of painting', *Applied Clay Science* **22** (2003) 223–236. www.iic.cas.cz/~grygar/ACS-earths.pdf (accessed 5 June 2010).
2. Barata, C., *Caracterização de materiais e de técnicas de policromia da escultura portuguesa sobre madeira de produção erudita e de produção popular da época Barroca*, Masters dissertation, Universidade de Lisboa - Faculdade de Ciências (2008).
3. Grygar, T., Hradilová, J., Hradil, D, Bezdička, P. and Bakardjieva, S., 'Analysis of earthy pigments in grounds of Baroque paintings', *Analytical and Bioanalytical Chemistry* **375**(8) (2003) 1154–1160. http://dx.doi.org/10.1007/s00216-002-1708-x (accessed 12 July 2011).
4. Turco, A., *Il doratore*, 2nd edn, Ulrico Hoepli Editore, Milan (1987).
5. Ribeiro, C., 'O Oratório Indo-Português do Museu de Évora; Estudo, Conservação e Restauro'. *Cenáculo - Boletim on line do Museu de Évora* **2** (2007) 2–15. http://museudevora.imc-ip.pt/Data/Documents/Cenaculo2/B2oratorioCR2007a.pdf (accessed 16 May 2010).
6. Tejeda, L.C., *El oro en hoja: aplicación y tratamiento sobre soportes móviles tradicionales, muro e resinas*, PhD Dissertation, Universidad Complutense de Madrid, Faculdad de Bellas Artes (2001). http://eprints.ucm.es/tesis/bba/ucm-t25606.pdf (accessed 14 January 2008).
7. Zamora, E.L., *Estudio de los materiales y procedimientos del dorado a través de las fuentes literarias antiguas: aplicación en las decoraciones de pinturas castellanas sobre tabla*, PhD dissertation, Departamento de Pintura-Restauración, Universidad Complutense de Madrid, Facultad de Bellas Artes (2007).
8. Toniolo, C.C.J.L., Bruni, S., Fermo, P., Casoli, A., Palla, G. and Bianchi, C.L., 'Gilded stuccoes of the Italian Baroque', *Studies in Conservation* **43** (1998) 201–208. www.jstor.org/stable/1506729 (accessed 5 June 2010).

Authors' addresses

- Patricia Mestre, Geobiotec (Geobiosciences, Geotechnologies and Geo Engineering Research Centre), Campus Universitário de Santiago, Aveiro, Portugal (patriciamestre@ua.pt)
- Fernando Rocha, Geobiotec and Geosciences Department of the University of Aveiro, Aveiro, Portugal
- João Coroado, Geobiotec and Geosciences Department of the University of Aveiro and Polytechnic Institute of Tomar, Art, Conservation and Restoration Department, Tomar, Portugal

Scanning electron microscopy and optical microscopic study of gilded tiles from Darb-i Imam Tomb in Isfahan, Iran

Moslem Mish Mast Nehi, Hamid reza Chaman and Mohammad Mortazavi

KEYWORDS gilded tiles, gold leaf, gilded technique, SEM-EDX, Iran

Introduction

Gilding is one of a number of techniques known worldwide in the arts. Extant gilded objects (from ancient Egypt to modern Japan) that belong to various periods in history are witness to this claim [1]. The use of gold leaf on a variety of materials such as stone, wood, metal, glass and gypsum may have been employed to produce objects resembling those made entirely of gold [1].

Bole (red clay) has been used to produce objects and is often accompanied by gilding. The principles of application of this material involve the following: first, the preparation of a ground with an application of bole in layers, which is moulded and carved to make a bas-relief ornament. This provides a smooth surface for later burnishing [2]. Secondly, a very thin layer of gold leaf is applied. The red layer (bole) prevents undesirable visual effects and enriches the colour of the gilding [2].

The use of gold leaf in Islamic tiles and objects has a long history as testified by objects found in Iran, Syria and Egypt (amongst others), so it is somewhat surprising that there has not been a comprehensive investigation of gilded techniques [3]. Islamic glazed ceramics decorated with gold leaf appear at the end of the twelfth century in Iran [3–6]. Among the most famous historic buildings using gilded tiles for their decoration are Takht-i Sulayman, Masjid-i Kabud of Tabriz and Historical Ensemble of Sheikh Safi al-Din Ardebili.

There are few historical sources describing recipes and techniques used in gilding glazed ceramics, but in one of the texts is a treatise written by Abu'l Qasim al-Kâshâni in 1301/700 [7, 8]. The technique used is the affixation of the gold leaf on a fired glazed ceramic, followed by firing [4–8]. An alternative technique, which has also been used (but is not mentioned), was gilding ceramics without any special firing after making the glazed support. In this technique, gold leaf was affixed to glazed ceramics with surfaces prepared in a similar way to bole gilding in painting, altarpieces and iconostases [2, 9, 10]. Gilded tiles from Darb-i Imam Tomb in Isfahan were produced using this lesser-known technique of gilding on glaze.

Darb-i Iman

The samples investigated in this study are from Darb-i Imam Tomb, a religious building built in the sixteenth–seventeenth century in Isfahan, Iran. The dados of the Northern Room are decorated with green hexagonal tiles with skilfully gilded delicate patterns.

Darb-i Imam Tomb was built in AD 1453 during the reign of Jahanshah Garagoyunlu; repairs and additions were carried out until the Qajar period (AD 1785–1925). In parts of the northern dado of the vestibule of this building (c.150 cm high) are green hexagonal gilded tiles with Cut-Tile Mosaic (*Moàrraq*) borders [11]. Between the green hexagonal tiles (average length 5 cm) are narrow blue tiles coloured using cobalt (Figure 1a). In the *Moàrraq* borders there are also brown tiles with traces of gold remaining.

Materials and methods

To avoid taking samples from undamaged tiles, experiments were carried out on 2–3 mm fragments of the edges of broken tiles upon which small quantities of gold leaf remained. Samples were embedded in resin and polished with silicon carbide abrasive paper (up to 1200 grit). Carbon coating of cross-sections was performed before scanning electron microscopy with energy dispersive X-ray microanalysis (SEM-EDX) investigation.

To study the raw materials and production technique of these gilded tiles, three complementary methods were used. Chemical tests were carried out by dipping gilded samples into distilled water and sulphuric acid (H_2SO_4) separately to differentiate between two different gilding techniques on ceramics: gold leaf that had undergone a firing, and bole gilding. Detachment of gold leaf as a result of dissolution binding in water or wet decomposition in sulphuric acid was used as an indication for distinguishing bole gilding from fired gilded tile. Because of the gold diffusion into the silica in the base of glaze (in the fired gilded type) [12] this could not be

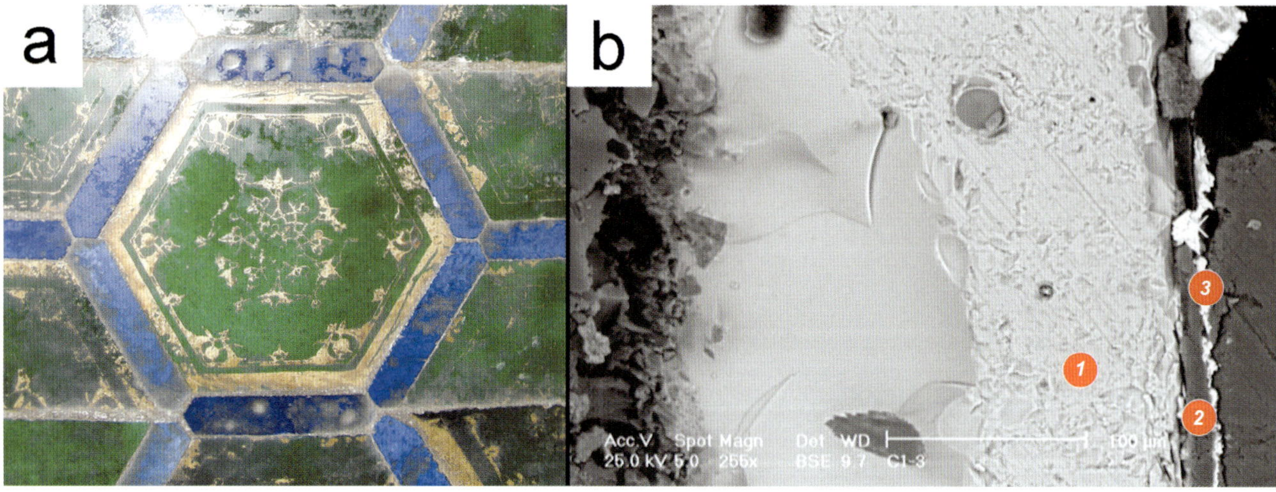

Figure 1 (a) Details of green hexagonal tiles showing gilded patterns. (b) BSE image of gilded tile cross-section showing three layers: (1) green glaze, (2) red layer and (3) gold leaf.

Table 1 SEM-EDX analysis of three different layers: glaze, red layer and gold leaf (elemental composition given in wt%).

Sample	Al	Si	Au	Cl	Ca	Fe	Cu	Zn	Pb	Na	K	Sn	Cr	Co
Glaze	2.5	22.4	1.2	–	1.7	1	3.8	–	60	0.4	4.3	2.3	0.2	0.3
Red layer	9.5	30.8	1.8	–	3.4	14	1.3	–	2.1	5.7	28.7	2.8	0.9	–
Gold leaf	–	0.8	83.2	2	–	0.6	12.8	0.7	–	–	–	–	–	–

separated through a chemical process. In addition, a spot test was carried out to identify protein using copper sulphate for the bole gilding type (Buiret test) [13].

SEM and optical microscopy (OM) were used for both surfaces and cross-sections. OM was carried out on a Japan (OGAWA SEIKI) MR-11 microscope. In order to determine the chemical composition of the samples, SEM-EDX analyses were undertaken with a Philips XL30 EM with an EDAX Si(Li) detector and Genesis software at the following measurement conditions: 25 kV, 30 μA, c.5000 cps, 60 s live-time, standardless ZAF quantification. Carbon coating of cross-sections was performed with a sputter coater BAL-TEC model SCD 050. A carbon film of 10 nm was produced in argon environment. Carbon was used for coating of samples instead of gold (to prevent the sample charging) because in EDX analysis the peaks from the gold coating can mask those of other elements present.

Results and discussion

The dissolution test using distilled water took several hours to detach the gold leaf from the glaze surface. This indicates the use of water-soluble glue (probably animal glue) as an adhesive, mixed with red bole. Where oxidation with sulphuric acid (H_2SO_4) was necessary to detach the gold leaf from the glazed surface it was assumed that the tiles were fired gilded. A spot test was carried out on red layers to verify the presence of proteins (probably animal glue).

The investigation of cross-sections with OM and SEM showed three different layers: glaze, red layer and gold (Figure 1b). OM images indicate a network of creases on the gold surfaces. Longitudinal bands can be seen in SEM images of the surface of the gold leaf, probably related to the hammering process for making gold leaf.

Over the green glaze a thin layer of red bole could be distinguished. In these tiles, the red layer is only 8–10 μm thick. It seems that the maker did not intend to use red bole to make relief decoration. Rather, it was the visual properties of a red layer under the gold leaf that was the main intention [2]. Red bole covered the areas that were to receive gold leaf similar to common bole gilding for painting and iconostases [2, 9, 10].

Elemental analysis using SEM-EDX of three different layers (Table 1) showed that a gold-copper (Au-Cu) alloy was used for gold leaf. There is a high iron (Fe) content in the red layer and lead-containing (Pb) green glaze with copper (Cu) as a chromophore. Because copper reduces the workability of gold, an appropriate heat treatment would have been necessary during the making of the gold leaf [14].

The use of lead (Pb) (60 wt%) as the flux in the green glaze would have resulted in a relatively low melting temperature. In addition, potassium (K) (4.3 wt%) acts as flux in glaze. The small amount of sodium (Na) indicates the use of an alkaline ash as flux [8] in combination with the Pb. The presence of high levels of Pb in the glaze would have reduced the firing temperature to c.700–900°C [15]. The presence of aluminium (Al), silicon (Si), Na and K as well as Fe (14 wt%) in the red layer implies the use of a composition of red clay with a significant amount of iron [2, 9, 10]; the most important material known to have these properties is red bole, which has a worldwide application in gilding [2, 9, 10, 16]. The small amount of gold, copper, lead and tin in the red layer probably comes from the glaze layer and gold leaf through preparation of the sample by polishing.

Conclusions

The technique of gilding can be divided into two different methods: first, affixation of the gold leaf by firing; secondly, the preparation of the glaze surface with soft red bole (iron-containing clay) mixed with animal glue, and then affixation of gold leaf onto these areas. Contrary to the common method of application of gold-leaf decoration on medieval Islamic glazed ceramics described by Abu'l Qasim al-Kâshâni, there was no firing process after applying the gold leaf on the tiles of the Darb-i Imam Tomb in Isfahan, Iran. Here, patterns were drawn on a fired glazed ceramic with red bole that was used to attach the gold leaf. Gold leaf was laid onto an adhesive layer of red bole in a similar manner to bole gilding (water gilding).

Acknowledgements

Every author is indebted to people who share ideas, concepts, and opinions, but most of all, to those who offer encouragement. In this light we would like to thank Abbas Abed Isfahani, Hesam Aslani, Hamed Sayadshahri and Parviz Holakoi.

References

1. Holakouei, P., *Technology and Pothology of the Laye-Chini Decorations (Pastiglia) of some Safavid Monuments in Isfahan (Ali-Qapu, Chihil-Sutun, and Hasht-Behesht)*, MA dissertation, Art University, Tehran, Iran (2008).
2. Sandu, I.C.A., Afonso, L.U., Murta, E. and De Sa, M.H., 'Gilding technique in religious art between East and West, 14th–18th centuries', *International Journal of Conservation Science* **1**(1) (2010) 47–62.
3. Porter, V., *Islamic Tiles*, The British Museum Press, London (1999).
4. Pacheco, C., Chapoulie, R., Dooryhee, E., Aucouturier, M., Bouquillon, A. and Makariou, S., 'Medieval Islamic gilding glazed ceramics: crossing analytical results in search of technological features', in *36th International Symposium on Archaeometry, Quebec, 2–6 May 2006*, ed. J.-F. Moreau, R.A.J. Chabot and A. Herzog Celat, Université Laval, Quebec, (2009), 269–278.
5. Pacheco, C., Chapoulie, R., Dooryhee, E. and Goudeau, P., 'Gold leaf decoration on medieval Islamic glazed ceramics: in search of technological features with XRD', *Zeitschrift für Kristallographie Supplement* **26** (2007), 317–323.
6. Pacheco, C., Chapoulie, R., Dooryhee, E., Aucouturier, M., Bouquillon, A., Makariou, S. and Miroudot, D., 'Gilded medieval Islamic glazed ceramics: production process and evolution in the Iranian world (12th–13th c.) and the Timurid Empire (14th–15th c.)', in *Vessels: Inside and Outside: Proceedings of EMAC'07, 9th European Meeting on Ancient Ceramics, 24–27 October 2007, Budapest*, Hungarian National Museum, Budapest (2009), 179–205.
7. Abu'l Qasim al-Kâshâni, *Arayis al-jawahirwanafayis al-atayib*, Persian edition, Tehran, Iran (1345 H).
8. Allan, J.W., 'Abu'l-Qasim's treatise on ceramics', *Iran* **11** (1973) 111–120.
9. Daniilia, S., Bikiaris, D., Burgio, L., Gavala, P. and Clrack, R.J.H., 'An extensive non-destructive and micro-spectroscopic study of two post-Byzantine overpainted icons of the 16th century', *Journal of Raman Spectroscopy* **33** (2002) 807–814.
10. Isar, N., 'Icons at the Willumsen Museum in Denmark: ethical issues on icon research and conservation', in *Icons: Approaches to Research, Conservation and Ethical Issues, International Meeting, Athens* (2006) (www.iconographyalmanac.com).
11. Golombek, L. and Wilber, D.N., *The Timurid Architecture of Iran and Turan*, translated from English to Farsi by K.A. Afsar and M.Y. Kaini, Cultural Heritage Organization, Tehran, Iran (1995).
12. Darque-Ceretti, E., Helary, D. and Aucouturier, M., 'An investigation of gold/ceramic and gold/glass interfaces', *Gold Bulletin* **35**(4) (2002) 118–129.
13. Stuart, B.H., *Analytical Techniques in Materials Conservation*, John Wiley & Sons Ltd, Chichester (2007).
14. Scott, D A., *Metallography and Microstructure of Ancient and Historic Metals*, The J. Paul Getty Museum in association with Archetype Publications, Singapore (1991).
15. Artioli, G., *Scientific Methods and Cultural Heritage: An Introduction to the Application of Materials Science to Archaeometry and Conservation Science*, Oxford University Press Inc., New York (2010).
16. Rapp, G., *Archaeomineralogy* (Natural Science in Archaeology), Springer-Verlag, Berlin/Heidelberg (2009).

Authors' addresses

- Moslem Mish Mast Nehi, Department of Art, Zabol University, Zabol, Iran (mmishmast@gmail.com)
- Hamid Chaman, Department of Art, Zabol University, Zabol, Iran
- Mohammad Mortazavi, Conservation of Cultural and Historic Properties, Art University of Isfahan, Isfahan, Iran

Evaluation of white rot deterioration in historic wood cellular structure by scanning electron microscopy and FTIR analysis

Mohsen Mohammadi Achachluei, Gholamreza Vatankhah and Aliakbar Enayati

KEYWORDS *Platanus orientalis*, white rot, *Trametes versicolor*, SEM, FTIR

Introduction

Biodeterioration caused by wood decay fungi is of great concern in the conservation of historic woods. Decay fungi can be classified as brown, white or soft rot according to their deterioration form in wood [1]. They strongly degrade wood structures and destroy wood aesthetically. White rot has a great capability to decompose the lignin of wood as well as polysaccharides contained therein [2, 3]. Activities of white rot fungi [4–9] and their properties [10, 11] have been extensively evaluated in the published literature, and it is clear that white rot fungi can affect all main structural components of wood including cellulose, hemicelluloses and lignin [12, 13]. However, many of these references do not discuss white rot in sound or archaeological woods. Due to its importance in conservation and its presence in some wooden parts of historic buildings in Iran, the cellular structure and components of oriental plane tree (*Platanus orientalis*) wood were assessed after exposure to *Trametes versicolor* (as a white rot).

Materials and methods

Wood samples were prepared from an historic door frame (about 200 years old) in Isfahan, Iran. For wood identification, a block of 2 cm^3 was sampled and soaked in 1:1 mixture of ethanol: glycerine, prior to sectioning. The samples were sectioned in transverse (TS), radial longitudinal (RLS) and tangential longitudinal sections (TLS) with a scalpel blade and were studied by an ERMA BM-8bi transmission light microscope.

The effect of the presence of *T. versicolor* was investigated according to [8]. After 16 weeks exposure, the following weight loss of each sample (eight wood blocks of 50 × 25 × 10 mm) was calculated:

$$\%W_L = [(W_O - W_A) / W_O] \times 100$$

W_O is the oven-dried weight before and W_A is the oven-dried weight after exposure to the fungus. A sample with mean percentage of weight loss was used for scanning electron microscopy (SEM) and Fourier transform infrared spectroscopy (FTIR) investigation. The TS was imaged with a Philips XL30 SEM. For FTIR analysis, spectra were measured by transmittance using a Nicolet 470 spectrometer and KBr pellet technique. All the spectra were measured in the range of 400–4000 cm^{-1} at a spectral resolution of 4cm^{-1} and 32 no. of scans.

Table 1 Mass loss data of samples after 16 weeks.

Sample	W_O (g)	W_A (g)	Mass loss%
C – 1	11.84	9.03	23.73
C – 2	10.38	6.78	34.68
C – 3	10.31	7.93	23.08
C – 4	10.88	7.86	27.76
C – 5	10.61	8.12	23.47
C – 6	10.41	7.6	26.99
C – 7	10.97	8.46	22.88
C – 8	12.8	10.9	14.84
Mean	11.02	8.33	24.41

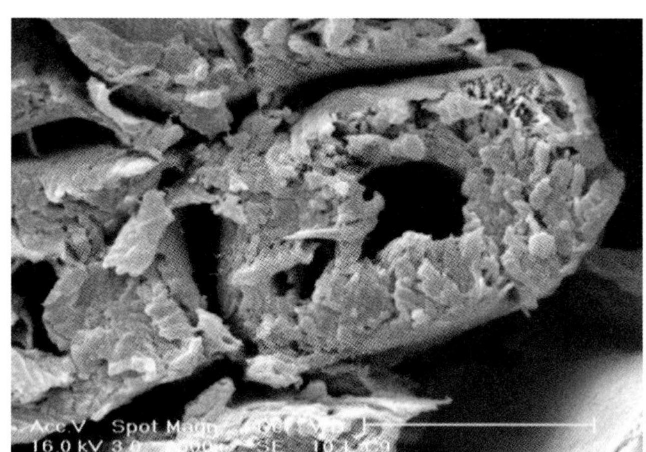

Figure 1 Degradation of the cell wall.

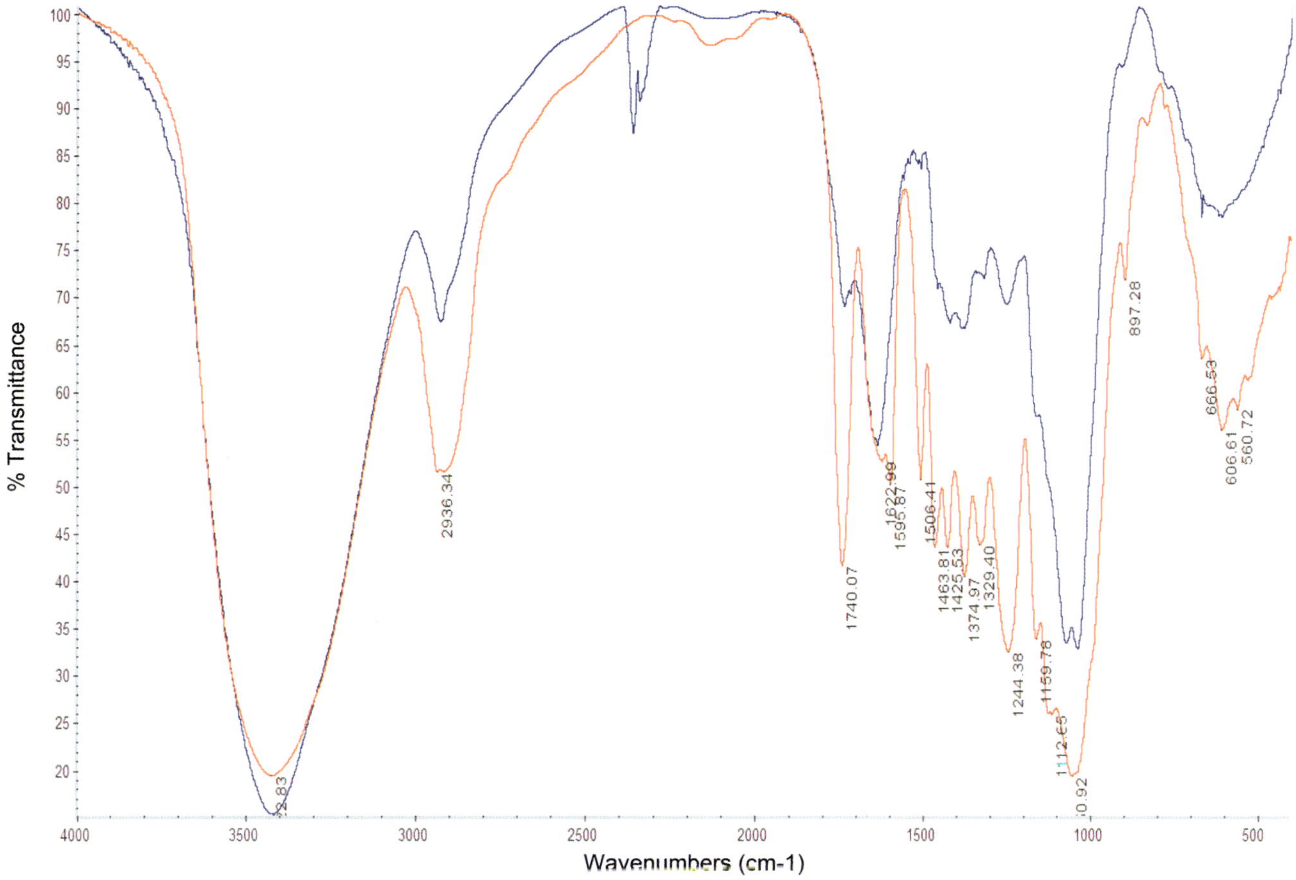

Figure 2 FTIR spectra showing undecayed (lower) and decayed (upper) samples.

Results and discussion

The TS of the wood showed a diffuse to semi-ring porous arrangement of vessels. The numerous vessels were generally in clusters in the earlywood and solitary, or in small irregular groups, in the latewood. Both the TS and TLS showed many large rays with an average width of 4 to 10 cells and often more than 2 mm high. Comparison between these sections and images of identified samples [14] indicated that the wood had anatomical characteristics of a *Plantanus* species and the location suggests that the wood is *P. orientalis*. Oriental plane tree wood has been used in many monuments in Isfahan.

Weight loss data after 16 weeks exposure to *T. versicolor* (Table 1) showed that the wood is susceptible to alteration and the fungus can diminish wood components extensively in a short time.

Before exposure, sound wood cells showed an integrity of wood structure and intact middle lamellae between cells. After exposure to the white rot fungus, several fractures were evident in cell walls. These fractures or erosion troughs indicated degradation of the secondary wall layers [14]. Mass loss observed for the samples after decay ranged from 14 to 34%. Moreover, the integrity of adjacent cells was destroyed and cells separated from each other (Figure 1). This resulted from decay of the middle lamella layer, attributed to the degradation of lignin which is mainly localised in the middle lamella [14].

FTIR spectra of different woods have been published and peaks in fingerprint region have been well defined [15–18]. Spectra of samples (Figure 2) showed strong hydrogen bonding of O–H stretching absorption and C–H stretching absorption around 3422 cm^{-1} and 2936 cm^{-1} respectively. After exposure to the wood of *T. versicolor*, the relative intensity of the band at 1740 cm^{-1} was extensively decreased. This band was assigned for C=O in xylans (polysaccharides) and its low intensity in decayed samples showed extensive degradation of hemicelluloses. An absorbed O–H and conjugated band at 1622 cm^{-1} and bands at 1595 cm^{-1} and 1506 cm^{-1} are assigned for aromatic skeletal vibration in lignin. In addition, the band at 1506 cm^{-1} is known as a lignin characteristic [19]. Decrease of their intensity in the decayed sample indicated structural degradation of lignin. These results confirmed SEM results that showed degradation of wood cells due to localised attack on lignin mainly in the middle lamella region. Decrease of C–H deformation at 1374 cm^{-1}, at 1159 cm^{-1} for C–O–C vibration, and at 1050 cm^{-1} for C–O stretch in cellulose and hemicelluloses and absence of C–H deformation at 897 cm^{-1} for cellulose indicated severe decomposition of cellulose polymer due to fungus activity. Absence of peaks at 1244 cm^{-1} for syringyl ring and C–O stretch in lignin and at 1329 cm^{-1} for CH vibration in syringyl derivatives was attributed to decomposition and successive decrease of lignin and adjacent hemicelluloses in the ultrastructure of wood.

Conclusions

Historic wood samples (c.200 years old) were used for the evaluation of white rot deterioration in wood samples identified as oriental plane (*Platanus orientalis*). After exposure to *T. versicolor*, mass loss of samples showed that the taxon is susceptible to white rot fungi. SEM studies showed cellular deterioration due to lignin degradation in the middle lamella layer. Cell wall erosion was localised in the S_2 layer. FTIR results and mass loss data confirm that structural degradation is clear in this wood due to the effect of the fungus. Degradation of cellulose and lignin (depolymerisation process) and their digestion by the fungus resulted in the decrease of mechanical strength of the wood. Due to decay of white rot, application of appropriate preservatives has a great importance in the conservation of historic structures made of oriental plane tree wood.

Acknowledgements

We would like to thank H. Purtaghi for assisting in sample preparation and M. Rezai for helpful collaboration in SEM studies.

References

1. Blanchette, R.A., 'A review of microbial deterioration found in archaeological wood from different environments', *International Biodeterioration & Biodegradation* **46** (2000) 189–204.
2. Unger, A., Schniewind, A.P. and Unger, W., *Conservation of Wood Artifacts*, Springer, New York (2001).
3. Blanchette, R.A., Otjen, L., Effland, M.J. and Eslyn, W.E., 'Changes in structural and chemical components of wood delignified by fungi', *Wood Science and Technology* **19** (1985) 35–46.
4. Enoki, A., Tanaka, H. and Fuse, G., 'Degradation of lignin-related compounds, pure cellulose, and wood components by white-rot and brown-rot fungi', *Holzforschung* **42** (1998) 85–93.
5. Blanchette, R.A., Cease, K.R., Abad, A.R., Koestler, R.J., Simpson, E. and Sams, G.K., 'An evaluation of different forms of deterioration found in archaeological wood', *International Biodeterioration* **28** (1991) 3–22.
6. Blanchette, R.A., Haight, J.E., Koestler, R.J., Hatchfield, P.B. and Arnold, D., 'Assessment of deterioration in archaeological wood from ancient Egypt', *Journal of the American Institute for Conservation* **33**(1) (1994) 55–70.
7. Blanchette, R.A., Krueger, E.W., Haight, J.E., Akhtar, M. and Akin, D.E., 'Cell wall alterations in loblolly pine wood decayed by the white-rot fungus, *Ceriporiopsis subvermispora*', *Journal of Biotechnology* **53** (1997) 203–213.
8. British Standard EN 113, *Wood Preservatives – Test Method for Determining the Protective Effectiveness against Wood Destroying Basidiomycetes – Determination of the Toxic Values* (1997).
9. Blanchette, R.A., 'Deterioration in historic and archaeological woods from terrestrial sites', in *Art, Biology, and Conservation: Biodeterioration of Works of Art*, ed. R.J. Koestler, V.H. Koestler, A.E. Charola and F.E. Nieto-Fernandez, The Metropolitan Museum of Art, New York (2003) 328–347.
10. Blanchette, R.A., Held, B.W., Jurgens, J.A. and Haight, J.E., 'Wood deterioration in Chacoan great houses of the southwestern United States', *Conservation and Management of Archaeological Sites* **6** (2004) 203–212.
11. Bj'ordal, C.G. and Nilsson, T., 'Waterlogged archaeological wood: a substrate for white rot fungi during drainage of wetlands', *International Biodeterioration & Biodegradation* **50** (2002) 17–23.
12. Jones, M. and Eaton, R., 'Conservation of ancient timbers from the sea', in *Conservation Science Heritage Materials*, ed. E. May and M. Jones, Royal Society of Chemistry, Cambridge (2006) 266–308.
13. Schmidt, O., *Wood and Tree Fungi Biology Damage Protection and Use*, Springer, Berlin (2006).
14. Schoch, W., Heller, I., Schweingruber, F.H. and Kienast, F., 'Wood anatomy of central European species' (2004) http://www.woodanatomy.ch/species.php?code=PLOR (accessed 5 January 2010).
15. Hedges, J.I., 'The chemistry of archaeological wood', in *Archaeological Wood Properties: Chemistry and Preservation*, ed. R.M. Rowell and R.J. Barbour, American Chemical Society, Washington, DC (1990) 111–140.
16. Schultz, T.P. and Glasser, W.G., 'Quantitative structural analysis of lignin by diffuse reflectance Fourier transforms spectrometry', *Holzforschung* **40** (1986) 37–44.
17. Owen, N.L. and Thomas, D.W., 'Infrared studies of "hard" and "soft" woods', *Applied Spectroscopy* **43** (1989) 451–455.
18. Faix, O., Bremer, J., Schmidit, O. and Stevanovic, T., 'Monitoring of chemical changes in white-rot degraded beech wood by pyrolysis-gas chromatography and Fourier transform infrared spectroscopy', *Journal of Analytical and Applied Pyrolysis* **21** (1991) 147–162.
19. Pandey, K.K. and Pitman, A.J., 'FTIR studies of the changes in wood chemistry following decay by brown-rot and white-rot fungi', *International Biodeterioration & Biodegradation* **52** (2003) 151–160.

Authors' addresses

- Mohsen Mohammadi Achachluei, Faculty of Conservation, Art University of Isfahan, Isfahan, Iran (mohsen.mohammadi@aui.ac.ir)
- Gholamreza Vatankhah, Faculty of Conservation, Art University of Isfahan, Isfahan, Iran
- Aliakbar Enayati, Department of Wood and Paper, Faculty of Natural Resources, University of Tehran, Karaj, Iran

The contribution of SEM-EBSD analysis to a microstructural interpretation of the elasticity of ancient fibulae springs

Luca Peruzzo, Daniela Ferro, Vania Virgili, Irene Calliari and Stefano Buson

KEYWORDS EBSD, archaeometallurgy, bronze, spring, elasticity

Introduction

Identification of the technical improvements that allowed the crucial progress in the production of straight pins and fibulae requires the study of metallurgical treatments. The technological advancement leading to this changeover was the manufacture of the spring; this confirms a direct correlation between the shape of the object and the working technique. A fibula is the result of four distinct manufacturing phases: fusion of the semi-manufactured product, hammering and polishing, the cold-work decoration of the arch and the clasp, and the folding of the spring that creates its ultimate shape [1]. These manufacturing processes influence the microstructure of the material used to create the fibula.

Materials and previous work

In order to recognise the different types of working, samples were selected from three different ancient Venetic bronze fibulae (seventh-sixth century BC) found in a necropolis near Este (Padova, north-east Italy), which have previously been analysed in cooperation with the Museo Nazionale Atestino (Este, Italy). The samples, taken from distinct parts of each of the three fibulae (spring, arch, clasp), show similar compositions (83–91 wt% copper, 6–13 wt% tin, 3–4 wt% lead, with traces of iron, nickel, zinc and cobalt) [2]. This composition is typical for cold-worked bronze.

Methodology

A crystallographic study based on the recent technique of scanning electron microscopy with electron backscattered diffraction (SEM-EBSD) was performed on metallographic cross-sections in order to detect diagnostic markers at sub-micron scale for the characterisation of the spring metal treatment. The instrument is a Camscan MX2500 with an HKL CHANNEL 5 system for EBSD analysis.

Electron backscattered diffraction is an SEM technique based on the acquisition of diffraction patterns from bulk samples, tilted at a high angle. Each pattern is generated with a spatial resolution <0.1 μm and is characteristic of the crystal structure and orientation from where it was generated. The resulting EBSD map permits a quantitative representation of the sample microstructure: the measure of the crystal orientation and the angular mismatch between grains, the examination of grain morphology, boundaries and twins, and the discrimination between different materials. The use of EBSD is not widespread in the field of cultural heritage and the lack of reference literature and comparison examples was one of the difficulties in applying the EBSD technique in this research work. Apart from a recent article on the identification of pigments and raw materials in historic glasses and ceramics [3], only a few EBSD-based studies focus on ancient metallurgical processes (see Northover in this vol.) [4–8].

Results

Automatic data collection allowed a wide variety of maps to be produced, of which only a few are presented in this paper. From all the results, we have chosen to report the inverse pole figure (IPF) orientation map (Figures 1 and 2). This kind of representation uses a basic red, green and blue (RGB) colour scheme, fitted to an inverse pole figure. For cubic phases, pure red, green, and blue are assigned to grains whose <100>, <110> or <111> axes, respectively, are parallel to the projection direction of the IPF (typically, the surface–normal direction) and intermediate orientations are coloured by an RGB mixture of the primary components. In the figures grain boundaries are defined as black contours, and the white areas indicate where the system was unable to index the pattern (i.e. holes or other phases). From these maps no preferred orientations are strongly evidenced in the samples.

Once the grains were reconstructed, a variety of grain size and shape parameters was easily measured (area, equivalent circle diameter, aspect ratio, etc.). In particular, the clasp

Table 1 Summary of the microstructural features observed by EBSD for the three samples.

Sample	Part	Grain size (average)	Shape	Twins	External – Internal
26815	clasp	22 μm	Rounded and equi-axed	Absent	Homogeneous
14942	spring	8 μm	Equi-axed with twins	Mostly in the larger grains	Increased grain size in the central area compared to external one
14989	arch	8 μm	Equi-axed with twins localized in the central area	Mostly in the larger grains	Increased grain size in the central area compared to external one

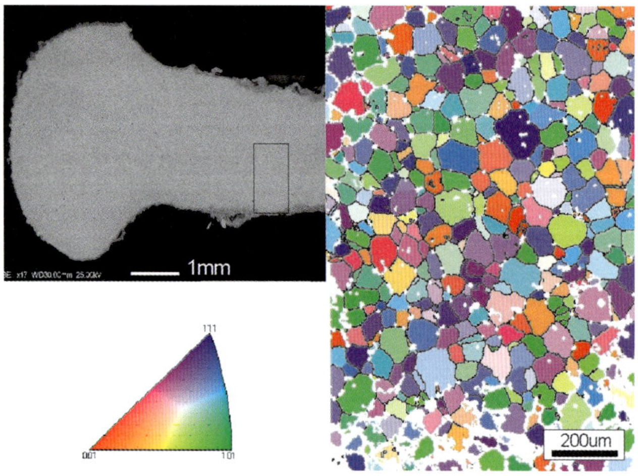

Figure 1 Clasp of fibula (sample 26815): BSE image and IPF map of the inset area.

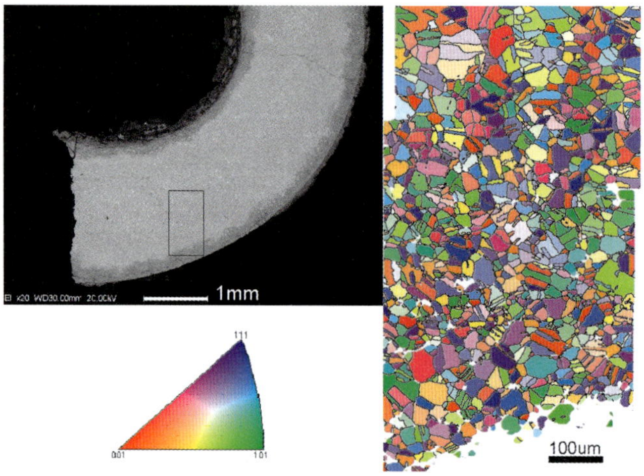

Figure 2 Spring of fibula (sample 14942): BSE image and IPF map of the inset area.

sample shows the highest grain size values, compared to the other two parts, which display a rounded, homogeneous and equi-axed shape, and an absence of twins.

Another type of map was used that distinguished fully recrystallised grains, deformed regions and substructured (recrystallised with subgrains) grains. The maps and respective histograms showed how the recrystallised fraction is nearly 31% in the clasp sample with an absence of deformed grains, while, inversely, this fraction is 8% in the spring sample with nearly 14% of deformed grains.

Many other angular measurements and calculations (e.g. misorientations, pole figures, etc.) were performed on these samples, but their interpretation is still doubtful, mostly due to the absence of known samples for comparison.

Discussion

The archaeological site where the fibulae were found is a necropolis in which cremation was practised. The fact that these artefacts have been subjected to fire/high temperatures seems to have resulted in a homogenisation of the alloy structure. Some differences in the structure of the metal however suggest that the fibulae have not undergone strong annealing processes. In fact, a closer inspection of the IPF orientation maps allows us to distinguish some microstructural differences between the spring, the arch and clasp.

The samples of the parts of the fibula whose functions require elastic properties, such as the spring and arch, indeed present specific and similar microstructures (only the IPF map related to the spring is reported, in Figure 2). They are characterised by a variation in grain size from the centre to the edge of the specimen: the central area is formed of recrystallised and twinned grains that become smaller towards the edge of the sample. This type of microstructure could be related to hot working followed by cold working. The composition and microstructure of the examined samples confirm that we are in the range of small deformations, which allow the use of spring several times without any appreciable degradation of its property.

Conclusions

In this initial approach to the study of elasticity treatments in ancient fibulae, the capabilities offered by EBSD were particularly useful in the investigation of structural features at a micron scale that would not be possible to evidence by means of other well-established metallographic methods. We therefore expect a greater use in the future of the EBSD technique in the reconstruction of ancient metallurgical processes and in the field of archaeological science.

References

1. Formigli, E. (ed.), *Fibulae: Dall'età del bronzo all'alto Medioevo tecnica e tipologia*, Quaderni archeologici, Sistema Musei Senesi, Firenze (2003).
2. Baldini, C., Buson, S., Calliari, I., Magrini, M., Ruta, A. and Variola, A., 'Studio archeometallurgico di fibule atestine', *Quaderni di Archeologia del Veneto* **20** (2004) 198–204.
3. Peruzzo, L., Fenzi, F. and Vigato, P.A., 'Electron backscatter diffraction (EBSD): a new technique for the identification of pigments and raw materials in historic glasses and ceramics', *Archaeometry* **53**(1) (2011) 178–193.
4. Peruzzo, L., Zaghis, F., Salviulo, G. and Molin, G., 'Use of electron backscatter diffraction (EBSD) for the characterization of ancient metallic artifacts: a preliminary investigation on bronze finds from Frattesina', *Epitome* **1** (2005) 32.

5. Wanhill, R.J.H., 'Embrittlement of ancient silver', *Journal of Failure Analysis and Prevention* **5**(1) (2005) 41–54.
6. Mapelli, C., Nicodemi, W. and Riva, R.F., 'Microstructural investigation on a medieval sword produced in the 12th century A.D.', *ISIJ International* **47** (2007) 1050–1057.
7. Wanhill, R.J.H., Hattenberg, T. and Northover, J.P., 'EBSD of corrosion, deformation and precipitation in the Gundestrup cauldron', in *Ligas Metálicas: Investigação e Conservação*, ed. A.C. Ferreiro da Silva and P. Menino Homem, Universidade do Porto, Porto (2008) 47–61.
8. Barnett, M. R., Sullivan, A. and Balasubramaniam, R., 'Electron backscattering diffraction analysis of an ancient wootz steel blade from central India', *Materials Characterization* **60** (2009) 252–260.

Authors' addresses

- Luca Peruzzo, National Research Council of Italy - Institute of Geosciences and Earth Resources, Padua, Italy (peruzzo@igg.cnr.it)
- Daniela Ferro, National Research Council of Italy - Institute of Nanostructured Materials, Rome, Italy (daniela.ferro@cnr.it)
- Vania Virgili, Department of Chemistry, 'Sapienza' University of Rome, Rome, Italy (vania.virgili@uniroma1.it)
- Irene Calliari, Department of Industrial Engineering, University of Padua, Padua, Italy (irene.calliari@unipd.it)
- Stefano Buson, Museo Nazionale Atestino, Este, Italy

Variable pressure scanning electron microscopy applied in the study of ancient manuscripts and inks

Flavia Pinzari and Marina Bicchieri

KEYWORDS SEM-VP-EDX, Raman, parchment, inks, manuscripts

Introduction

Variable pressure (VP) scanning electron microscopy (SEM) represents a micro-invasive methodology that can be successfully applied in the study of ancient manuscripts, inks, illuminated parchments and other writing supports. This method makes it possible to analyse degradation phenomena, study constituent materials and manufacturing processes, evaluate restoration practices, and describe the microscopic effects of chemical treatments. Energy dispersive X-ray microanalysis (EDX) of organic materials can subject samples to greater invasive impact than simple SEM observation [1]. Paper and parchment can deteriorate if their exposure to ionising radiation is prolonged. For example, the water content of paper is normally ≥5%; prolonged radiation could vaporise such moisture, thereby modifying the material. These limitations aside, VP-SEM coupled with EDX has proved (over many decades) to be an excellent technique in archaeological and cultural heritage studies. Backscattered electron detectors (BSE), in particular, represent invaluable tools for the visualisation of chemical composition, by contrasting areas with different average atomic numbers, which creates highly contrasted images when organic and inorganic materials are present together, such as in the case of metallic inks, fillers, and microorganisms and/or impurities distribution on parchment and paper surfaces.

Materials and methods

VP-SEM and EDX were applied to six fragments of ancient manuscripts discovered at Sana'a in Yemen. The analyses were carried out using a Carl Zeiss EVO 50 fitted with both BSE and secondary electron (SE) detectors in VP mode. SEM observations were made at 20 kV accelerating voltage using a tungsten filament. EDX INCA Energy 250 was used for microanalyses.

A Renishaw In-Via Reflex Raman microscope equipped with a diode laser adjusted to 785 nm (output power 300 mW; power applied to sample 5 mW) was used to support the SEM-EDX analysis; backscattered light was dispersed by a 1200 line/mm grating, and the signal detected by a Peltier cooled (−70°C) deep depletion charge-coupled device (CCD, 578 × 384 pixel) optimised for near-infrared and ultraviolet wavelengths.

The fragments investigated during this study originated from a series of discoveries (starting in 1973) made at the Sana'a Mosque in Yemen following damage sustained to its roof. A major cache of ancient texts was discovered behind the mosque's ceiling panels [2]. The manuscripts, which are known throughout the world as 'Sana'a's Manuscripts', proved to be of great importance to Arab history and Muslim culture. The works attracted considerable interest among Orientalists and were afforded World Heritage status by UNESCO in 1976.

A preliminary historical assessment of the fragments that were found more recently (2007) led to the hypothesis that they represent parts of full volumes, presumably dating to the seventh and eighth centuries AD and, for the most part, containing religious texts. The fragments were in very poor condition and displayed various types of damage such as missing fragments (gnawed by rodents), the conspicuous presence of stains, charred portions, and severely creased areas.

Results

The scientific study began with an investigation combining attenuated total reflectance and Fourier transform infrared spectroscopy (ATR-FTIR) with Raman spectroscopy and X-ray fluorescence (XRF) analyses of the six fragments (three of parchment, three of paper). These techniques allowed the identification of the glues present in the paper fragments (gelatine) and the composition of black (iron-gall and iron-tannic inks), red (cinnabar) and yellow (orpiment) graphic media, as well as a characterisation of the conservation status of the supports. Particularly interesting was a comparison made between the manufacturing techniques used for the black inks, as well as the study of a possible superimposition of the different graphic media, which suggested a succession of writing episodes that, in some cases, appeared to have been used to clarify the meaning of the text.

Due to the considerable differences between the ages of the various fragments, the basic materials (papers and parchment), and the manufacturing techniques and writings,

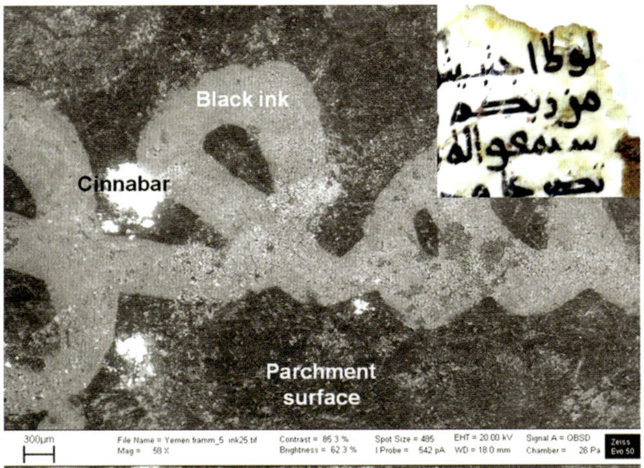

Figure 1 SEM-BSE image at low magnification. The highly different average atomic numbers between organic and inorganic materials (inks and impurities) allowed the clear observation of ion distribution on the surface of the parchment and in the writing.

an investigation of the inks and chemical properties was undertaken using VP-SEM. SEM BSE images revealed the distribution of inks and impurities on the surface of the papers and parchments (Figure 1) and the overlapping of different ink strata. EDX analysis detected the presence of orpiment and cinnabar, which were widely used in ancient Middle Eastern manuscripts, and the presence of both copper and iron black inks, with different metal concentrations and metallic particle sizes.

Tannin-based inks were frequently used in the creation of documents from antiquity to the twentieth century [3]. In addition to extensively studied iron-gall tannic inks, ink recipes based on sumac or myrobalan have been known at least since the age of the Roman Empire [4]. Even though no clear and complete recipes are known yet, it is widely agreed that in the ancient Jewish and Arabian traditions copper and iron sulphates were used in combination with various plant extracts (such as pomegranate and acacia) [4]. The study of a diverse group of manuscript fragments of Arabian origin could be of great value in defining the techniques used in the preparation of inks over the centuries. It must be said that tannins and mineral salts play different roles in parchment-supported documents; they can be found in either the substrate or writing media, or sometimes in both. The attribution of specific chemical signals to inks, impurities or manufacturing processes can be achieved by observing the topographical distribution of such elements over a sample surface at microscopical level by means of EDX-BSE-SEM imaging. For example, several samples appeared to have very curious tubular particles distributed across their surfaces in discoloured/brown areas. SEM revealed and documented that fungal bioleaching processes had induced a significant biogenesis of calcite, monohydrate and dihydrate calcium oxalate crystals through the leaching of the calcium-based materials (chiefly lime) originally used in the manufacture of the parchment.

Alongside the extensive literature that addresses the ability of some fungi (and, in particular, the lichen-forming fungi) to utilise metallic ions present in rocks and soils to form oxalates [5], the results of this research present evidence and documentation of the occurrence of this phenomenon on parchment surfaces. Mineralised fungal hyphae have been reported in limestone and calcareous soils originating from a range of localities [5]. Gadd [5] reported that fungi are able to re-precipitate secondary carbonates and biomineralise hyphae containing both calcite and whewellite. Oxalic acid excretion and the formation of calcium oxalate results in the dissolution of limestone or other calcium-rich matrices, so that the solution becomes enriched with free carbonate. Moreover, it has been experimentally proved that chitin, the chief component of fungal hyphal walls, is a substrate on which calcite readily nucleates and grows [6]. The presence of biogenic structures only on some of the fragments that were found together suggested that the biodeterioration event occurred before the burial of the materials behind the mosque's ceiling panels.

Conclusions

SEM-EDX analysis, confirmed by Raman spectroscopy, allowed the precise identification of pigments, iron-gall inks and iron-tannic inks, and the detection of different distributions of calcite and oxalates, which depended on previous fungal activity and growth. The presence of fungal traces dating back to before the burial in the mosque indicates a possible event in the past that caused the damage to the manuscripts and their final disposal.

The comparison made of the manufacturing techniques used in the black inks, as well as the study of the possible overlap of different graphic media, suggested a series of different stages in the writing process, i.e. the addition of vowels to the skeleton, using red dots in order to clarify the reading of the text. Further studies of 'Sanaa's Manuscripts' and a close collaboration with palaeographers would bring to light new knowledge on ancient Islamic writing techniques.

References

1. Goldstein, J., Newbury, D., Joy, D., Lyman, C., Echlin, P., Lifshin, E., Sawyer, L. and Michael, J., *Scanning Electron Microscopy and X-Ray Microanalysis*, 3rd edn, Kluwer Academic, New York, (2003).
2. Batori, A., Casetti Brach, C., Bicchieri, M., Di Bella, M. and Pinzari, F., *Yemen: i frammenti ritrovati nella Grande Moschea di Sana'a*, Salone dell'Arte e del Restauro e della Conservazione dei Beni Culturali ed Ambientali – XVI Edizione FerraraFiere, Edizioni MP Mirabilia srl, Italy (2009).
3. Carvalho, D.N., *Forty Centuries of Ink*, Dodo Press, Gloucester (2007) (1st edn 1904).
4. Hagerman, A.E., *Tannin Handbook*, Miami University, Oxford (1995).
5. Gadd, G.M., 'Fungal production of citric and oxalic acid: importance in metal speciation, physiology and biogeochemical processes', *Advances in Microbial Physiology* **41** (1999) 47–92.
6. Manoli, F., Koutsopoulos, E. and Dalas, E., 'Crystallization of calcite on chitin', *Journal of Crystal Growth* **182** (1997) 116–124.

Authors' addresses

- Flavia Pinzari, Istituto Centrale per il Restauro e la Conservazione del Patrimonio Archivistico e Librario, Rome, Italy (flavia.pinzari@beniculturali.it)
- Marina Bicchieri, Istituto Centrale per il Restauro e la Conservazione del Patrimonio Archivistico e Librario, Rome, Italy

Scanning electron microscopy and ceramic technology: crucibles from late prehistoric Scotland

Daniel Sahlén

KEYWORDS crucibles, Scotland, late prehistoric, ceramic technology, metallurgy

Introduction

Studies employing scanning electron microscopy (SEM) and energy dispersive X-ray analysis (EDX) in the study of crucibles and other metalworking ceramics have been important for our methodological use of the SEM in research on archaeological ceramics [1]. Early studies of metalworking ceramics mainly examined changes in ceramic microstructure following vitrification, highlighting metallurgical processes and use [2]. These issues have continued to play a major role in the application of SEM for the study of metalworking ceramics, but recent studies have shown the possibility for further progression [3]. The aim of this article is to discuss the ceramic technology of late prehistoric crucibles from Scotland, drawing on results from my PhD research (University of Glasgow) on late prehistoric moulds and crucibles from Scotland [4]. Here follows a summary of the results focusing on the investigation using SEM/SEM-EDX; a more detailed publication is forthcoming [5].

The current study discusses ceramic materials from later prehistoric Scotland; a period from the Late Bronze Age, through the Iron Age, to the Early Historic period (c.1000 BC–AD 800). This period is often termed the 'long Iron Age' [6], due to the homogeneous nature of materials and social developments in part of Scotland at this time. Following the nature of the material recovered, the focus here is on the Late Bronze Age (LBA), the Middle Iron Age (MIA) and the Late Iron Age/Early Historic period (LIA/EH). My PhD investigated four sets of materials: moulds, crucibles, pottery and clays; the current study focuses on the analysis of crucibles. Crucibles are vessels used for high-temperature processing of various materials, such as metals, glass and pigments [7]. Scottish late prehistoric crucibles were employed mainly for casting non-ferrous metals, primarily copper alloys [8]. Casting of precious metals is first known from the later part of the MIA and the LIA/EH periods.

The SEM and EDX analyses and imaging were performed on selected samples to support petrographic analysis. Backscattered electron (BSE) images of polished blocks and polished thin-sections were analysed to study the microstructure of the material, assessing firing/heating temperatures and vitrification processes, and the preparation of the ceramic material. This work was primarily carried out at the National Museums of Scotland (NMS) Analytical Research section on uncoated polished thin-sections and blocks in en-vac using a CAMSCAN MX2500. Additional samples were analysed as fractured surfaces and carbon-coated polished blocks at the University of Glasgow, School of Geographical and Earth Sciences using a field emission environmental Quanta 200F SEM. Microanalysis with SEM-EDX was used to study the compositional relationship between different groups of ceramics, and among ceramics from particular sites. Analyses were performed at NMS with a Norum Vantage spectrometry package, at 20 kV and 35 mm working distance (which is the standard at NMS), ×500 magnification and 4 × 4 μm spot size [4]. Analyses were taken at three random spots across the section.

Results and discussion

Resources, preparation and use

Studies of pottery and other ceramics from prehistoric Britain have shown that, to a large extent, local clay sources and tempering materials were used [9], a conclusion supported by the current research [5, 10]. The study of provenance was not a central goal, but a key question was to assess if different ceramic materials were made from different clays and to what degree function affected the choice of clay used. There is an indication that craftworkers during the MIA/LIA period in some cases selected different clays for the production of different ceramic materials [5]. Following a study by Martinón-Torres and Rehren [11], the refractory properties of the crucibles were assessed against the properties of other

Table 1 Comparison of largest grain size (in mm) in Scottish late prehistoric crucibles.

	LBA (n=10)	MIA (n=22)	LIA/EH (n=7)
$\bar{x}$	1.6	0.9	1
Min	0.5	0.2	0.2
Max	3.3	2	2.2

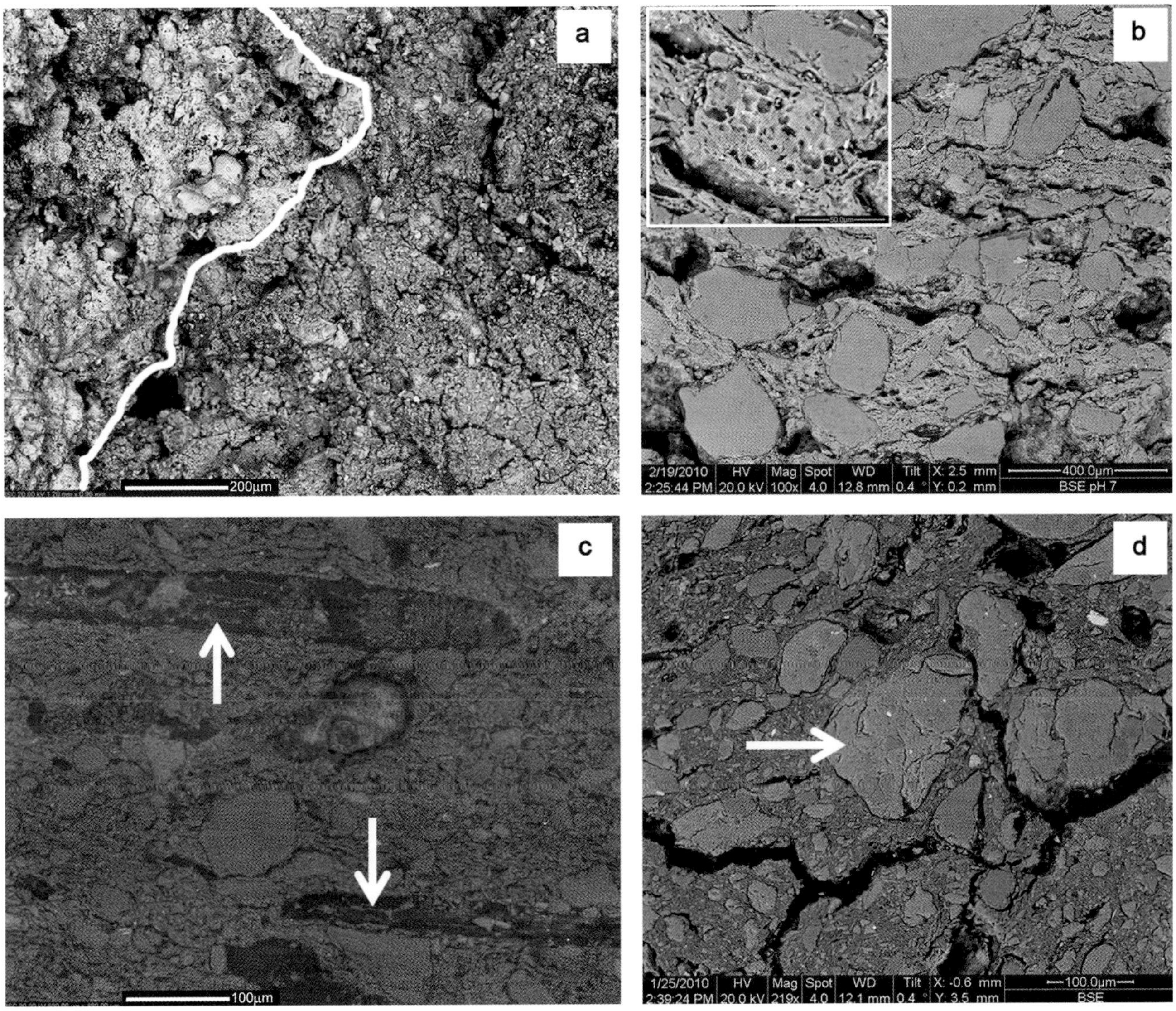

Figure 1 Selected BSE images of late prehistoric crucibles: (a) difference in vitrification in the LBA crucible – the area to the left of the line is more heavily vitrified and partly bloated (scale bar: 200 µm); (b) MIA crucible, extensive vitrification throughout the sherd (scale bar: 400 µm) – the inserted image shows an area with higher magnification (scale bar: 50 µm); (c) linear voids (marked with arrows) from fibrous temper in MIA crucible from Culduthel (scale bar: 100 µm); (d) grog inclusions (marked with an arrow) in MIA crucible from Traprain Law (scale bar: 100 µm).

ceramics by comparing the content of alumina with alkali and alkali earth oxides (defined here as the refractory index: RI). This assessment showed that in most cases the refractoriness was not indicative of the selection of particular clays, but individual crucibles from some sites showed higher refractory properties following this model. A clear relationship between heavily vitrified samples and higher RI value was seen, and it is suggested that this increase in the RI is due to alteration of the clay's composition following vitrification rather than the use of alumina-rich clay [5].

SEM analysis in combination with polarising light microscopy (PLM) was successful in assessing the preparation of the ceramic material. LBA crucibles were commonly tempered with mixed larger local rock fragments (Table 1), similar to materials used for pottery, whilst MIA crucibles were to a higher degree tempered with quartz and quartz-rich inclusions [4]; a change that improved the thermal and mechanical properties of the material. Production in the LIA/EH was more varied. Crucibles from some MIA sites show the use of specific types of temper (Figure 1c and Figure 1d), such as the fibrous material seen in crucibles from Culduthel (Inverness) and grog temper seen in one crucible fabric from Traprain Law (East Lothian). This indicates local practices of material preparation and technological solutions, and the presence of different craftworkers/traditions.

Analyses of BSE images of the three sets of material highlighted differences in use and heating. LBA crucibles show generally uneven vitrification with bloating pores (Figure 1a), while MIA/LIA crucibles are often more thoroughly vitrified (Figure 1b). This suggests different heating processes:

- LBA crucibles were heated in an open fire with a top temperature for a short period.
- MIA and LIA/EH crucibles were predominantly heated in a more controlled environment where the top temperature was maintained for a longer period. The crucibles were possibly reused for several castings.

As far as could be assessed, the crucibles from the LBA had been heated from above, with vitrification and slagging on the rim and inside of the vessel. The MIA and LIA/EH crucibles have mainly been heated from below. When in the Iron Age this technological change took place in Scotland has not been possible to establish.

Conclusions

This study has examined the ceramic technology of crucibles in late prehistoric Scotland, a perspective that has been unusual in Scottish studies, where classification and chronology have been more in focus. It has been possible, in line with studies from other areas, to confirm that the SEM in combination with petrography is a powerful tool for the study of ceramic technology, investigating material preparation, heating processes, and the use of specific clays and materials.

The study of crucibles has indicated a general development from the LBA to the MIA and LIA/EH period towards a more specialised technology, with some selection of particular resources and specific preparation methods. This development can be explained as technological advances of the crucibles that improved their mechanical and thermal properties, leading to more successful castings and allowing the vessel to be reused to a greater extent. The latter is also clear from the vitrification of MIA and LIA/EH crucibles showing evidence of a more prolonged use. Archaeological evidence of the production supports this conclusion, indicating a more specialised and extensive production in the MIA and LIA/EH periods [4].

References

1. Tite, M.S., 'The impact of electron microscopy on ceramic studies', in *New Developments in Archaeological Science*, ed. A.M Pollard, Oxford University Press, Oxford (1992) 111–131.
2. Tite, M.S., Freestone, I.C., Meeks, N.D. and Craddock, P.T., 'The examination of refractory ceramics from metal-production and metalworking sites', in *The Archaeologist and the Laboratory*, ed. P. Philips, CBA Research Report 58, London (1985) 50–55.
3. Hein, A. and Kilikoglou, V., 'Modelling of thermal behavior of ancient metallurgical ceramics', *Journal of the American Ceramic Society* **90** (2007) 878–884.
4. Sahlén, E.D., *Ceramic Technology and Technological Traditions: Manufacture of Metalworking Ceramics in Late Prehistoric Scotland*, unpublished PhD dissertation, University of Glasgow (2011).
5. Sahlén, D. 'Selected with care? – The technology of crucibles in late prehistoric Scotland. A petrographic and chemical assessment', *Journal of Archaeological Science* (submitted).
6. Harding, D., *The Iron Age in Northern Britain: Celts and Romans, Natives and Invaders*, Routledge, London (2004).
7. Bayley, J. and Rehren, T., 'Towards a functional and typological classification of crucibles', in *Metals and Mines: Studies in Archaeometallurgy*, ed. S. La Niece, D. Hook and P.T. Craddock, Archetype Publications, London (2007) 46–55.
8. Heald, A., *Non-ferrous Metalworking in Iron Age Scotland, 700 BC–AD800*, unpublished PhD dissertation, University of Edinburgh (2005).
9. Morris, E.L., 'Iron Age production and exchange', in *The Iron Age in Britain and Ireland: Recent Trends*, ed. T.C. Champion and J. Collis, J.R. Collis Publication, Sheffield (1996) 41–65.
10. Sahlén, D., 'Ceramic traditions – the evidence from clay sampling at two late prehistoric sites: Birnie (Moray) and Traprain Law (East Lothian), Scotland', *The Old Potter's Almanack* **17**(1) (2012) 2–9.
11. Martinón-Torres, M. and Rehren, Th., 'Post-medieval crucible production and distribution: a study of materials and materialities', *Archaeometry* **51** (2009) 49–74.

Author's address

Daniel Sahlén, Department of Archaeology, University of Glasgow, Glasgow, UK (sahlen.d@gmail.com)

Scanning electron microscopy study of Chinese jade working technology: comparing excavated Bronze Age artefacts with jades in the British Museum

Margaret Sax, Li Boqian, Nigel Meeks and Qin Ling

KEYWORDS scanning electron microscopy, nephrite jade, China, Bronze Age, technology

Introduction

The imaging power of scanning electron microscopy (SEM) has been demonstrated for investigating the technology of ancient hard stone carvings in the collections of the British Museum (BM), London: principally, the engraving of Mesopotamian quartz cylinder seals, c.3000–400 BC, and the working of nephrite jade in China since c.3500 BC until the present [1, 2]. The use of a range of non-rotary and rotary tools and techniques can be recognised from the characteristic morphology of tool marks preserved in artefacts.

The methodology is based on the preparation and SEM examination of detailed silicone mouldings of the carved features of artefacts for comparison with moulded experimental standards. Small mouldings are made of features of interest using Coltène President Jet Plus light body, a dental polyvinylsiloxane resin, and mounted on aluminium stubs, 30 mm in diameter, for SEM observation. This method of moulding was adopted to avoid any risk of damage to objects posed by direct observation in the high vacuum chamber of an SEM, and proved to be essential for examining deeply carved parts of artefacts where important evidence for technique is often preserved.

In a collaborative study between the Department of Conservation and Scientific Research at the British Museum and the School of Archaeology and Museology at Peking University, Beijing, a similar approach was adopted for examining 30 nephrite jades from the burial assemblages of the Jin Marquisate excavated at Tianma-Qucun, Shanxi province. The jades were dated according to the sequence of the tombs and stylistic considerations, mainly to the Western Zhou dynasty, 1050–771 BC and the preceding Shang dynasty, c.1600–1050 BC [3]. The results of this study are summarised in this article and compared with the evidence of 21 British Museum jades thought on stylistic grounds to be of similar Bronze Age date [4, 5, 6].

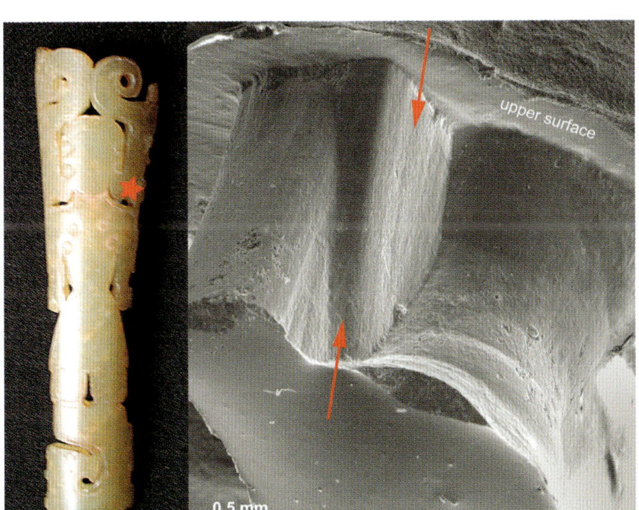

Figure 1 Curved ornament M8:203, early eighth century BC, Western/Eastern Zhou dynasty (92 mm high). On the right, the SE SEM image (20 Kv, 39 mm) shows a mould of the decorative hole, marked with an asterisk on the jade: the hole was worked using small rigid straight files and drills.

Use of high vacuum (HV) and variable pressure (VP) SEM

The moulded details of most jades from both the Jin Marquis cemetery and the BM were examined with the high vacuum SEM (JEOL 840) at the BM. The moulds were coated with a thin conducting layer of gold to facilitate examination, generally using a secondary electron (SE) detector (SE, Figures 1 and 2). Difficulties were however encountered with the cracking of gold coatings and consequent charging effects, particularly on polished silicone surfaces.

The subsequent acquisition of the VP-SEM (Hitachi S-3700N) by the BM offered several advantages. First, the ability to examine uncoated moulds; optimum viewing conditions are usually achieved using the environmental secondary electron detector (ESED). Secondly, the mechanism controlling the manoeuvrability of the stage allows optimum conditions of tilt, rotation and working distance to be achieved for moulded samples with respect to detectors,

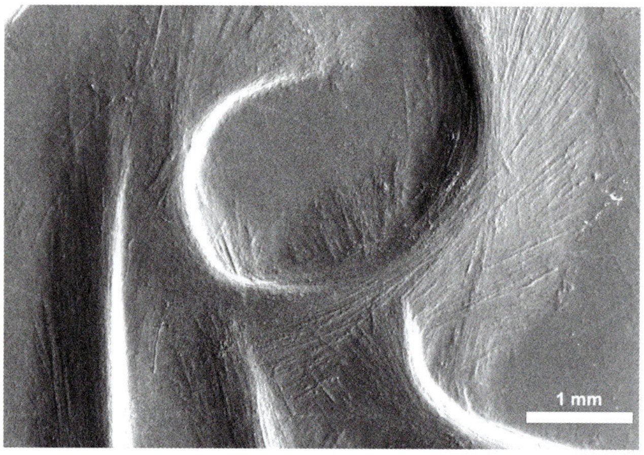

Figure 2 SE SEM image of a mould of tool marks in ornament M63:161, tenth century BC, Western Zhou dynasty, consistent with the use of pointed tools for incising.

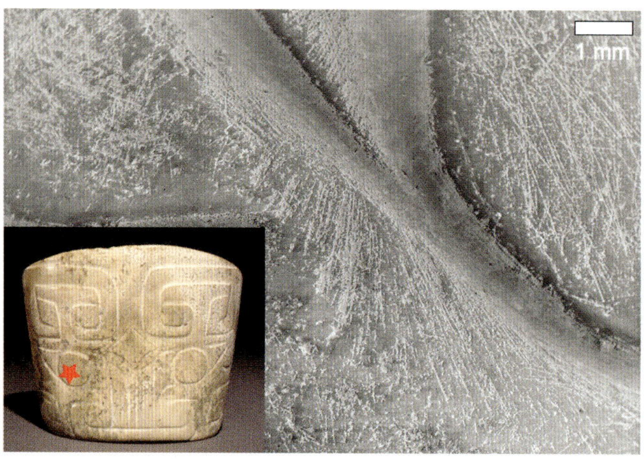

Figure 3 Inset: hoof-shaped ornament BM 1945,1017.1, originally worked c.3250 BC, Hongshan culture (68 mm high). Hitachi VP-SEM image (ESED, 20 kV, 39 mm, 30 Pa) of tool marks (annotated on the jade with an asterisk): the design was worked in the Bronze Age using pointed tools.

providing maximum three-dimensional contrast of features. A third advantage of VP-SEM is the ability to examine artefacts up to 200 mm diameter directly at a chamber pressure of 30 Pa (Figure 3).

Technology of Jin Marquis jades

SEM examination of the moulded tool marks from the Jin Marquis jades provided evidence for the extensive use of non-rotary rigid saws and thicker files for various stages of shaping and smoothing artefacts. These tools have long, straight working edges or broader surfaces and are applied in a backwards and forwards motion, producing features with essentially linear characteristics. Clear evidence for the use of such tools is apparent in a narrow decorative hole through the curved ornament (Figure 1). On the right of Figure 1, the SEM micrograph shows the broader end of a mould of the hole; the two perpendicular flat surfaces arrowed are the sides of a cut made with a small straight rigid file, c.0.5 mm thick.

Handheld pointed tools were commonly used for incising features. Figure 2 shows a mould of the design worked in another ornament; numerous linear grooves protrude upwards on the mould and diverge beyond the ends and sides of curved features. Single grooves occur in multiple orientations, reflecting the use of handheld pointed tools for scoring the jade in series of linear strokes. Previous carving tests showed that narrow grooves can be cut in nephrite without loose abrasive using single strokes of pointed tools composed of a suitably tough and hard material, such as flint or chert, corundum and also diamond [2].

Solid drills were employed for working narrow perforations; they had straight, c.1.5–2.0 mm wide, or tapered sides as shown by the biconical profile on the far right side of Figure 1. Evidence was also found for the use of a tubular drill, 50 mm in diameter, and lathes on which jades were mounted and rotated for working artefacts with circular sides.

Technology of BM jades

The results of examining the Jin Marquis jades complemented the evidence of BM jades, confirming the Bronze Age origin of most carvings and the decoration of a Neolithic ornament in the form of a hoof (inset Figure 3) during the Bronze Age. The Neolithic jade is extensively weathered and has friable surfaces, precluding moulding. The upper right corner of the VP-SEM micrograph (Figure 3) shows details of a raised circular feature; encompassing it are narrow recessed features. On the left, the sloping surfaces are marked by single diverging grooves, characteristic of the handheld pointed tools commonly employed for incising jades during the Shang and Western Zhou dynasties (Figure 2). The decorative features of jades of later date show that it was not until the Tang dynasty, AD 618–906, that consistent use began to be made of lathe-mounted disc-shaped tools (or wheels) for incising [7].

The BM carvings were acquired from collections formed in the nineteenth century and early twentieth century and provide supplementary evidence for small tubular drills, 5–6 mm in diameter, during the Bronze Age.

Conclusions

Using SEM for examining the carved features of nephrite jades recovered from the Jin Marquis cemetery provided detailed information about the non-rotary and rotary tools and techniques applied for several stages of carving during the Bronze Age. The results complemented the evidence of nephrite jades in the collections of the BM. VP-SEM has greatly facilitated the study of ancient lapidary technology, allowing both uncoated mouldings and artefacts to be examined.

Acknowledgements

We thank Antony Simpson for processing the images, Carol Michaelson and Sascha Priewe at the British Museum and Ji Kunzhang at Shanxi Provincial Institute of Archaeology, Houma, for contributing to this study. Margaret Sax is most grateful to Michael S. L. Liu, c/o Latham & Watkins, 35F One Exchange Square, Central, Hong Kong, for funding her travel to Beijing.

Material and supplier

Coltène President Jet Plus: available from Wright Cottrell (www.wright-cottrell.com/)

References

1. Sax, M., Meeks, N.D. and Collon, D., 'The introduction of the lapidary wheel in Mesopotamia', *Antiquity* **74**(284) (2000) 380–387.
2. Sax, M., Meeks, N.D., Michaelson, C. and Middleton A.P., 'The identification of carving techniques on Chinese jade', *Journal of Archaeological Science* **31**(10) (2004) 1413–1428.
3. Department of Archaeology at Peking University and Shanxi Institute of Archaeology, excavation reports of the Jin State cemetery at the site of Tianma-Qucun, *WENWU* **3** (1993), **1** and **8** (1994), **7** (1995), **8** (2001).
4. Sax, M., 'The technology of jades excavated at the Jin Marquis cemetery, Tianma-Qucun, Shanxi province: recognition of tools and techniques', in *Ancient Civilization*, Peking University (forthcoming 2012).
5. Sax, M. and Ji Kunzhang, 'The technology of jades excavated at the Jin Marquis cemetery, Tianma-Qucun, Shanxi province; recognition of tools and techniques', *Journal of Archaeological Science* (forthcoming).
6. Rawson, J., *Chinese Jade from the Neolithic to the Qing*. British Museum Press, London (1995).
7. Sax, M., Meeks, N.D., Ambers, J. and Michaelson, C., 'The introduction of rotary incising wheels for working jade in China', in *Scientific Research on the Sculptural Arts of Asia, Proceedings of the Third Forbes Symposium at the Freer Gallery of Art*, eds. J.G. Douglas, P. Jett and J. Winter, Archetype Publications, London (2007) 20–25.

Authors' addresses

- Margaret Sax, Department of Conservation and Scientific Research, The British Museum, London, UK (msax@thebritishmuseum.ac.uk)
- Li Boqian, School of Archaeology and Museology, Peking University, Beijing, China
- Nigel Meeks, Department of Conservation and Scientific Research, The British Museum, London, U.K.
- Qin Ling, School of Archaeology and Museology, Peking University, Beijing, China

A petrographic study of the anthropomorphic stelae from the megalithic area of Saint-Martin-de Corléans (Aosta, northern Italy)

Margherita Serra, Lorenzo Appolonia, Alessandro Borghi, Stefano De Leo and Valentina Rubinetto

KEYWORDS petrography, anthropomorphic stelae, prehistory, Aosta Valley, archaeometry

Introduction

The prehistoric site of Saint-Martin-de-Corléans, discovered in 1969 [1], is located on the alluvial sediments of the River Dora Baltea [2] in the western outskirts of the city of Aosta (north-west Italy).

The site displays evidence of articulated history that highlights the cultural importance of this monumental area in the context of European prehistory, relating to the last 5000 years, from the transitional period between the Neolithic and the Bronze Age (Eneolithic), to the present day. The site, which was initially devoted to the cult of the living rather than to the dead, has been divided into five main structural stages, dating back from 3000 BC to about 2000 BC.

Anthropomorphic stelae, about two metres in height and about one metre wide, with a depth of just a few tens of centimetres, have been attributed to a period between the late Neolithic and the Eneolithic. Throughout the history of the site, anthropomorphic stelae were systematically reused for the construction of different types of megalithic tombs, determining significant variations in the original alignments [3]. Many analogies, in terms of style and iconography [4], link some of the Eneolithic stelae to those found in Petit Chasseur's archaeological site near Sion (Switzerland).

Methodology

This work consisted of the petrographic characterisation of the lithology used for the construction of the stelae. The examination included 47 thin-sections of the stelae and 41 thin-sections of stone reference samples collected from nine different outcrops within the Aosta Valley (Figure 1). The research aimed to define the original quarry sites exploited within the megalithic area. Samples were analysed by both optical microscopy (OM) and scanning electron microscopy (SEM) (Cambridge Stereoscan S360). SEM analyses were conducted using backscattered electron imaging mode. An energy dispersive X-ray spectrometer (EDX) equipped with a pentafet detector (Oxford Instruments) was used to determine the chemical composition of the most discriminative minerals. Natural oxides and silicates (Astimex Scientific Limited) were analysed as standards. The acceleration voltage was set at 15 kV and counting time was 60 seconds. Analyses were quantified using the MINSORT program [5]. Analytical spectra were processed by INCA 300 software (Oxford Instruments).

Results

Geological setting

The archaeological site of Saint-Martin-de-Corléans is located in the geological unit known as the Piedmont Zone, extensively outcropping along the Aosta Valley (Figure 1). This unit is further divided into an Internal Unit, dominated by rocks of oceanic crust metamorphosed during the Alpine orogeny under eclogitic facies conditions, and an External Unit that consists mainly of metasediments now transformed into calcschists and marbles under blueschist facies conditions. In proximity to the investigated area, there are also other relevant geological units, corresponding to the Great Saint Bernard Nappe, the Gran Paradiso Massif and the Dent Blanche Nappe. They consist of several varieties of continental crustal rocks distinguished on the basis of their metamorphic evolution and paleogeographic provenance.

Petrographic characterisation

Petrographically, the analysed stelae reflect the complex geological setting of the Aosta Valley, showing strong compositional heterogeneities. About 81% correspond to different varieties of marble that are difficult to distinguish on a macroscopic scale. The most abundant materials belong to a large

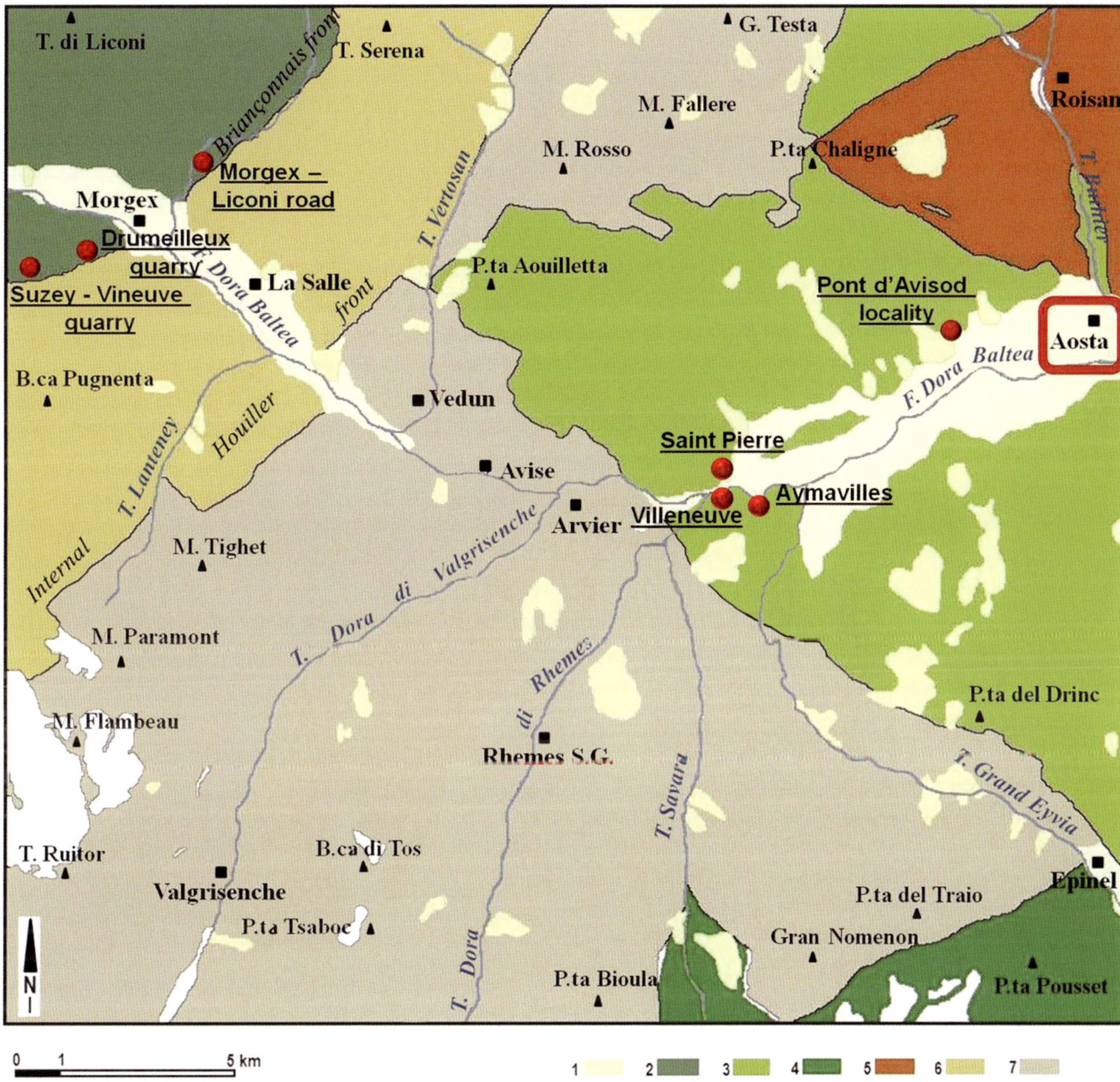

Figure 1 Schematic geological map of the central sector of Aosta Valley, modified by [7]. 1= Quaternary deposits; 2 = Sion-Courmayeur geological unit; 3 = External Piedmont Zone, 4 = Internal Piedmont Zone; 5 = Dent Blanche Nappe 6 = Houiller unit 7 = Great Saint Bernard Nappe.

class of foliated marbles and calcschists. They consist mainly of calcite (40–80 vol.%), with minor quartz, white mica, chlorite, epidote, albite, rutile and graphite. Pseudomorphs after lawsonite (a high pressure, low temperature metamorphic mineral) characterise these lithologies. SEM-EDX analysis showed that the composition of white micas correspond to Na (paragonite) and K (muscovite) end members. Muscovite, recalculated on the basis of 11 oxygen atoms per formula unit is characterised by Si between 3.00 and 3.50 atoms per formula unit displaying a strong compositional zoning, which implies a growth under high pressure conditions. These stelae show significant mineralogical and compositional similarities with quarry samples collected in the External Piedmont Zone, which outcrops in close proximity to the Megalithic area (Figure 1). The analyses of representative chlorite confirm the good correlation between the quarry samples and the stelae, matching ripidolite composition. According to Hey's classification diagram [6], chlorites show Fe_{tot} / (Mg + Fe_{tot}) ratio between 0.35 and 0.50, with a Si content between 5.25 and 5.65 atoms per unit formula (recalculated on 28 oxygen atoms per formula unit).

Another abundant rock used for the stelae is the grey marble known as Bardiglio, which comes from quarries located near the town of Villeneuve (Figure 1). This marble is easily recognised by the alternating fine-grained dolomite and calcite layers, as well as the lawsonite pseudomorphs and, according to EDX results, the coexistence of three di-octahedral micas (muscovite, margarite and paragonite). The characteristic paragenesis of this marble is completed by phlogopite, a magnesium-rich end member of the biotite solid solution series. Also chlorite is magnesium-enriched, plotting in the field of clinochlore.

Nine of the stelae were carved using an isotropic marble interlayered with foliated micaceous zones. This marble is microscopically characterised by a medium to fine grain size with a marked preferred orientation of the micas and albite poikiloblasts. The modal minerals are calcite (60–80 vol.%), quartz, phengitic mica, Mg-chlorite, epidote-zoisite, albite, graphite. Garnet porphyroblasts, with Almandine 39%, Spessartine 31%, Grossular 28% and Pyrope 2% composition, and lawsonite pseudomorphoses are also present. Analogue features were observed in the marbles from the Sion-Courmayeur geological unit (Figure 1). Reference samples were collected in the Suzey and Liconi quarries, near the village of Morgex. It is particularly noticeable that there is a good correlation between the composition of the di-octahedral micas (muscovite and paragonite) of the stelae samples and that of the reference quarry samples. The distance required to transport the carved material from the outcrops to the Megalithic area would have been more than 30 km.

Less frequent orthogneisses, micaschists and prasinites have been interpreted as coming from erratic boulders commonly distributed around the archaeological site. In particular, the mineralogical assemblage of gneisses refers to the ortho-derivates of Dent Blanche or Great Saint Bernard Nappe. Micaschists were identified as garnet-chloritoid-bearing para-derivates from the Gran Paradiso massif, whilst prasinites, corresponding to tabular beds of greenschist facies metabasalts of oceanic origin, can be clearly related to the External Piedmont Zone.

Conclusions

Most of the late Neolithic-Eneolithic stelae uncovered in the archaeological site of Saint-Martin-de-Corlèans (Aosta, Italy) were carved from local stones (foliated marble and calcschists), outcropping in the close surroundings; also prasinites, interlayered with calcschists of the External Piedmont Zone, which outcrop near the Megalithic area. Bardiglio marble outcrops 12 km south-west of the archaeological site. Sporadic orthogneisses and micaschists from further geological units were interpreted, according to field observations, as erratic boulders transported by Quaternary glaciers. Marbles with centimetre scale micaceous zones seem most likely to have come from the Sion-Courmayeur geological unit, but currently there is no evidence as to whether they were carved elsewhere and transported or transported as blocks to be carved. As the Sion-Courmayeur Zone extends in continuity from the Aosta Valley to Swiss Valais, this evidence can be interpreted as a corroboration of the strong archaeological affinities with the area of Petit Chausseur, supporting the hypothesis of cultural exchanges over the Great Saint Bernard Pass before the Early Copper Age.

References

1. Mezzena, F., 'Le stele antropomorfe dell'arco alpino (Considerazioni ed appunti)', in *L'arte Preistorica dell'Italia settentrionale*, ed. A. Aspes, Catalogo esposizione, Verona (1978) 49–59.
2. De Leo, S., 'Caratterizzazione petrografica dei materiali costituenti le stele antropomorfe dell'area megalitica di Saint-Martin-de-Corléans in Aosta (Néolithique)', in *Colloque International sur les alpes dans l'antiquité: La pierre in milieu alpine*, ed. D. Daudry, Champsec/Val de Bagnes/Valais-Suisse (2006) 33–40.
3. Mezzena, F., 'Le stele antropomorfe dell'area megalitica di Aosta', in *Dei di Pietra: La grande statuaria antropomorfa del III millennio a.C.*, ed. A. Renda, F. Mezzena and A.M. Belley, Skira, Milano (1998) 90–121.
4. Bornaz, L., Porporato, C., Rinaudo, F., De Gattis, G. and Focareta, R., '3D high accuracy survey and modelling of one of Aosta's anthropomorphic stelae', *Proceedings of the 21st CIPA Symposium AntiCIPAting the Future of the Cultural Past: International Archives of the Photogrammetry, Remote Sensing and Spatial Information Sciences*, XXXVI-5/C53, Athens (2007) 76–181.
5. Petrakakis, K. and Dietrich, H., 'MINSORT: a program for the processing and archivation of microprobe analyses of silicate and oxide minerals'. *N. Jb. Miner. Mh.* **8** (1985) 379–384.
6. Hey M.H., 'A new review of chlorites', *Mineralogical Magazine* **30** (1954) 277–292.
7. Malusà, M.G., *Post-metamorphic Evolution of the Western Alps: Kinematic Constraints from a Multidisciplinary Approach (Mesostructural Analysis, Fission-track Dating, Fluid Inclusion Analysis)*, unpublished PhD thesis, University of Torino (2004)

Authors' addresses

- Margherita Serra, Dipartimento di Scienze Mineralogiche e Petrologiche – Università degli Studi di Torino, Turin, Italy
- Lorenzo Appolonia, Soprintendenza per i Beni e le Attività Culturali della Regione autonoma della Valle d'Aosta, Aosta, Italy (l.appolonia@regione.vda.it)
- Alessandro Borghi, Dipartimento di Scienze Mineralogiche e Petrologiche – Università degli Studi di Torino, Turin, Italy (alessandro.borghi@unito.it)
- Stefano De Leo, Soprintendenza per i Beni e le Attività Culturali della Regione autonoma della Valle d'Aosta, Aosta, Italy
- Valentina Rubinetto, Dipartimento di Scienze Mineralogiche e Petrologiche – Università degli Studi di Torino, Turin, Italy (valentina.rubinetto@libero.it)

Scanning electron microscopy investigation of the Nuzi frits

Andrew Shortland, Katherine Eremin, Marc Walton, Susanna Kirk, Patrick Degryse and Joseph Greene

KEYWORDS Late Bronze Age, Near East, frits, Egyptian blue, SEM

Introduction

The Hurrian city of Nuzi lies 13 km from the city of Kirkuk in modern Iraq. It was intermittently occupied from around 5000 BC right through to AD 400 [1]. However, it is perhaps best known for its Late Bronze Age remains, which terminate in a destruction layer known as Stratum II [1]. The site was excavated from 1925 to 1931 by a series of American groups, culminating in the leadership of Richard Starr from the Fogg Museum, Harvard. The excavated remains were split between Harvard University, the University of Pennsylvania and the Baghdad Museum. The majority of the finds in the United States are held at the Semitic Museum, Harvard University. The collection studied here was that now stored at Harvard [1, 2] and includes ceramics, metals, vitreous materials and organics. A major re-examination of the material culture of Nuzi is ongoing by an international team based in Europe and the USA [3]. This aims to characterise all the material types from the site, and to use the latest scientific techniques to shed further light on the manufacture, provenance, use and preservation of the material.

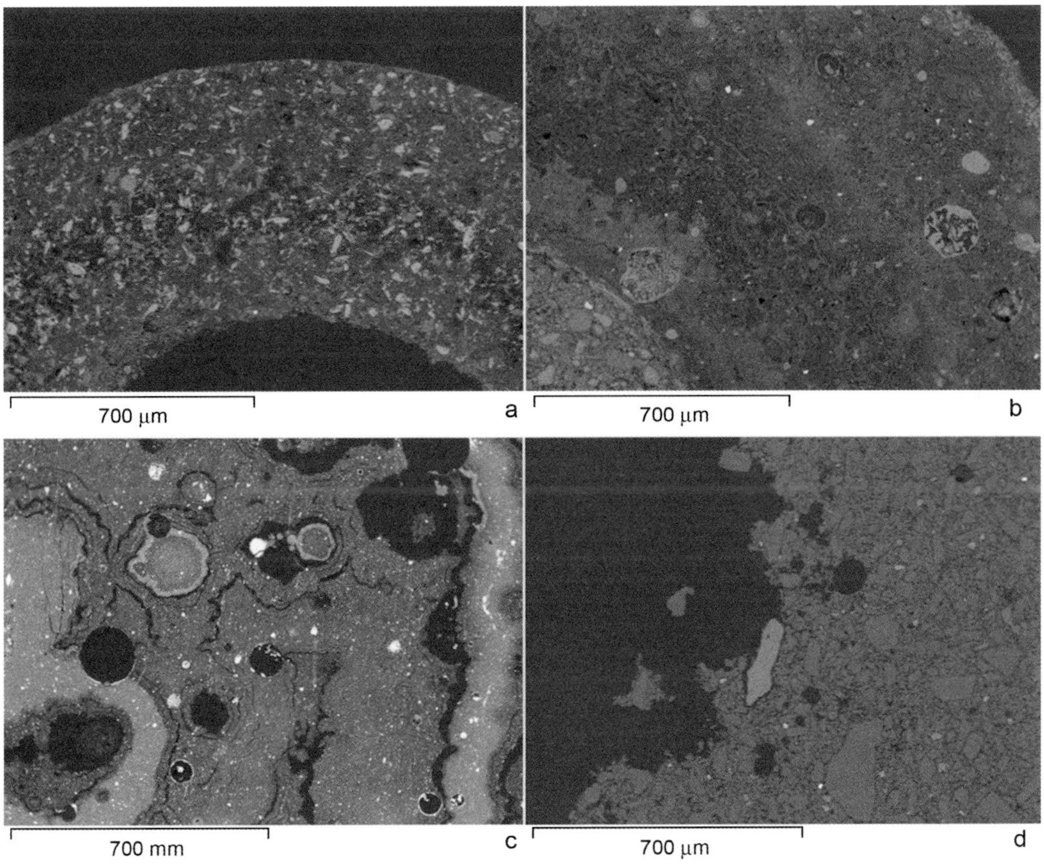

Figure 1 SEM images of frit beads from Nuzi: (a) blue; (b) red; (c) and (d) white or cream.

Stratum II and its vitreous materials

Stratum II marks a destruction layer in which the city of Nuzi was almost completely destroyed, dated to around 1350 BC [4]. It has some of the best preserved and most numerous finds of vitreous materials from the entire Near East. Many thousands of objects, mostly beads, but including tiles, vessels, amulets, figurines, etc. were uncovered in a range of materials including glass, faience, glazed ceramic and frit [1, 5]. This paper describes in detail the group of materials loosely characterised as frits. They are important in that they are quite numerous across the site, and have the possibility of shedding light on the manufacturing technique used in the other vitreous materials and their links to metalworking and ceramic manufacture.

The frits

Whilst excavating Nuzi, thousands of beads were discovered, many of which were described initially as 'composition' and more often as 'frit' in later publications [1, 5, 6]. This implies a sintered quartz material, perhaps associated with the production of glass. Excavators identified several colours including blues, yellow, white, red and creams. The exact nature of these frits and how they were produced remained uncertain, but their widespread occurrence at Nuzi makes them an important part of the bead assemblage. A few earlier studies have suggested various possible origins for the frits and other materials [5–7], but this is the first major re-examination for several decades of the frits in their entirety.

Methodology

Sampling of the frits was permitted by the Semitic Museum, Harvard University, and over 50 beads were sectioned and mounted in resin blocks for polishing. These were then examined in a JEOL scanning electron microscope (SEM), both at the Natural History Museum, London and at the Museum of Fine Arts in Boston, at 20 kV and high vacuum (HV). An additional 12 beads, which could not be sampled, were examined in the JEOL SEM non-destructively using the controlled pressure mode.

Results

Beads described in the excavation records as 'frits' or 'composition' were identified in a range of colours and examined individually to determine their internal structure.

Blue frits

Fine-grained blue frits in colours ranging from pale turquoise to mid-blue are the most common of all the frits at Nuzi. They are used in the production of beads, but also in rare lumps, perhaps pigments and a single vessel. SEM images of cross-sections show the presence of quartz, cuprorivaite and usually a glassy phase (Figure 1a). This structure and composition is typical of the pigment Egyptian blue, which is found widely throughout the Late Bronze Age world of the Eastern Mediterranean. In most of the examples from Nuzi, the glassy phase is heavily weathered and compositionally altered, which limits the compositional information that can be obtained.

Dark blue frits

Rare examples of blue frits that are significantly darker in colour and coarser grained than the abundant mid-blue frits described above have been identified. In the SEM, these can be seen to consist of coarser quartz and cuprorivaite with very little glass. Some cuprorivaite grains are up to 100 μm in length giving a strong colour. The dark blue frits are therefore closely linked to the other blues and can also be classed as Egyptian blue.

Red frits

Red frits are comparatively rare and largely restricted to small beads. These show the typically semi-parallel weathering lamellae that are very common in glass at Nuzi. Figure 1b shows some core material (bottom left) but other than that, only products associated with the weathering of glass. Some of the red frits sampled are therefore weathered glass, which may originally have been colorant free. Some other beads have quartz cores similar to the cream bead shown in Figure 1d, but with additional hematite, and appear to be red faience. These red faience beads occur both as rounded and globular beads.

Yellow and cream frits

There are significant numbers of pale yellow and cream beads described as composition or frit by the excavators. In the SEM these can also be seen to be weathered glass (Figure 1c) some originally yellow with lead antimonate opacifier; others contain copper so were blue translucent glasses. Some of the very pale to white frits consist almost entirely of sintered crushed quartz (Figure 1d), and seem to be the weathered quartz cores. Variable levels of copper can be detected by SEM-EDX in the majority of the white examples studied, indicating that these were originally blue faience that had been heavily altered. These include a wide variety of bead typologies, including globular.

Black beads

The black beads studied were found to contain sintered crushed quartz with no separate glaze layer visible with SEM, despite a separate layer occasionally being present visually. In most examples any glassy phase had completely weathered out but in a few cases small remnants were present. On analysis with SEM-EDX the glassy phase was found to contain silica, potash, soda and aluminium with copper, manganese and iron also present as colorants. Any remnants of this glassy phase were usually found at the outer margins of the beads rather than throughout the body, perhaps suggesting that they are weathered faience rather than frits.

Conclusions

This study of the frits found that they could all be assigned to other classes of vitreous materials commonly found at Nuzi and elsewhere in the Late Bronze Age. Most abundant was Egyptian blue, but severely weathered glass and faience is also present and can produce a range of colours, not just white, creams and buffs, but also reds and blacks. They fit, therefore, within the production processes of these materials rather than forming a separate group of their own.

Acknowledgements

We would like to thank the Electron Microscopy and Microanalysis Division, Mineralogy Department, at the Natural History Museum, London; Richard Newman at the Museum of Fine Arts, Boston; Adam Aja and James Armstrong, Semitic Museum at Harvard University.

References

1. Starr, R.F.S., *Nuzi: report on the excavations at Yorgan Tepa and Kirkuk, Iraq: conducted by Harvard University in conjunction with the American Schools of Oriental Research and the University Museum of Philadelphia, 1927–1931, Volume I Text.* Cambridge, MA, Harvard University Press (1939).
2. Starr, R.F.S., *Nuzi: report on the excavations at Yorgan Tepa and Kirkuk, Iraq: conducted by Harvard University in conjunction with the American Schools of Oriental Research and the University Museum of Philadelphia, 1927–1931, Volume II Plates and Plans.* Cambridge MA, Harvard University Press (1937).
3. Shortland, A.J., Eremin, K., Kirk, S. and Armstrong, J., 'Reassessing Bronze Age manufacturing technologies at Nuzi', in *Materials Issues in Art and Archaeology VIII*, eds. P.B. Vandiver, B. McCarthy, R.H. Tykot, J.L. Ruvalcaba-Sil and F. Casadio, Materials Research Society Symposium Proceedings 1047, Warrendale, PA (2008) 217–232.
4. Moorey, P.R.S., *Ancient Mesopotamian Materials and Industries*, Clarendon Press, Oxford (1994) 196.
5. Vandiver, P., 'Mid second millennium BC soda-lime-silicate technology at Nuzi (Iraq)', in *Early Pyrotechnology: The Evolution of the First Fire-using Industries*, eds. T.A. Wertime and S.F. Wertime, Smithsonian Institution Press, Washington, DC (1982) 73–92.
6. Vandiver, P., 'Glass technology at the mid-second millennium BC Hurrian site of Nuzi', *Journal of Glass Studies* **25** (1983) 239–247.
7. Tite, M.S., 'Characterisation of early vitreous materials', *Archaeometry* **29** (1987) 21–34.

Authors' addresses

- Andrew Shortland, Centre for Archaeological and Forensic Analysis, DEAS/CDS, Cranfield University, Shrivenham, UK (a.shortland@cranfield.ac.uk)
- Katherine Eremin, Harvard Art Museums, Harvard University, Cambridge, MA, USA
- Marc Walton, The Getty Institute of Conservation, Getty Centre, Los Angeles, USA
- Susanna Kirk, Department of Conservation and Research, National Museums Scotland, Edinburgh, UK
- Patrick Degryse, Centre for Archaeological Sciences, Katholieke Universiteit Leuven, Belgium
- Joseph Greene, Harvard Semitic Museum, Harvard University, Cambridge, MA, USA

Use of low vacuum scanning electron microscopy to study the morphology and degradation of organic materials

Michelle Taube, Anna-Grethe Rischel and Maj Ringgaard

KEYWORDS low vacuum scanning electron microscopy, waterlogged wood, wool, paper, archaeological objects

Introduction

Many published studies have shown that low vacuum (LV) or variable pressure (VP) scanning electron microscopy (SEM) allows the examination of organic materials without the application of a conductive coating, which can interfere with imaging and elemental analysis. With the addition of a backscattered electron (BSE) detector, variations in composition can be identified visually. At the National Museum of Denmark, LV-SEM is used regularly to investigate organic materials and we present here the analysis of paper, textile and wood (both dried and waterlogged). In addition to determining the state of preservation, analysis of paper and textile fibres *in situ* provides information about manufacturing methods, which can include the use of fillers and pigments in paper, or the spinning technique in textiles. With wooden samples, the condition of the cells and/or the use of a consolidant are of primary interest.

Methods

The JEOL JSM-5310LV SEM, equipped with an Oxford Instruments energy dispersive X-ray (EDX) detector, at the School of Conservation in Copenhagen was used for our investigations. Imaging was performed in low vacuum mode, typically at a pressure of 24 Pa (except with the cold stage) and an accelerating voltage of 15 or 20 kV. All images were collected with a BSE detector set to give compositional information. The detector was positioned slightly off the beam axis (shadow position), so topographic information could also be obtained. Most specimens can be attached to a stub with carbon tape and the images can be recorded directly without further sample preparation. Usually wooden samples are fractured mechanically and images acquired on the freshly cleaved surface because other preparation methods, such as cutting with a blade, tend to obscure the cell structure. A Deben Coolstage accessory allowed wet materials to be investigated in the LV-SEM in their original condition. With the cold stage, materials are freeze-dried and images are recorded within the SEM chamber without the structural changes that might otherwise occur with changes in pressure and environment (e.g. collapse of the cells in waterlogged wood when allowed to air-dry). To use the cold stage, a piece of wet material is placed on the stage and cooled until ice forms on the surface. The sample chamber is then evacuated. The pressure in the chamber is not fixed and the sample freeze-dries as the ice sublimates directly to water vapour.

Results

Textiles

In a project related to the condition and colouring of archaeological textile finds, we investigated new wool that was coloured and then aged [1]. To create the samples, test swatches were systematically mordanted and sometimes dyed using traditional methods. The samples were then aged in soil under wet,

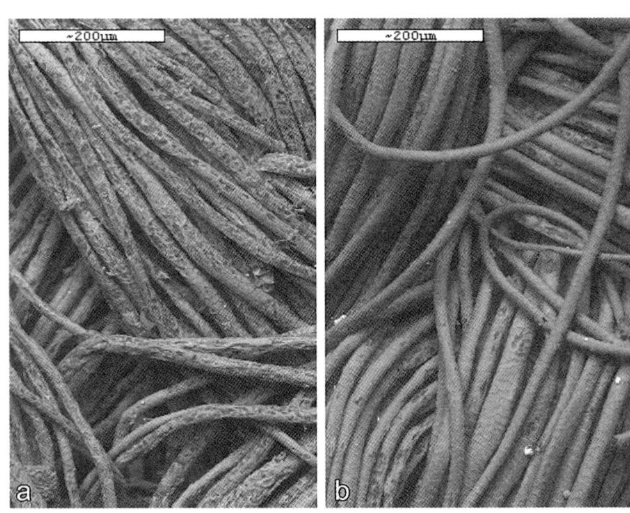

Figure 1 Electron micrographs of two wool samples after ageing for eight months. Both were mordanted with alum. The sample in Figure 1b was dyed with madder after mordanting.

anaerobic conditions. The textiles were rinsed gently after excavation and then freeze-dried at atmospheric pressure.

The SEM investigation was performed on small pieces of dry fabric, both aged and unaged. The extent of degradation was judged by the integrity of the wool scales before and after colouring and ageing. Opaque areas seen with an optical microscope could be identified as cavities using SEM and were most likely due to the activity of microbes. Loss of scales was also easy to identify in the micrographs. Figure 1 shows two micrographs of samples that were aged for eight months. Both samples were mordanted with alum (potassium aluminium sulphate). The sample on the right (Figure 1b) was dyed with madder after mordanting. The sample shown in Figure 1a is heavily deteriorated; the surfaces of the fibres are pitted and only a few scales remain. The madder dye was more protective than the other treatments tested in the study.

Paper

Paper fibre types are often identified using transmitted light optical microscopy, but SEM can provide information about the construction of the paper as a whole. In addition to investigating the splitting and twisting of the fibres within the matrix, any inorganic fillers, processing remains and pigment particles can be identified and analysed. In one case, small pieces removed from the edges of paper monetary notes in the Museum's Royal Coin and Medal Collection were analysed to determine the method of manufacture since paper money needs to be durable and difficult to counterfeit [2]. SEM was used to study and record images of the visible ends of the fibres. Based on their susceptibility to splitting, the fibres were tentatively identified as hemp and linen from recycled fabric. Inorganic materials, visible as bright areas in the images, were also identified and analysed. In particular, smalt was found throughout the paper. Smalt is a blue glass and was used as a pigment or a whitening agent.

Wood samples

Many different consolidation treatments have been used in an attempt to preserve the integrity of waterlogged wood when it is removed from the archaeological environment [3]. Without a consolidant, the thin cell walls of deteriorated wood will collapse as the water evaporates. The wood is kept in water during impregnation with the consolidant and is then dried. Wood treated by several different consolidation methods has been investigated with SEM. The state of degradation of the wood and sometimes the location of the consolidation medium within the structure can be seen. Some consolidants, such as polyethylene glycol (PEG), are organic and are therefore difficult to see within the wood structure even in BSE images.

Figure 2 shows a BSE micrograph of an alum-impregnated archaeological wood sample from Jelling, Denmark. Alum was used as a consolidant for many years, but is no longer widely used because it tends to absorb water and crystallise in humid environments with an increase in volume, which can break apart the remaining cell structure [3]. The alum has formed cylindrical plugs within the cells, both axially and radially. The condition of the wood after burial can also be determined from the image; the cell walls are very thin suggesting that the wood is quite deteriorated.

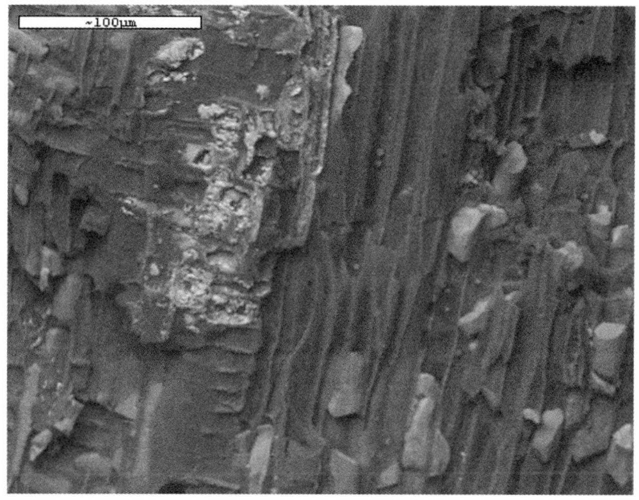

Figure 2 Micrograph of alum-impregnated archaeological wood. Cylindrical plugs of alum can be seen along the cell direction and within the perpendicular rays (top left).

Wet organic materials

As mentioned above, wet organic materials can be studied in the SEM on the cold stage in their as-found condition [4]. The cell structure of waterlogged wood can then be investigated without the addition of a consolidant, such as alum or PEG. The freeze-drying that occurs during the SEM investigation preserves the cell structure without collapse but with some alteration. In addition, hyphae from fungi are often preserved and may be visible above the surface of the wood.

Conclusions

The various case studies presented here demonstrate that LV-SEM is a powerful technique for the investigation of the morphology, state of deterioration and method of manufacture of wet and dry organic samples. Both imaging and analysis can be accomplished with little sample preparation and without the use of conductive coatings.

Acknowledgements

We wish to thank Jørn Bredal and Jettie van Lanschot at the School of Conservation, Royal Academy of Fine Arts, Copenhagen for instruction and guidance on the SEM.

References

1. Ringgaard, M. and Scharff, A.B., 'The impact of dyes and natural pigmentation of wool on the preservation of archaeological textiles', in *North European Symposium for Archaeological Textiles X,*

Copenhagen, May 2008, ed. E.B. Andersson Strand et al., Oxbow Books, Oxford (2009) 221–225.
2. Rischel, A.-G., 'The interaction between Eastern and Western papermaking', in *IPH Congress Book Vol. 17, 29th International Congress of Paper Historians, Stockholm, May 2008*, ed. J.-E. Levlin, IPH (2008) 55–64.
3. Rowell, R.M. and Barbour, R.J. (eds), *Archaeological Wood: Properties, Chemistry, and Preservation*, American Chemical Society, Washington, DC (1990).
4. Jensen, P., Jørgensen, G. and Schnell, U., 'Dynamic LV-SEM analyses of freeze drying processes for waterlogged wood', in *Proceedings of the 8th ICOM-CC Working Group on Wet Organic Archaeological Materials, Stockholm 2001*, ed. P. Hoffmann et al., ICOM-CC WOAM, Bremerhaven (2002) 319–331.

Authors' addresses

- Michelle Taube, Department of Conservation, National Museum of Denmark, Copenhagen, Denmark (michelle.taube@natmus.dk)
- Anna-Grethe Rischel, Department of Conservation, National Museum of Denmark, Copenhagen, Denmark (anna-grethe.rischel@natmus.dk)
- Maj Ringgaard, Department of Conservation, National Museum of Denmark, Copenhagen, Denmark (maj.ringgaard@natmus.dk)

SEM-EDX technological studies of some remarkable Early Bronze Age gold artefacts from Bulgaria

Svetla Tsaneva, Martin Hristov, Victoria Karatsanova and Zdravko Tsintsov

KEYWORDS SEM-EDX, gold, technology, Early Bronze Age, sintering

Introduction

The finds studied were recovered from over 20 ritual structures located in Balinov Gorun, Karlovo Region, Bulgaria. These structures were excavated in 2004–2007 by a team from the National Museum of History led by Martin Hristov. Most date from the second half of the third millennium BC, i.e. the Early Bronze Age III (EBA III), but a few are dated to the Iron Age (first millennium BC). This study discusses the EBA III finds.

The structures excavated (so far) yielded over 21,000 small gold objects including beads, string separators and spirals, as well as several gold, electrum and bronze daggers, a lidded silver box and pottery. The only known parallels to the gold finds from Dubene can be found at the EBA sites of Troy and Poliochni.

Methodology and results

The majority of the pieces analysed (thus far) by scanning electron microscopy with energy dispersive X-ray analysis (SEM-EDX) are elements of complex adornments: necklaces and other jewellery. Some that were associated with remains of organic materials were also subjected to SEM. The surfaces of about 15,000 small gold pieces were initially studied using optical microscopy. Over 500 of these were chosen for SEM observations and most were analysed by EDX (Philips SEM-515 fitted with an EDAX PV 9100 and run at 20–25 kV and 0.5 nA). The SEM image of one of the many almost identical EBA small gold beads from Dubene is shown in Figure 1.

The results suggest the use of hot sintering (powder metallurgy) as a basic manufacturing technology. Structures of EBA gold finds from Dubene were compared to structures of placer gold from a deposit near Burgas and of a modern cast gold drop [1]. The structures observed in single sintered gold particles that are still discernible in some of the EBA gold objects (Figure 2) are of the 'cut through a cabbage' type. This structure is typical for alluvial gold [2], although here the folded layers have become thicker and the structure has become denser through heating and plastic deformation, which additionally distorted the 'cut through a cabbage' structure. SEM images of the flattened apertures in the initial structure of sintered gold particles clearly show the direction of the applied forces in the shaping by plastic deformation.

Some areas of the complex string separators made of gold leaf display grained structures. These areas are slightly concave and have not been flattened and polished sufficiently to hide the grained structure of the gold leaf. When they were

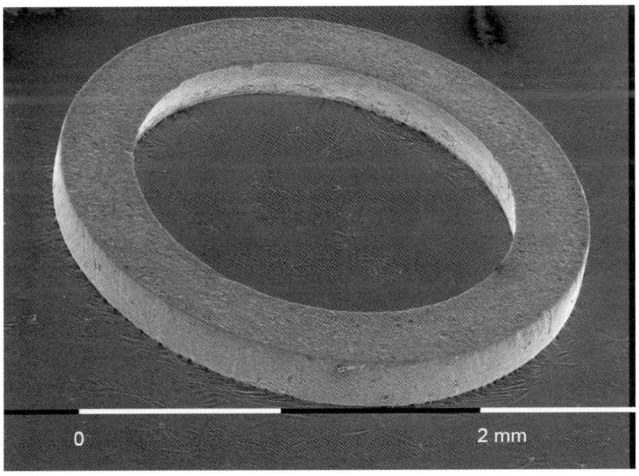

Figure 1 EBA gold bead from Dubene (×50).

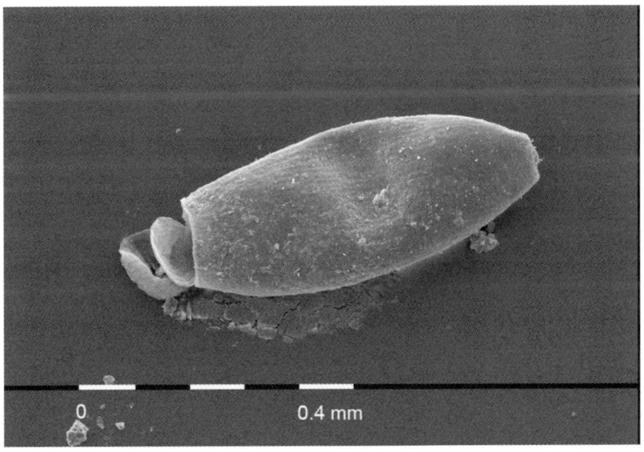

Figure 2 SEM image of a bulrush seed-head found in a closed pottery vessel (×115).

compared to another gold piece from Dubene showing the grained structure of embrittled gold significant differences were observed. The embrittled grains exhibit straight grain boundaries, while those in the gold leaf from Dubene show uneven shape and size, and their boundaries are rounded. EDX data indicated that the composition of the grains in the embrittled structure is almost identical, whilst that of the grains in the gold leaf differs, as shown by the analysis of 17 neighbouring grains from a small area. The silver content varied from about 3 wt% to almost 24 wt%; four grains showed traces of iron (0.2–0.3 wt%) and four contained copper in the range 0.10.01–0.5 wt%.

The instrument used is (unfortunately) very old and the mapping could not reveal the different particles in this structure due to the fact that the image was acquired by a single pass of the electron beam – the only option available on this equipment.

Mineral inclusions of various sizes (10 to almost 100 μm) and deeply locked in the gold matrix were recorded in almost all the gold pieces studied. Backscattered electron (BSE) images revealed the inclusions better. These seem to be either fine mineral particles that had originally been present in the alluvial gold dust to be sintered, or the particles had been in the moulds and were embedded into the surface of the items during their pressing into shape. The rounded edges of these inclusions suggest that they originated from the alluvial gold particles. EDX analysis of some of these inclusions identified them as feldspars and quartz.

Although the beads and the string separators were obviously intended to be part of a necklace, they did not show any traces of wear. Beads and thread remains, however, were preserved in the holes of several string separators. Some of these threads were mounted on a holder, coated with gold and placed in the SEM. The SEM image showed that the thread is composed of parallel, untwisted cotton fibres.

Another relatively intriguing material was discovered in one of the structures. All items, including gold ones, which were found in a pottery vessel coupled with an inverted pot, were covered by hundreds of very small (c. 0.3 × 1 mm) light yellow, transparent, pod-like capsules. Samples of these were coated with gold and identified in the SEM as bulrush seed-heads.

Conclusions

SEM proved to be an excellent tool for this technological study of archaeological materials and especially of gold. However, general statistics – and cluster analysis in particular – proved to be unsuitable tools in this instance as the pieces are composed of minute gold particles agglomerated by diffusion, each of them probably composed of even smaller particles that were cold-welded to one another during transportation. Conventional provenance studies are also inapplicable with such inhomogeneous pieces. The huge amount of gold finds suggests a local alluvial deposit that had formed by several primary sources in the mountain, but there were no clearly differentiated clusters to indicate their possible number.

SEM images greatly facilitated identification of the minute organic remains that provide valuable information. EBA cotton was identified for the first time from the territory of present-day Bulgaria. The presence of numerous bulrush seed-heads in this closed pottery vessel also suggests that this originally fluffy material may have been used to protect the items.

The emergence of sintering at such an early date should not be surprising as people already had thousands of years' experience in pottery making, which is based on the same principles. There had been an even longer experience in testing various colour earths in the production of pigments. The deposits of finest alluvial gold look very much like fine coloured clay [1], but the finest gold particles cannot form deposits by gravitation. They need flotation, which could have been ensured by rotting vegetation, thus forming pockets of 'coloured' earth in the deposits of finest clays. Even nowadays, sintering is regarded as a simple method to produce numerous almost identical objects, items of special quality and of complicated and very precise shape [1].

Acknowledgements

We thank Elena Marinova, Department of Geosciences, Leuven Catholic University, Belgium, for her identification of the bulrush seed-heads. This study is part of the ТК-СКИН 4/2007 Project financed by the National Science Fund, Ministry of Education, Science and Youth, Republic of Bulgaria

References

1. Tsintsov Z., Hristov M., Karatsanova, V. and Tsaneva, S., 'Preliminary results from the study of Early Bronze Age golden artefacts from ritual structures by the village of Dubene, Karlovo district, south Bulgaria', *Archaeologia Bulgarica* **13**(3) (2009) 7–21.
2. Raub, Ch.J., 'The metallurgy of gold and silver in prehistoric times', in *Prehistoric Gold in Europe: Mines, Metallurgy and Manufacture*, ed. G. Morteani and J.P. Northover, Kluwer Academic Publishers, Dordrecht and Boston (1995) 243–259.

Authors' addresses

- Svetla Tsaneva, National Museum of History (NMH), Sofia, Bulgaria (svetla_tsaneva@abv.bg)
- Martin Hristov, National Museum of History (NMH), Sofia, Bulgaria
- Victoria Karatsanova, National Museum of History (NMH), Sofia, Bulgaria
- Zdravko Tsintsov, Institute of Mineralogy and Crystallography, Bulgarian Academy of Sciences (IMC-BAS), Bulgaria

The formation of complex crusts in oil paints containing lead white and smalt: dissolution, depletion, diffusion and deposition

Annelies van Loon, Petria Noble and Jaap J. Boon

KEYWORDS oil paint, Rembrandt, surface whitening, lead white, smalt

Introduction

The formation of crusts and whitish deposits on the surface of oil paintings is a widely observed problem. It can be visually very disfiguring and raises many questions during conservation treatment. What is the composition of these crusts? Are they part of the original paint build-up? How are they formed? Investigation of paint cross-sections with the scanning electron microscope in combination with energy dispersive X-ray analysis (SEM-EDX) is the key to forming a better understanding of this type of deterioration.

Analysis of a large number of seventeenth-century oil paintings has shown the presence of complex crusts, where several chemical and physical processes are taking place [1]. A very interesting group are those deposits/crusts rich in lead, potassium and sulphur that form as a result of deterioration of paints containing lead white, a basic lead carbonate, and smalt, a finely ground cobalt potash glass. In addition to smalt, alum, used in the production of organic lake pigments or as an ingredient in lining adhesives, as well as earth pigments, are other possible sources of potassium [2, 3]. Similar crusts have been found on the surface of historic (high potassium/low lime) glass and other objects [4].

This paper describes the formation of surface crusts on a late Rembrandt painting *Homer* (dated 1663) in the collection of the Royal Picture Gallery Mauritshuis (The Hague).

Methods

In this study, the composition of the crust was analysed using light microscopy (LM), SEM-EDX, Fourier transform infrared spectroscopy (FTIR), Raman spectroscopy and direct temperature-resolved mass spectrometry (DTMS). SEM-EDX analysis was performed on a FEI XL30 SFEG high vacuum electron microscope with the EDAX EDX system (for further details see [1]). The SEM backscattered electron (BSE) images in particular provide valuable insight into the composition, morphology and origin of the surface crusts and clearly illustrate the different steps in the reaction mechanism: dissolution, depletion, diffusion and deposition. The compositional contrast between lead, which is a high atomic weight element and therefore highly scattering and appears bright in the BSE image, and the low atomic weight components present in the paint (low scattering and appear dark) is particularly revealing in the investigation of lead-rich crusts [5].

Results and discussion

Crust formation

In Rembrandt's *Homer*, which was extensively studied during its treatment in 2004–2006, a whitish, insoluble crust was found to cover most of the dark passages in the painting: the garment, the cap and the background. Based on LM observations alone, the crust was first interpreted as residual varnish or broken up faded glaze of organic lake. In the affected areas, a coarse smalt, mixed with bone black, lakes and earths, is present in the paint in unusually large proportions giving it texture and translucency, in addition to colour. The preparatory layers in this painting consist of a double ground: a thick orangey lower ground containing chalk and earth pigment, followed by a grey-brown upper ground comprising lead white and a little chalk, earth and black pigment.

Dissolution of lead white

Traditional Dutch stack process lead white has a typical particle distribution of large lumps and smaller crystals that fill the open spaces between the lumps [6]. Previous studies have shown that lead white can be affected by free fatty acids released from the oil network (by hydrolysis of the ester bonds of the cross-linked triglyceride moieties during ageing) to form mobile lead soaps [7]. Here SEM BSE images are instrumental in revealing changes in particle morphology and distribution

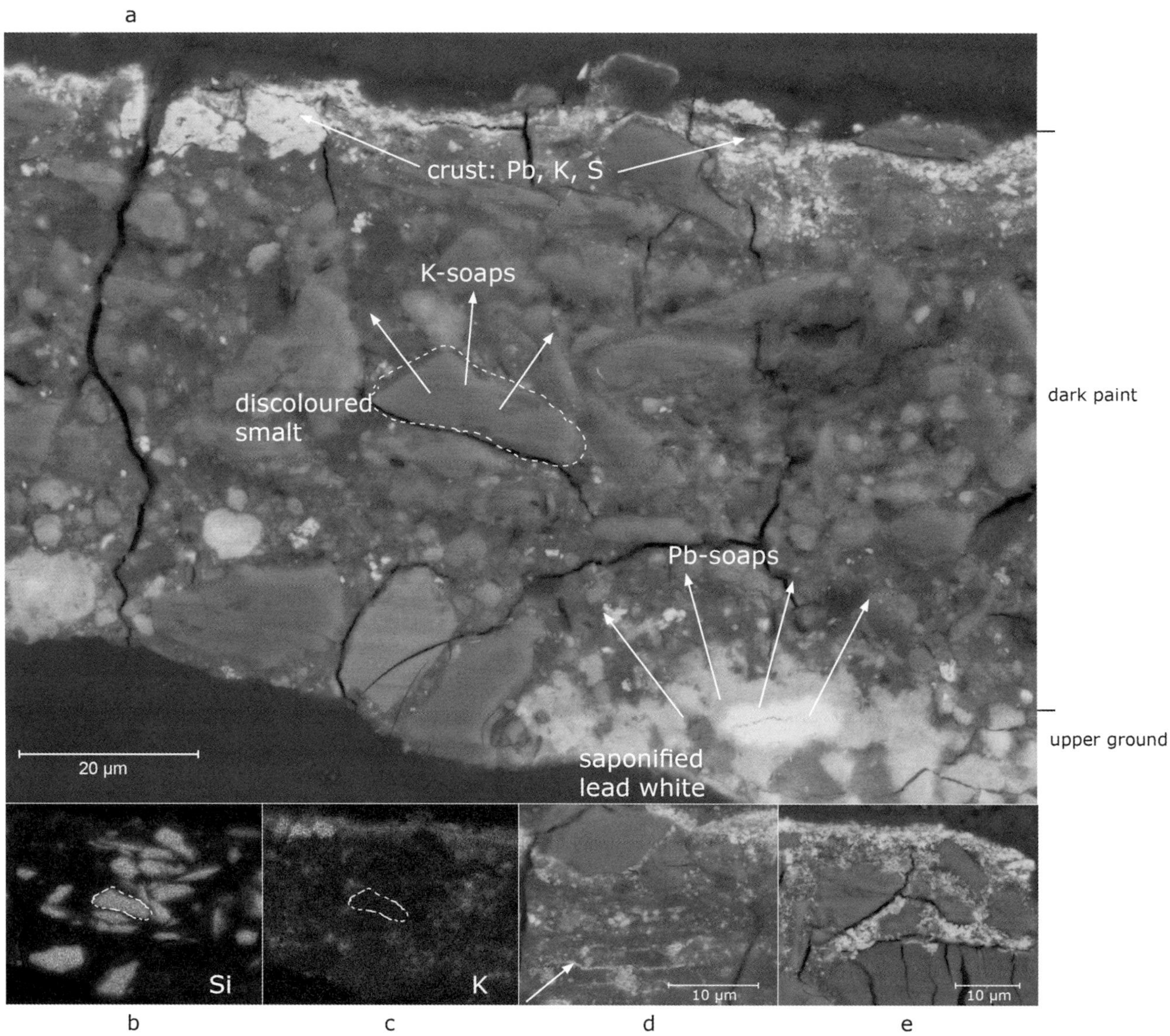

Figure 1 SEM images of paint cross-sections from dark passages in Rembrandt's *Homer* (1663), oil on canvas, Royal Picture Gallery Mauritshuis (MH584), The Hague. (a) BSE image showing the dissolution and saponification of lead white in the upper ground and deposition of highly scattering Pb-K-S-rich deposits at the paint surface (sample MH584_X05). (b) and (c) Corresponding Si and K maps showing the depletion of potassium from the discoloured smalt (sample MH584_X05). (d) BSE image showing horizontal lead-rich bands (see arrow) and many small dispersed highly scattering lead-rich particles throughout the dark paint (sample MH584_X35). (e) BSE image showing detail of the surface crust with highly scattering small dispersed particles integrated in the paint structure and surrounding the angular smalt particles (sample MH584_X56).

of lead white. In Figure 1a, the light grey, amorphous lead-rich areas in the upper ground differ significantly from the white, highly scattering intact lead white lump. Saponified areas appear darker (less scattering) than intact lead white, due to their low degree of crystallinity and higher organic content. Figure 1a also shows that all the smaller particles have already reacted away. In this sample, as in other samples from this painting, the presence of lead soaps in these areas in the upper ground has been confirmed by FTIR and DTMS analyses [1].

Depletion of potassium in smalt

The characteristic angular shape of smalt (c.20 μm in diameter) in the dark paint layers is easily recognised in the BSE images. EDX detected silicon as the main element, with lower concentrations of potassium, cobalt, iron, nickel, arsenic and bismuth in the particles. The iron, nickel, arsenic and bismuth are typical impurities from the cobalt ore used in the seventeenth century. The percentage of cobalt (3–4 wt% CoO) used to colour the glass, compares well with other seventeenth-century smalts (see Spring in this volume). In all cross-sections studied, the very low levels of potassium in the smalt particles (<1 wt%) demonstrate that the smalt has completely discoloured. In the deterioration process, potassium is depleted from the smalt by interaction with the oil, reacting with free fatty acids from the oil to form mobile potassium soaps [8]. The depletion of potassium from the smalt is made visible through EDX mapping by comparison of the silicon and potassium maps (Figure 1b and c).

Diffusion and migration of lead and potassium

The lead and potassium soaps that have formed in the paint are highly mobile. This is clearly shown in the SEM BSE images and EDX elemental maps that indicate that lead and potassium, possibly in the form of soaps, have migrated through the paint layers towards the paint surface. These transport processes are considered to be driven by gradients in temperature and moisture [7, 1]. The presence of potassium in the paint matrix between the smalt particles and in large quantities at the surface is revealed by the potassium map (Figure 1c). Furthermore, horizontal light grey bands of finely dispersed lead-rich particles, visible in the upper dark paints in the BSE images, are interpreted as migration fronts of lead originating from the saponified upper ground (Figure 1d). The lack of distinct interface between the ground and the upper dark paint layer is notable (Figure 1a). Here the light grey saponified areas of lead white in the ground can be seen to diffuse into the dark paint layer above.

Deposition and mineralisation of lead/potassium/sulphur crusts

In these samples, lead and potassium were seen to deposit and mineralise to form salts at the paint surface and around the smalt particles (Figure 1a and e). EDX demonstrated the abundance of lead, potassium and sulphur in these deposits. Sulphates, among other products, were identified using FTIR and Raman [1]. The source of sulphur/sulphate is thought to originate from the atmosphere. The lead-potassium-sulphur-rich deposits appear as light grey, finely dispersed crystals and large chunks in the BSE images, and contrast with the angular smalt particles in the paint layer, which show less scattering and appear dark grey (Figure 1a and e). The SEM images clearly demonstrate that, in this painting, the deposits that make up the crust are relatively thick (c.10 μm) and intimately bound with the paint. This has implications for conservation treatment since its removal would damage the original paint. In this case, the final stages of cleaning proceeded under the stereomicroscope, after which the light scattering caused by the surface crust was reduced by using a low molecular weight varnish to saturate the surface.

Conclusions

In paint samples from affected areas, SEM-EDX analysis demonstrated the depletion of potassium from the discoloured smalt and the severe saponification and dissolution of lead white in the upper ground. The subsequent migration of the lead and potassium towards the paint surface, possibly in the form of soaps, was made visible in the BSE images. The deposits are thought to undergo further reactions with the atmosphere to form insoluble complex salt mixtures rich in lead, potassium and sulphur. SEM (in combination with other techniques) provided a powerful tool for understanding the nature of surface deposits and the complex ageing processes that occur at, and below, the paint surface.

Acknowledgements

The research took place as part of the Open Laboratory agreement (2002–2007) between the FOM Institute AMOLF in Amsterdam and the Mauritshuis.

References

1. Van Loon, A., *Colour Changes and Chemical Reactivity in Seventeenth-Century Oil Paintings*, PhD dissertation, University of Amsterdam, Molart Series (14), AMOLF, Amsterdam (2008) 119–204.
2. Noble, P. and Van Loon, A., 'Rembrandt's *Simeon's Song of Praise*, 1631: pictorial devices in the service of spatial illusion', *ArtMatters Netherlands Technical Studies in Art* **4** (2007) 19–36.
3. Boon, J.J. and Oberthaler, E., 'Mechanical weakness and chemical reactivity observed in the paint structure and surface of *The Art of Painting* by Vermeer', in *Vermeer: The Art of Painting*, exh. cat., Kunsthistorisches Museum, Vienna (2010) 328–335.
4. Sterpenich, J., 'Cristallochimie des produits d'altération des vitreaux médiévaux: application au vieillissement des déchets vitrifiés', *Bulletin of Engineering, Geology and the Environment* **61** (2002) 179–193.
5. Keune, K., Van Loon, A. and Boon, J.J., 'SEM backscattered-electron images of paint cross-sections as information source for the presence of the lead white pigment and lead-related degradation and migration phenomena in oil paintings', *Microscopy and Microanalysis* **17**(5) (2011) 696–701.
6. Groen, K., 'Investigation of the use of the binding medium by Rembrandt: chemical analysis and rheology', *Zeitschrift für Kunsttechnologie und Konservierung* **11**(2) (1997) 207–227.
7. Boon, J.J., Hoogland, F. and Keune, K., 'Chemical processes in aged oil paints affecting metal soap migration and aggregation', *AIC Paintings Specialty Group Postprints Vol. 19*, comp. H. Mar Parkin, American Institute of Conservation, Washington, DC (2007) 16–23.
8. Spring, M., Higgitt, C. and Saunders, D., 'Investigation of pigment-medium processes in oil paint containing degraded smalt', *National Gallery Technical Bulletin* **26** (2005) 56–70.

Authors' addresses

- Annelies van Loon, Royal Picture Gallery Mauritshuis, The Hague, The Netherlands (vanloon.a@mauritshuis.nl)
- Petria Noble, Royal Picture Gallery Mauritshuis, The Hague, The Netherlands (noble.p@mauritshuis.nl)
- Jaap J. Boon, FOM Institute AMOLF, Amsterdam, The Netherlands (boon@amolf.nl)

Uncovering painted surfaces in historic buildings: a comparison of paint removal techniques using light and scanning electron microscopic surface imaging

Edwin Verweij

KEYWORDS surface textures architectural paint research

Introduction

The process of examining and documenting the painted colour schemes in historic houses is often referred to as architectural paint research (APR). This research field, strongly developed in building conservation since the 1990s, has resulted in proposals to improve working methods and techniques [1–3]. In order to reveal the painted history in an interior, a paint window or uncovering is made on the architectural elements. Additionally, paint cross-sections are taken for the interpretation of the layered paint build-up that can, at the same time, be used to analyse pigments used in the paint composition.

In general, we distinguish between three different methods for revealing underlying paint layers: mechanical, thermal and chemical. Comparing these removal techniques enables us to quantify how much damage is done to a surface. The uncovered surfaces are assessed with light microscopy (LM) techniques and imaged with scanning electron microscopy (SEM) [4]. This showed that each method resulted in a different paint surface. Both texture and colour are affected therefore the overall visual appearance is changed.

Materials

The interior of the former Logement van Amsterdam, now part of the Dutch parliament buildings in The Hague, was built in the 1740s and has been restored recently [5]. This required local replacement of parts of the original oak wainscoting that could be used as test panels. They contained the complete history from the first paint coats applied in the mid-eighteenth century until the end of the twentieth century. In total, 23 layers were identified in the cross-sections, including ground layers, fillers and topcoats. After exposing each layer separately, we concluded that they represented 10 different colour schemes. The original colour scheme, from the first period in the 1740s, was found to be a dark brown monochrome oil paint painted on top of an orangey-brown ground coat. This colour scheme was repeatedly exposed by the use of the three different paint removal techniques listed above.

Methods

Mechanical paint removal was performed with surgical scalpel blades and a stereomicroscope. Thermal paint removal involved using scalpels in addition to a heated airflow (30–35°C). A commercial paint stripper was used as a chemical paint removal method. The paint stripper was removed from the surface partly with scalpel blades and partly with cotton swabs and a low aromatic hydrocarbon solvent (see 'Materials and suppliers' below).

Macro photography and LM were used to examine the texture, colour and gloss of the surfaces. Normal light imaging of the surface textures was performed with a stereomicroscope (Zeiss Opmi pico, focal length 200 mm) with a digital camera (Canon EOS 5D mark II) and an LCD light source (Schott KL 1500 at 3000 K). Subsequently, the paint surfaces were imaged in a variable pressure scanning electron microscope (VP-SEM) in backscatter (BSE) mode at 15 kV for both compositional contrast and textural assessments. SEM imaging was performed on a VP-SEM (Hitachi S-3600N with Quantax 400 software). BSE images of the surfaces were taken at 15 kV in 10 Pa vacuum giving compositional contrasts combined with surface information. Owing to the size of the chamber of the VP-SEM the wainscoting parts had to be sawn into smaller pieces and their thickness reduced to 4 mm.

Results

The surface of a dark brown painted eighteenth-century colour scheme was compared after the use of the three different uncovering methods: mechanical, thermal and chemical.

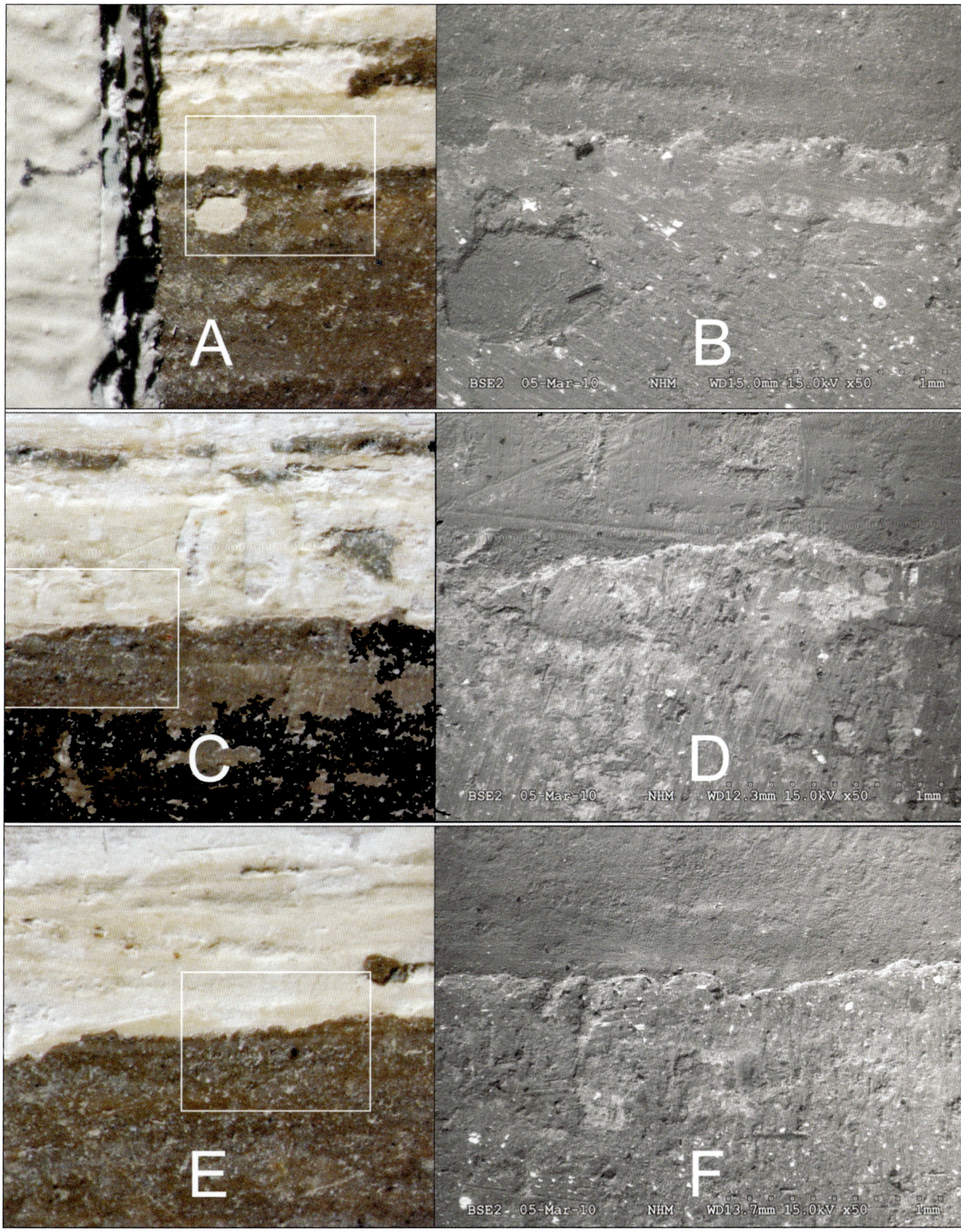

Figure 1 Comparing three uncovering methods in use in architectural paint research. Mechanical exposed surface (A) with a smooth surface and SEM-BSE image (B) showing the evenness of the surface following the texture of brushstrokes with minor linear scratch marks. Thermal exposed surface (C) with softened and flattened surface and SEM-BSE image (D) showing overall directional scratch marks. Chemical exposed surface (E) with greyish discoloration and SEM-BSE image (F) showing a roughened surface, overall tiny pinholes and omnidirectional scratch marks on a flattened texture.

Mechanical

The painted brushstroke seems to be well preserved after uncovering. The force applied when using scalpel blades sometimes results in local paint loss, owing to poor adhesion to the ground layers because of cleaning and/or sanding before the overpaint was applied.

Minor surface cuts can be observed as small scratch marks as a result of the scalpel blade cutting into the paint surface. The method is relatively time-consuming but all paint layers can be exposed.

Thermal

The heated airflow softens the overlying paint layers. Depending on the binding medium, the edges between the ground and topcoat are less distinct and partly discoloured. Depending on age, condition and the binding medium used in the overpaints the effect of this method varies – it can either improve or delay the separation of paint layers. It tended to diminish friction and reduce the force needed to separate the layers. The surface was generally more easily pressed and flattened owing to the softening effect of the heated airflow.

Chemical

Due to the solvent's power, a clear distinction between the different layers tends to disappear; surface edges seem to be less clear. Paint loss was minimised as a result of diminished friction. The effectiveness of the solvent can be slowed down by using a low aromatic hydrocarbon solvent on cotton swabs. After uncovering, a greyish-white colour tinge formed on the paint surface, influencing the hue. The paint surface looks evenly exposed but on closer inspection it has a peach-like skin, with a roughened and matted appearance resulting in a change of colour. This method can be relatively fast depending on the age and the binding medium used in the overpaints.

Discussion and conclusions

The visual appearance of exposed paint layers was assessed for texture, colour and gloss. With the use of SEM imaging we could examine surface defects to an extent that could not be matched by conventional LM. All three techniques seemed to affect the surface textures of the paints. As a result, light is being reflected in a different way, seemingly producing another colour. Chemical paint removal seemed to have the most impact compared with the thermal and mechanical techniques. This implies that the painted history of buildings and interiors can be misinterpreted depending on the technique applied.

A valid interpretation of the painted decorative history can be made more carefully by using a more refined uncovering method. Limiting the damage improves the final quality of the exposed paint surfaces. This will result in a more accurate interpretation of the colour schemes that can be found in the built cultural heritage. It is hoped that by showing the consequences of using different paint removal techniques, a more cautious approach will be practised within architectural paint research.

Acknowledgements

We would like to thank Hans-Jørgen Berg, Department of Technical and Scientific Conservation, National History Museum, University of Oslo, Norway for his help with the SEM imaging. This work was performed at the Norwegian Institute for Cultural Heritage Research (NIKU) in Oslo, Norway, as part of the Strategic Institute Programme (SIP) for the 'Conservation of Cultural Heritage – Careful Management of Changes' (PRECARE) project under the title: 'Across cross-sections: previous and current colour schemes'.

Materials and suppliers

- The paint stripper Duxola was obtained from Krefting & Co. AS, Nesveien 15, P.B. 4, 1305 Haslum, Norway (www.krefting.no/). It contains dichloromethane and isotridecanol polyglycolether. Excess paint stripper was removed by using cotton swabs with Exxsol D60, a low aromatic hydrocarbon solvent.
- The scalpel blades were obtained from Swann-Morton Limited, Owlerton Green, Sheffield S6 2BJ, UK (www.swann-morton.com/). Both blade sizes 10 and 15 were used to uncover the paint layers.

References

1. Hughes, H., *Layers of Understanding: Setting Standards for Architectural Paint Research*, Donhead/English Heritage, London (2002).
2. Hughes, H., 'The potential of architectural paint research in building analysis and conservation', in *Paint Research in Building Conservation*, eds. L. Brenghøi, H. Hughes, J. Lindbom, T. Olstad and E. Verweij, Archetype Publications, London (2006) 3–9.
3. Solberg, K., 'Data on strata: NIKU's practice for the documentation of architectural paint research', in *Paint Research in Building Conservation*, eds. L. Brenghøi, H. Hughes, J. Lindbom, T. Olstad and E. Verweij, Archetype Publications, London (2006) 10–15.
4. Berg, K.J.v.d., Daudin, M., Joosten, I, Wei, B., Morrison, R. and Brunstock, A., 'A comparison of light microscopy techniques with scanning electron microscopy for imaging the surface cleaning of paintings', in *9th International Conference on NDT of Art, 25–30 May 2008, Jerusalem*. http://212.8.206.21/article/art2008/papers/033VandenBerg.pdf (accessed 2010)
5. Polman, M. and Verweij, E., *Logement van de Heren van Amsterdam, s'-Gravenhage. Rapport Kleurverkenning*, unpublished report, Amsterdam (2003).

Author's address

Edwin Verweij, Conservation Department, The Norwegian Institute for Cultural Heritage Research (NIKU), Oslo, Norway (edwin.verweij@niku.no)